Advances in Intelligent Systems and Computing

Volume 740

Series editor

Janusz Kacprzyk, Polish Academy of Sciences, Warsaw, Poland
e-mail: kacprzyk@ibspan.waw.pl

The series "Advances in Intelligent Systems and Computing" contains publications on theory, applications, and design methods of Intelligent Systems and Intelligent Computing. Virtually all disciplines such as engineering, natural sciences, computer and information science, ICT, economics, business, e-commerce, environment, healthcare, life science are covered. The list of topics spans all the areas of modern intelligent systems and computing such as: computational intelligence, soft computing including neural networks, fuzzy systems, evolutionary computing and the fusion of these paradigms, social intelligence, ambient intelligence, computational neuroscience, artificial life, virtual worlds and society, cognitive science and systems, Perception and Vision, DNA and immune based systems, self-organizing and adaptive systems, e-Learning and teaching, human-centered and human-centric computing, recommender systems, intelligent control, robotics and mechatronics including human-machine teaming, knowledge-based paradigms, learning paradigms, machine ethics, intelligent data analysis, knowledge management, intelligent agents, intelligent decision making and support, intelligent network security, trust management, interactive entertainment, Web intelligence and multimedia.

The publications within "Advances in Intelligent Systems and Computing" are primarily proceedings of important conferences, symposia and congresses. They cover significant recent developments in the field, both of a foundational and applicable character. An important characteristic feature of the series is the short publication time and world-wide distribution. This permits a rapid and broad dissemination of research results.

Advisory Board

Chairman

Nikhil R. Pal, Indian Statistical Institute, Kolkata, India
e-mail: nikhil@isical.ac.in

Members

Rafael Bello Perez, Universidad Central "Marta Abreu" de Las Villas, Santa Clara, Cuba
e-mail: rbellop@uclv.edu.cu

Emilio S. Corchado, University of Salamanca, Salamanca, Spain
e-mail: escorchado@usal.es

Hani Hagras, University of Essex, Colchester, UK
e-mail: hani@essex.ac.uk

László T. Kóczy, Széchenyi István University, Győr, Hungary
e-mail: koczy@sze.hu

Vladik Kreinovich, University of Texas at El Paso, El Paso, USA
e-mail: vladik@utep.edu

Chin-Teng Lin, National Chiao Tung University, Hsinchu, Taiwan
e-mail: ctlin@mail.nctu.edu.tw

Jie Lu, University of Technology, Sydney, Australia
e-mail: Jie.Lu@uts.edu.au

Patricia Melin, Tijuana Institute of Technology, Tijuana, Mexico
e-mail: epmelin@hafsamx.org

Nadia Nedjah, State University of Rio de Janeiro, Rio de Janeiro, Brazil
e-mail: nadia@eng.uerj.br

Ngoc Thanh Nguyen, Wroclaw University of Technology, Wroclaw, Poland
e-mail: Ngoc-Thanh.Nguyen@pwr.edu.pl

Jun Wang, The Chinese University of Hong Kong, Shatin, Hong Kong
e-mail: jwang@mae.cuhk.edu.hk

More information about this series at http://www.springer.com/series/11156

Jugal Kalita · Valentina Emilia Balas
Samarjeet Borah · Ratika Pradhan
Editors

Recent Developments in Machine Learning and Data Analytics

IC3 2018

 Springer

Editors
Jugal Kalita
College of Engineering and Applied Science
University of Colorado Colorado Springs
Colorado Springs, CO, USA

Samarjeet Borah
Department of Computer Applications
Sikkim Manipal University
Sikkim, India

Valentina Emilia Balas
Automation and Applied Informatics
Aurel Vlaicu University of Arad
Arad, Romania

Ratika Pradhan
Department of Computer Applications
Sikkim Manipal University
Sikkim, India

ISSN 2194-5357   ISSN 2194-5365   (electronic)
Advances in Intelligent Systems and Computing
ISBN 978-981-13-1279-3   ISBN 978-981-13-1280-9   (eBook)
https://doi.org/10.1007/978-981-13-1280-9

Library of Congress Control Number: 2018945899

This Springer imprint is published by the registered company Springer Nature Singapore Pte Ltd.
The registered company address is: 152 Beach Road, #21-01/04 Gateway East, Singapore 189721, Singapore

Preface

Recent Developments in Machine Learning and Data Analytics is a collection of research findings of the Second International Conference on Computing and Communication. The conference is centered upon the theme of machine learning and data analytics.

The works incorporated in this volume can roughly be divided into three parts, namely data analytics, natural language processing, and soft computing. The following section contains a brief information about the various contributions to this volume.

In the first paper, Aski et al. provide an architectural overview of IoT-enabled ubiquitous healthcare data acquisition and monitoring system for personal and medical usage powered by cloud application. The next one is also an IoT-based paper where Ishita Chakraborty, Anannya Chakraborty, and Prodipto Das discuss sensor selection and data fusion approach for IoT applications. An overview of Hadoop MapReduce, Spark, and scalable graph processing architecture is provided by Talan et al. in their paper. On the other hand, Hore et al. discuss a machine intelligence-based approach to analyze social trend toward girl child in India. Analyzing class-imbalanced data is found to be a difficult task always. In the next paper, an improvement in boosting method for class-imbalanced datasets is discussed by Kumar et al. In their paper, Ambika Choudhury and Deepak Gupta provide a survey on medical diagnosis of diabetes using machine learning techniques. Another classification-related issue on diabetes data is presented by Santosh Kumar Majhi in his research work. Findings on the usefulness of big data technologies for statistical homicide dataset are discussed by Askew et al. The next research work discusses a journal recommendation system through content-based filtering approach. Another content-based filtering and collaborative filtering technique for movie recommendation is provided by Bharti et al in their research work.

The next 13 contributions are from the domain of natural language processing. The first work of such kind discusses a word-sense disambiguation for Assamese language. In addition to this, other two works are found for Assamese language where Choudhury et al. present a context-sensitive spelling checker for Assamese language and Mirzanur Rahman and Shikhar Kumar Sarma discuss a hybrid approach to analyze the morphology of an Assamese word. The next work presents an aptitude question paper generator and answer verification system. Ghosh et al. discuss affinity maturation of homophones in a word-level speech recognition in their work. This follows a discussion on feature map reduction in CNN for hand-written digit recognition. Multilingual text localization from camera-captured images is presented by Dutta et al. The technique is based on foreground homo-geneity analysis. Jajoo et al. propose script identification from camera-captured multiscript scene text components in their research work. Again, Khan et al. present a distance transform-based stroke feature descriptor for text–non-text classification. The volume also includes a contribution on emotion mining. This follows two works on Nepali language, where Thapa et al. discuss a finger spelling recognition for Nepali sign language and Yajnik et al. present a work on parsing in Nepali language.

A number of contributions are found which can roughly be categorized under the domain of soft computing. Mishra et al. discuss a BFS-NB hybrid model in intrusion detection system, whereas Saikia et al. propose an effective alert corre-lation method in their research work. An application of ensemble random forest classifier for detecting financial statement manipulation of listed Indian companies is discussed by Hiral Patel and co-authors. Another security-related paper is dis-cussed on dynamic shifting genetic non-adjacent form elliptic curve Diffie–Hellman key exchange procedure for IoT heterogeneous network. This follows few classification-related works, where Vijaya et al. discuss fuzzy clustering with ensemble classification techniques to improve the customer churn prediction in telecommunication sector and Ahmed et al. propose a technique to remove the bottleneck of FP tree. Additional works include elephant herding algorithm, improved K-NN algorithm through class discernibility and cohesiveness, a reduction-level fusion of PCA and random projection for high-dimensional data, a stable clustering algorithm for mobile ad hoc networks (MANETs), and interval-valued complex fuzzy concept lattice and its granular decomposition. Umesh Gupta and Deepak Gupta discuss their findings on twin-bounded support vector machine based on L2-norm. A work to perform natural scene labeling using neural networks is presented by Das et al. Kalvapalli et al present their findings on a city-scale transportation system using XGBoost. A selfish controlled scheme in an opportunistic mobile communication network is presented by Moirangthem Tiken Singh and Surajit Borkotokey. Few quality works on image processing techniques are also included in this volume. The first part of these papers discusses a fusion-based underwater image enhancement using weight map techniques. The next work proposes an algorithm for automatic segmentation of pancreas histo-logical images for glucose intolerance identification. An edge detection technique using ACO with PSO for the noisy image is discussed by Aditya Gautam and Mantosh Biswas in their research work. Another work discusses improved con-volutional neural networks for hyperspectral image classification. Mohanraj et al. present a neural network-based approach for face recognition. A method on auto-mated vision inspection system for cylindrical metallic components is proposed by

Govindaraj and co-authors in their research work. The volume also includes a research work on gene selection of microarray datasets. A case study on geo-statistical modeling of remote sensing data for forest carbon estimation is presented by Kumar et al. Finally, it includes a study of DC–DC converters with MPPT for stand-alone solar water pumping.

IC3 2018 represents a global forum for research on computational approaches to learning. It includes mostly the current works and research findings from various research laboratories, universities, and institutions and may lead to the development of market-demanded products. The works report substantive results on a wide range of learning methods applied to a variety of learning problems. It provides solid support via empirical studies, theoretical analysis, or comparison to psychological phenomena. The volume includes works to show how to apply learning methods to solve important application problems as well as how machine learning research is conducted.

The volume editors are very thankful to all the authors, contributors, reviewers, and the publisher for making this effort a successful one.

Colorado Springs, USA   Jugal Kalita
Arad, Romania   Valentina Emilia Balas
Sikkim, India   Samarjeet Borah
Sikkim, India   Ratika Pradhan

Contents

About the Editors

Jugal Kalita is Professor in the Department of Computer Science at the University of Colorado, Colorado Springs, CO, USA. He completed MS and Ph.D. from the University of Pennsylvania, USA; M.Sc. from the University of Saskatchewan, Canada; and B.Tech. from the Indian Institute of Technology Kharagpur, India. His areas of research are artificial intelligence, bioinformatics, information retrieval, and computer security. He has over 185 journal/conference papers, 2 technical reports, 5 edited volumes, and 4 books to his credit. He is Editor-in-Chief of Posoowa, non-resident Assamese monthly magazine and edited 75+ issues. He is also Editor of Asomi, yearly magazine of Assam Society of America. He is also involved in several non-profit organizations, mostly geared toward the southeastern Himalayan foothills region.

Valentina Emilia Balas is currently Full Professor in the Department of Automatics and Applied Software at the Faculty of Engineering, "Aurel Vlaicu" University of Arad, Romania. She holds a Ph.D. in applied electronics and telecommunications from Polytechnic University of Timisoara. She is the author of more than 250 research papers in refereed journals and international conferences. Her areas of research interests are intelligent systems, fuzzy control, soft computing, smart sensors, information fusion, modeling, and simulation. She is Editor-in-Chief of *International Journal of Advanced Intelligence Paradigms* (IJAIP) and *International Journal of Computational Systems Engineering* (IJCSysE), Editorial board member of several national and international journals, and an evaluator expert for national and international projects. She is Director of Intelligent Systems Research Centre in Aurel Vlaicu University of Arad. She served as General Chair of the International Workshop Soft Computing and Applications (SOFA) in seven editions 2005–2016 held in Romania and Hungary. She participated in many international conferences as Organizer, Honorary Chair, Session Chair and Member in Steering, Advisory or International Program Committees. She is Member of EUSFLAT and SIAM and Senior Member of IEEE, Member in TC—Fuzzy Systems (IEEE CIS), Member in TC—Emergent Technologies (IEEE CIS), and Member in TC—Soft Computing (IEEE SMCS). She was Vice-President

(Awards) of IFSA International Fuzzy Systems Association Council (2013–2015) and is Joint Secretary of the Governing Council of Forum for Interdisciplinary Mathematics (FIM)—A Multidisciplinary Academic Body, India.

Samarjeet Borah is currently working as Professor in the Department of Computer Applications, Sikkim Manipal University (SMU), Sikkim, India. Dr. Borah handles various academics, research and administrative activities. He is also involved in curriculum development activities, board of studies, doctoral research committee, IT infrastructure management etc. along with various administrative activities under SMU. Dr. Borah is involved with three funded projects in the capacity of Principal Investigator/Co-principal Investigator. The projects are sponsored by, AICTE (Govt. of India), DST-CSRI (Govt. of India) and Dr. TMA Pai Endowment Fund. Out of which one is completed and two are undergoing. He is associated with ACM (CSTA), IAENG and IACSIT. Dr. Borah organized various national and international conferences in SMU. Some of these events include ISRO Sponsored Training Programme on Remote Sensing & GIS, NCWBCB 2014, NER-WNLP 2014, IC3-2016, ICACCP 2017, IC3-2018 etc. Dr. Borah is involved with various journals of repute in the capacity of Editor/Guest Editor such as IJSE, IJHISI, IJGHPC, IJIM, IJVCSN, JISys, IJIPT, IJDS etc.

Dr. Ratika Pradhan is working as Professor in the Department of Computer Science and Engineering, Sikkim Manipal Institute of Technology since July 1999. She received Ph.D. from Sikkim Manipal University (SMU) in 2011 and M.E. (CSE) from Jadavpur University, Jadavpur, in 2004. Her areas of research interest are digital image processing, remote sensing, and GIS. She has published 25 journal papers, and 5 conference papers.

IoT Enabled Ubiquitous Healthcare Data Acquisition and Monitoring System for Personal and Medical Usage Powered by Cloud Application: An Architectural Overview

Vidhaydhar J. Aski, Shubham Sanjay Sonawane and Ujjwal Soni

Abstract Modern lifestyle, swift adoption of fast-food diet and various environmental changes causing chronic life-threatening diseases stimulates a real necessity of advancements in built-in Internet technology for remote healthcare. Proposed work discourses an IoT paradigm comprising of Wireless Health Sensors (WHS) that allows us to observe essentially continuous monitorable and/or should be monitored biometric parameters like pulse rate, pulmonary functional quality, blood pressure, body temperature, electro cardio activity, etc., which in turn helps us for self-evaluation and control future severity by predictive analysis via smart healthcare systems in a long run diagnostic procedure by a medical practitioner. The work addresses the development of an Arduino-based all-in-one cost-effective, miniaturized Wireless Intelligent Embedded Healthcare Device (WIEHD) that can provide home-based health services to the patient. The sensors are connected to Arduino to track the status which is interfaced to a display as well as Wi-Fi connection in order to transmit collected data and receive user requests. The parameters can be monitored from a smartphone application. This system is based on a cloud platform and keeps track of the device data on daily basis. This data is shared with doctors through a website where the doctor can analyze the condition of the patient and provide further details online and intimate patient about future severity well in time.

Keywords WHS (wireless health sensors) · Arduino
WIEHD (wireless intelligent embedded healthcare device) · NodeMCU
IoT · Cloud · Database · Wi-Fi

V. J. Aski (✉) · S. S. Sonawane · U. Soni
Manipal University Jaipur, Jaipur, Rajasthan, India
e-mail: vidyadharjinnappa.aski@jaipur.manipal.edu

S. S. Sonawane
e-mail: shubhamsanjaysonawane@muj.manipal.edu

U. Soni
e-mail: ujjwalsoni@muj.manipal.edu

© Springer Nature Singapore Pte Ltd. 2019
J. Kalita et al. (eds.), *Recent Developments in Machine Learning and Data Analytics*,
Advances in Intelligent Systems and Computing 740,
https://doi.org/10.1007/978-981-13-1280-9_1

1 Introduction

Domestic and self-remote healthcare plays a vital role when it comes to the frequent ups and downs in the health of elderly and patients having the chronic illness which needs recurrent health examination. The application spectrum shall be extended to a wide variety of people having numerous medical backgrounds like differently abled people, diabetic, asthmatic, cancerous people, etc. need their daily health status for further precautions. This opens up huge opportunities for advancements in healthcare and medical devices. The system we propose is purely meant for home sitting applications but regardless of that, the same technology can be used in medical industry as well. Although similar technologies have emerged, there is still space for better, open-source, and low-cost solution.

The system is designed in such a way that it is capable of measuring various human body factors. Various sensors are used to determine the health structure of the patient. Basic sensors like DS18B20 [1] digital temperature sensor, MLX90614 [2], noncontact infrared temperature sensor, pulse rate sensor, ECG module, spirometry sensor, and more can be used to collect primary body factors. Complex sensors like those used in body composition meters can also become part of the system to provide specific data like fat, water, bone, and muscle percentage. More accurate sensors can even determine weight, body age, BMI, and resting metabolism. These sensors are to be connected to the Arduino Uno or nanomicrocontroller. Arduino is to be programmed in such a way that it is not to be limited to the number of sensors. The Arduino will mainly act as a sensor manager and help the system to collect the data for further processing. The collected data is further sent to NodeMCU [3]. NodeMCU is another microcontroller with inbuilt ESP8266 [3] Wi-Fi module and MCU-based Arduino architecture. NodeMCU will receive the data from Arduino via serial communication established between them. NodeMCU will act in master mode and will then be used to command Arduino for the sensor data that has been requested by the user. The input to the system or NodeMCU will be over intranet where MCU will act as a server to provide a response to the user requests.

This system is based on a cloud platform and hence the data is to be uploaded on to the cloud. When performing this task, MCU goes to client mode and requests the cloud for upload. The system keeps track of the daily data collected in the past. This data is then shared with doctors through a website where the doctor can analyze the condition of the patient and provide further details online. Analytical platforms like Watson and Matlab can be used to perform data analysis on to the information collected by the system over the cloud. Patients and their respective doctor can communicate with one another on this platform and can go back and view the past data any time they desire. If the system detects any abrupt abnormalities in patient's any of the health condition like heart rate or body temperature, the system automatically notifies the user about the patient's status over their Android phones or web portal, and also shows details of the patient in real time over the Internet. The system uses ubiquitous data accessing method in IoT-based information system [3] for management of huge data collected on a daily basis. This will make the system more efficient with big data management allowing the user to share information with medical experts on a large scale.

Patient's health parameter data stored in the cloud makes it even more beneficial than maintaining the records in physical format or digital memory storage devices. There are chances that these devices can get corrupt and data might be lost, whereas the cloud storage is more reliable and does have minimal chances of data loss. Thus, this cloud and IoT-based health tracking system effectively uses the Internet to monitor patient's health status and save lives on time.

2 Methodology

Following methods are used in creating this overall research plan.

2.1 Sensors

The system will consist of mainly three health monitoring sensors: temperature sensor, pulse rate, and ECG sensor. But the design is to be made in such a way that the system will not be limited to few sensors. Depending on the patient's need, the system will provide options for the body status they are looking for. Other important factors that can be added to the system are the body composition sensors which will provide extensive insight of body and visceral fat, skeletal muscle level, BMI, resting metabolism, and more.

Sensors like DS18B20 [4] digital temperature sensor, SEN-11574 pulse rate sensor, AD8232 heart ECG monitoring sensor, and spirometry sensor are to be used in the system. DS18B20 contains a unique 64-bit code stored in ROM [4]. The least significant 8 bits of the ROM code contains the DS18B20's 1-Wire family code: 28 h. The next 48 bits contain a unique serial number. The most significant 8 bits contain a cyclic redundancy check (CRC) byte that is calculated from the first 56 bits of the ROM code [4].

SEN-11574 is easy to use pulse rate sensor [5]. The power of SEN-11574 ranges from 3v to 5v [5]. It consists of optical heartbeat sensor circuit, an amplification circuit, and noise cancellation circuit [5] all embedded on the single chip.

An analog sensor requires signal conditioning circuit to interface with MCU for better precision, likewise for ECG sensor AD2832 acts as an internal signal conditioning circuit and many other potential bio-parameters [6]. This AD8232 can also act as high-pass filter that removes low-frequency motion objects. This filter has got large gain as it is tightly coupled with amplifier.

2.2 Microcontroller

The sensors will collect the analog and digital input for the microcontroller. The system will use Arduino nano or Uno for processing the data collected by the sensors. Arduino will be programmed in such a way that it should be capable of taking as many body factors as possible into consideration. This will make the system not limited to few sensors. Arduino will mainly be programmed for handling various sensors. As there is no limitation to the sensors we are using, one microcontroller is fully assigned to perform the hand over the task of data collection.

The data will then be sent to NodeMCU, another microcontroller with Wi-Fi connectivity for further processing via hardware serial communication established between both controllers as shown in Fig. 1. NodeMCU has a built-in ESP8266 module. The module supports standard IEEE 802.11 agreement and complete TCP/IP protocol stack [3]. The module can be added to an existing networking device or built-in a separate network controller as NodeMCU [3]. ESP8266 is high integration wireless SOCs, designed for space- and power-constrained mobile platform designers. It provides unsurpassed ability to embed Wi-Fi capabilities within other systems or to function as a standalone application, with the lowest cost, and minimal space requirement [3]. NodeMCU will be our primary microcontroller. It will communicate with Arduino to fetch the data over a hardware serial communication established between them. Arduino will be in slave mode, while NodeMCU will be in master mode. NodeMCU will also be interfaced with a display which will provide output in real time.

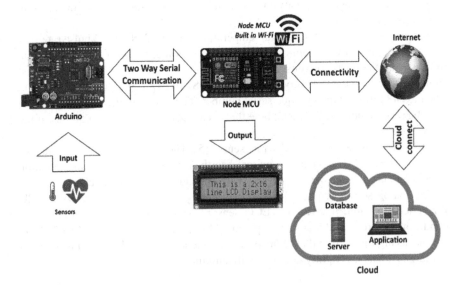

Fig. 1 System architecture

NodeMCU is a low-power device capable of operating on 3.3v input, while Arduino takes up to 5v of power. This makes the system more power efficient.

2.3 Internet Connectivity

Internet connectivity plays an important role when it comes to IoT platforms. We will use the ESP8266 Wi-Fi module inbuilt in our NodeMCU. It will deal with all the Internet requests sent by the client. The node will be connected to the home Wi-Fi network. Since the system can be controlled by Android or Web application, the request has to be sent to the system. Same goes for requests that system sends to cloud as in Fig. 1.

When receiving requests, NodeMCU will go into a server mode, providing a response to the requests. The response is nothing but an acknowledgement for the user to use requested sensor and storing collected information in the cloud. NodeMCU will command Arduino for the provision of the data that is to be uploaded to the cloud. When uploading data to the cloud, NodeMCU will be in client mode and request the cloud server to upload collected data to its database. Most of the request from a user to the system will be over the intranet, and hence connectivity will only be limited to local area network when the node is in server mode. As for client mode, MCU will switch over to the Internet to access the cloud.

2.4 Cloud

Information collected by the system will be stored in a database located on cloud virtual machine. The database will contain all the biodata of the patient. It will also have information of the doctors. The web application and the online server that drives the system will also be part of the cloud virtual machine as in Fig. 1. Cloud platforms like IBM Bluemix [7] can be ideal for this task. IBM Bluemix is open cloud platform that offers mobile and web developer access to IBM and other software for integration, security, transactions, and other key functions [7]. The IBM IoT service lets our apps communicate with and retrieve data collected by the system [7]. Bluemix's recipes make it easy to get devices connected to the Internet of Things cloud [7]. The web and Android application can then use real-time and REST APIs to communicate with devices and use the data that sensors had been collecting [7].

2.5 Application Development

Depending on the requirement, an Android or web application is to be developed which shall contain both patients and doctors portal. The patient can select their respective doctor(s) so that they can share their daily health status. The website will contain all the past information of the patient, so that doctor or patient can check and analyze the data any time they want.

Various tools like IBM Watson or Matlab can be used in the portal to provide analysis and visualization of the data that is collected by our system time to time. This will provide better user experience and will also help doctors determine current health condition of the patient.

3 Cloud Platform and Database Management

With the usage of the medical devices being high, collecting a huge amount of sensor data from n number of users and a load of data on the cloud platform demands technologies depending upon big data, hence resulting in a requirement for data management on multitenant basis rather than conventional distributed data acquiring modules.

Multilayer data storing architecture module is used for increased efficiency directly in terms of data fetching and data storing speed. The bottom data layer is a cause of tenant data storage layer, which allows us to store collective tenant databases. Next layer is used as controlled data access layer. This constitutes different mechanisms for controlling resources in order to form effective distributed healthcare model. Among all the layers, top layer being the business layer. Data interpretations, controlling operations, and business operations sharing resources between various data formats will occur in the business layer that coordinates shared data as well as interpreted data. Two separate data interoperable protocols are used for creating isolated nature between shared data and interpreted data. In healthcare applications, patient data is used as an extremely big dataset in hospitals and MQTT protocols provide relatively good encryption to datasets for transporting data in remote places. Because various hospitals provide different cloud infrastructures for patient's database, it is always better to have a protocol that deals with extracting data from different databases by maintaining isolation between each infrastructure [8].

4 Implementation

The sensors which are to be used must be connected to the Arduino Uno depending on the analog and digital pins. Temperature sensor is like DS18B20 digital temperature sensor. The DQ pin of like DS18B20 is connected to any of the Arduino digital pins. It is also connected to 4.5k register which is then interfaced with VCC of the sensor input and 5v Arduino output as shown in Fig. 2. The GND of the sensor goes to the GND of Arduino. Similarly, analog output of the SEN-11574 pulse rate sensor goes to any one of the Arduino analog pins with an input voltage of 5v.

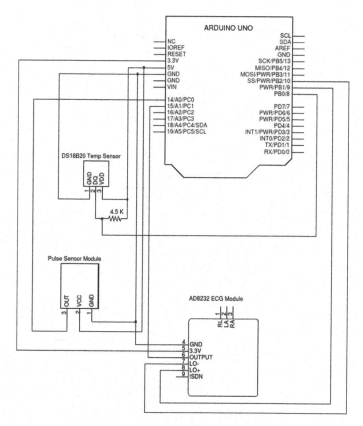

Fig. 2 Sensor interfaced circuit diagram

AD8232 [9] bioelectrical signal acquisition development module and the sensor are most important of all as it will be used for electrocardiography, i.e., ECG visualization. It has got the efficient structure to excerpt, strengthening signal by means of amplification and removal of less biopotential signals. In the case of noisy signal condition occurrence like those designed by distant electrode placement [9], this also enables to have extremely low-power ADC embedded in the microcontroller to obtain output signal power more easily [9]. AD8232 is low-power 3.3v device. It will get its input from Arduino 3.3v pin. An analog output of AD8232 will be connected to an analog pin of Arduino, whereas LO− and LO+ will be going to digital pins of Arduino as shown in Fig. 2 (Fig. 3).

The 6 analog pins and 13 digital pins of Arduino Uno being more than enough for above implementation, many more sensor in future can be added to the system using 4051 Multiplexer. The 4051 is an 8-channel analog multiplexer/DE multiplexer with one common and 8 I/O pins. The common pin is connected to any of the analog pins, whereas sensors are connected to 8 channels. 4051 also have three digital pins—A, B, and C; these pins together will determine which channel is to be used by providing

Fig. 3 Arduino sensor
manager flowchart

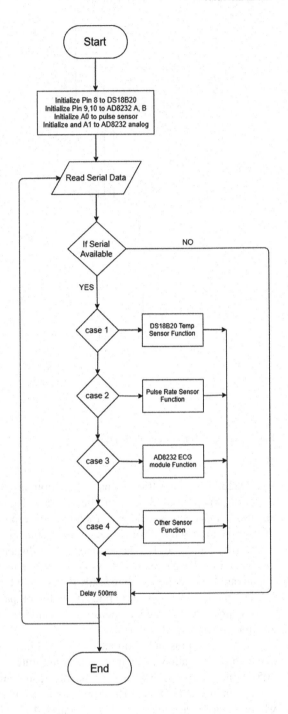

Fig. 4 4051 multiplexer
with Arduino Uno

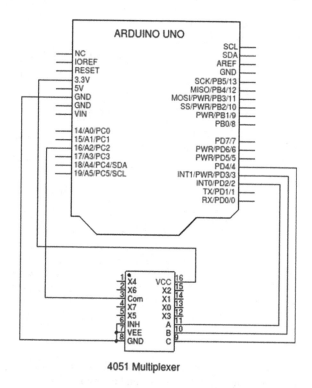

4051 Multiplexer

three-digit binary number from 0 to 7 as of respective channel pin of 4051. Pins A, B, and C will be connected to digital pins of Arduino as shown in Fig. 4.

The Arduino is internally programmed to handle all the sensors at the same time. A programmable switch is written in such a way that depending upon what is the request for sensor available on the hardware serial, nothing but a request from NodeMCU in form of an integer. This commands Arduino to call for a function specific to the requested sensor. Each function of sensors will have its own program depending upon its characteristics. The flowchart of the sensor is shown in Fig. 3.

This is only to assure that the user has a home or personal Wi-Fi and is supposed to provide Wi-Fi name and password to the system. Once the system is provided with the details, it will then store this info on to its memory and start accessing the Internet. Once the connection is established, the users are free to use the system to its full potential. This task only has to be performed once. The Wi-Fi details are to store in the memory of NodeMCU. NodeMCU has the memory of 128 KB and flash storage of around 4 MB which is more than enough for this task. This memory can also be used to temporarily store the data fetched when the system is in offline mode or has experienced any problems with the connectivity (Fig. 5).

Once an Internet connection is available, the system will upload the data on to the cloud. The flowchart of NodeMCU is shown in Fig. 6.

Fig. 5 Arduino Uno and
NodeMCU interfacing

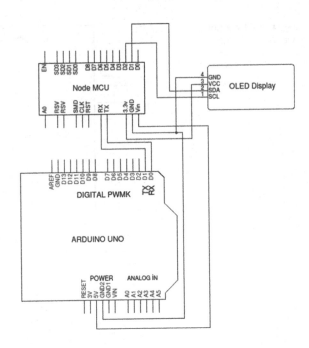

An application programming interface will be designed in order to provide information to the available doctors and hospital. Since we are using three-layer database management architecture [8], both shared and isolated databases have to be managed. The shared database will contain all the patient's daily health and body status. Doctor- and hospital-related information also are to be stored in the shared database. Patients and doctors personal information will be stored in an isolated database. The web application will be developed using JSP technology and for storage purpose, we will be using MongoDB document-oriented database which is ideal for our three-layer architecture.

The analytical tool is a vital feature of the system that can be used for predictive measures. Specifically, IBM Watson analytics is used as an analytical tool. This is an essential cloud service platform which enables us to create various regressive models and design supervised training techniques for the purpose of data analytics [9]. Watson analytical tool allows working with extremely simple and common milieu schemes such as .csv, .xsl, .xslx, etc. As a use case of this platform, it provides an environment and asserts an offset value of data quality automatically after uploading datasets. Along with this, many other factors are also taken into consideration such as it requires less number of records, filed completeness, and various other qualitative parameters. We created a data acquisition system for collecting biomedical parameters such as ECG, body temperature, pulse, and ECG, among which ECG data is very crucial and is extremely necessary to feed Watson with collected ECG data on daily basis for prediction of abnormalities in the functioning of heart.

Fig. 6 NodeMCU request
handle flowchart

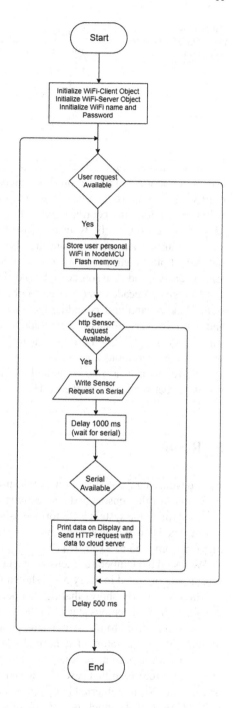

Table 1 Results obtained by using DS18b20 digital temperature sensor

Time (s)	Sensor result (°C)
0	27.4
5	28.8
10	29.5
15	29.8
20	30.6

The electrocardiogram is a graphical representation of activities that occur after every contraction and relaxation. This can be detailed and observed using electrodes of ECG sensors, and data processing unit (microcontroller) will take care of mapping voltage variations in accordance with the various records observed through ECG sensors placed on the body in response to the heart function. Basically, cardiac activity can be measured through the observations of voltage differences in some predefined placement positions of electrodes on the skin [9]. ECG is a very basic, less expensive, and vital source of diagnostic information in abnormalities of cardiac function since decades and is used to observe electrical activity of cardiovascular stem. Reading and understanding ECG signal was a real challenge for an engineer and can refer various medical journals to study on ECG waves. Varieties of ECG variations cause and intuit different health statuses of the heart. It acquires signature waves whose variations may indicate different problems of the heart. These changes are referred to as heart rate vulnerability (HRV); this generally refers to time length variations between two heartbeats. HRV dataset [9] can be used in the Watson.

5 Result

After developing the system using above implementation, we can test the system for the outputs of the sensors and connectivity.

A test run of the system is performed using DS18B20 one wire digital temperature sensor. The DS18B20 is very precise sensors even capable of measuring temperature ranging from −55 to +125 °C [1].

We tested the temperature sensor on a person with normal body temperature. The sensor data obtained in every 5 s is shown in Table 1.

The data clearly shows that the temperature gradually increases till the certain point and then it becomes stable. In this case, the data is stable at 31.9 °C. Since the sensor is connected to the palm of the patient, the temperature shown in Table 1 is normal palm temperature. The normal skin temperature of the person is said to be 32–33 °C, while normal external body temperature is 34 °C, i.e., 2 °C more than that of skin. Therefore, in this case, the normal body temperature is 31.9 + 2 = 33.9 °C or 93.02 °F. Normal internal body temperature most commonly referred by doctors is 37 °C or 100 °F which is 3 °C more than that of internal body temperature in this case, that is, 33.9 + 3 = 36.9 °C or 98.42 °F, which is close enough to normal

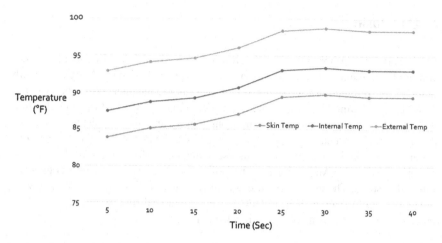

Fig. 7 Graphical presentation of data obtained from DS18B20 temperature sensor in degree Fahrenheit

body temperature. The output provided by the system will be in degree Fahrenheit. Internal body temperature will be the factor taken into the consideration but external and skin temperature will also be used for better analysis as shown in Fig. 7.

It can be observed that the sensor provides the required data accurately after at least 30 s. Hence, Arduino is programed for a delay of 30 s after the sensor is requested by the user, to provide more precise body temperature. The resolution of DS18B20 is also set to max for better results. Similarly, results are to be found in data obtained from pulse sensor and ECG sensor.

6 Future Scope

Currently, the system is developed by keeping domestic applications in mind but it can be expanded to the medical industry as well. The system can be integrated into the medical devices in hospitals and data can be fetched from there and then to the cloud. This will allow us to perform analytics on better scale to get more accurate and realistic results. Same results can also be shared with the hospital database using given three-layer model. IBM Watson and its IoT features can be utilized to full extent as it develops itself, making our Internet of Things, Internet that thinks. As per domestic applications of the system are concerned, modern day home automation features can be added to our system as well. The system can be configured with popular AI like Google Home's Google Assistant [10] and Amazon Echo's Alexa [11] to provide more immersive and living experience. As for sensor, with growing sensor technology and fields of application of sensors and sensor networks [12], more medical sensors like blood pressure, insulin and sugar monitor, RBC, WBC, and platelet monitors can also become part of the system as well.

7 Conclusion

In this paper, we proposed an architecture of a healthcare system capable of performing basic medical operations for analysis and visualization of body and health factors. The architecture was designed with two microcontrollers, Arduino Uno and NodeMCU. Arduino Uno was set up to manage sensor network of the model, while NodeMCU was responsible for Internet connectivity and user request handling. We also developed flowcharts of the programs that are to be installed in these microcontrollers. We studied the three-layer database management for ubiquitous data access. Visualization of the Android and web application was also taken into consideration in this study. The cloud structure and various analytical tools like Watson were understood. Observation of results of the sensors in real time was done for better understanding of the data collection mechanism.

Acknowledgements *Ethical approval and informed consent* All procedures performed in studies involving human participants were in accordance with the ethical standards of the institutional and/or national research committee and with the 1964 Helsinki declaration and its later amendments or comparable ethical standards. This article does not contain any studies with animals performed by any of the authors, and informed consent was obtained from all individual participants included in the study.

References

1. Nie, S., Cheng, Y., Dai, Y.: Characteristic analysis of DS18B20 temperature sensor in the high-voltage transmission lines' dynamic capacity increase. Energy Power Eng. **5**, 557–560 (2013)
2. Vempali, M.D., Tanveer Alam, K., Punyasheshudu, D., Ramamurthy, B.: Design and development of digital proximity type IR thermometer based on Arduino unoR3. Int. J. Innovative Res. Sci. Eng. Technol. **5**(2) (2016)
3. Oh, T.-G., Yim, C.-H., Kim, G.-S.: Esp8266 Wi-Fi module for monitoring system application. Global J. Eng. Sci. Res. (2017)
4. DS18B20 Programmable Resolution 1-Wire Digital Thermometer, Data Sheet, Maxim Integrated Products, Inc. datasheets.maximintegrated.com
5. SEN-11574 pulse sensor introduction. Heart Beat Pulse Sensor Interfacing with Pic Microcontroller. http://www.instructables.com/id/Simple-DIY-Pulse-Sensor/. Viewed in January 2018
6. Single-Lead.: Heart Rate Monitor Front End, Data Sheet AD8232. www.analog.com/AD8232
7. Stifani, R.: IBM Bluemix the cloud platform for creating and delivering applications. Int. Tech. Support Organ. (2015)
8. Xu, B., Da Xu, L., Senior Member IEEE, Cai, H., Xie, C., Hu, J., Bu, F.: Ubiquitous data accessing method in IoT-based information system for emergency medical services. IEEE Trans. Ind. Inf. **10**(2) (2014)
9. Guidi, G., Miniati, R., Mazzola, M., Iadanza, E.: Analytics-as-a-Service System for Heart Failure Early Detection. Department of Information Engineering, Unversità degli Studi di Firenze, v. S. Marta, 3-50139 Firenze, Italy (July 2016)

10. IoT.: Home Automation with Android Things and the Google Assistant. http://nilhcem.com/a ndroid-things/google-assistant-smart-home. 12 June 2017

11. Kaundinya, A.S., Atreyas, N.S.P., Srinivas, S., Kehav, V., Naveen Kumar, M.R.: Voice enabled home automation using amazon echo. Int. Res. J. Eng. Technol. (IRJET) **04**(08) (2017)

12. Smart Sensor Networks: Technologies and Applications for Green Growth. Organisation for Economic Co-Operation and Development (December 2009)

Sensor Selection and Data Fusion Approach for IoT Applications

Ishita Chakraborty, Anannya Chakraborty and Prodipto Das

Abstract The wireless sensors which are used in IoT applications are battery powered, which imposes extreme energy constraints on their operations. Therefore, a fundamental challenge lies in designing a sensor network which can maximize its lifetime. To address this issue in WSNs, data aggregation techniques have come up as a novel approach that minimizes the number of transmissions in the network, which in turn can optimize the network and increase its lifetime. In this paper, we have discussed a simple hierarchical data aggregation method for energy optimized data transmission in wireless sensor networks.

Keywords Wireless sensor network · Data fusion · IoT · Sensor selection
Path optimization

1 Introduction

In the current years of development of wireless technology and micro-electromechanical systems, development of small, cost-efficient sensors with the ability of sensing and wireless communication is paving a huge path in the domain of wireless communication. This amalgam of wireless technology with sensors has given rise to the wireless sensor network communication. WSNs are deployed in various domains like military surveillance, environmental monitoring, civil applications, etc. Wireless sensor networks have been heavily used for acquiring environmental data such as humidity, temperature, sound, pollution, and so on which are later used

I. Chakraborty (✉) · A. Chakraborty
Department of Computer Science and Engineering, Royal Global University,
Guwahati, India
e-mail: ishitachakraborty18@gmail.com

A. Chakraborty
e-mail: 155anannya.2@gmail.com

P. Das
Department of Computer Science, Assam University, Silchar, India

© Springer Nature Singapore Pte Ltd. 2019
J. Kalita et al. (eds.), *Recent Developments in Machine Learning and Data Analytics*,
Advances in Intelligent Systems and Computing 740,
https://doi.org/10.1007/978-981-13-1280-9_2

I. Chakraborty et al.

Fig. 1 Sensor network
architecture [1]

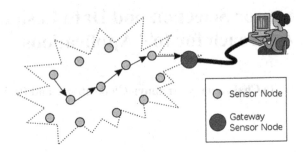

for statistical analysis by various other fields of research. Along with the advancement in wireless networking, a new paradigm of information is paving its way, that is, none other than the Internet of Things (IoT). A huge number of devices are connected by IoT and is expected to cross a large number not less than 50 billion.

Wireless sensor networks are ad hoc in nature, in the sense that they can be formed spontaneously and they rely entirely on wireless communication for sensing data from the environment that can be transmitted wirelessly among various nodes. An interesting notion of the sensor networks is that they are sometimes also referred to as "dust network" which comes from the relevance of the small sized sensors [1]. Smart dust is a UC Berkley project and the goal of which is to build a self-contained, millimeter-scale sensing platform for large distributed sensor networks [2]. This project is sponsored by DARPA (Defence Advanced Research Project Agency). Wireless sensor networks are spatially distributed networks made of "nodes" which may count from a few to hundreds and not less than thousands. The sensor nodes are also referred to as motes. Each mote in a sensor network is none other than a computer, that is, processing collected data and transmitting it to other motes. These motes must have the characteristics of being programmable, self-organizing in case of an ad hoc network as well as self-repairing whenever a network failure occurs [3]. One major problem of sensor nodes is that they are battery operated; hence, they bear a limited lifetime. As the design of a sensor node is made such that it can be deployed and reused many times and also be made cost-efficient, these are some of the limitations of a WSN which must be addressed while designing a wireless sensor network. Figure 1 shows the basic skeleton of a sensor network.

The architecture of a WSN specifically requires both hardware and software platforms. Operating systems for wireless sensor networks are generally less complex in nature. TinyOS is one such example of an operating system used in WSNs. It is based on event-driven programming model and not a multithread model [4]. Application areas of wireless sensor network include a lot of areas such as habitat monitoring, fire detection, traffic control, etc.

1.1 IoT Application of WSN

The term IoT was coined by the US National Intelligence Council (NIC) in 2008. The increasing number of IoT applications is to reproduce useful information that is gathered by the environment wherein WSNs play a major part [5].

In an IoT environment, the amount of data that is produced in a very short time span is huge. And to make this data accurate and useful is a major task. Elimination of redundant information from the huge mass of data acquired and extracting only the useful ones is a critical task. However, this can be achieved by the process of data fusion where large volumes of data are combined to get more accurate and potentially viable data, thus optimizing the network resources [5]. Multi-sensor data fusion approach performs fusion of data gathered from multiple sensors and producing globally acceptable information which cannot be produced by a single source [5].

1.2 Data Fusion in Wireless Sensor Network

Wireless sensor network is an autonomous network which is primarily made up of a number of sensor nodes which communicate among themselves wirelessly for transferring sensed data. But it is already known that sensor nodes are of limited battery power, so the overall network has limited energy. The fundamental issue in wireless sensor network is to process the collected data in an energy efficient manner, thus increasing the QoS of the network [6].

Since the sensor nodes are deployed over large geographical area, and if each and every sensor starts transmitting their data to the sink individually, then it would take a lot of time and energy and also generate inconsistent result. It may so happen that the sensor mote dies halfway while transmitting as they are battery operated and have limited energy [7]. This is another issue of WSNs which has to be looked upon. One solution could be collectively sending the data packets to the destination node by fusing them through selective intermediate nodes.

Data fusion is the process of systematically collecting sensed data from multiple sensor nodes, processing them at an intermediate node, and again retransmitting the processed information to the destination or sink node. It is the process by which information from multiple sources are fused to get an aggregated information. An intermediate fusion node in the network also ensures that less amount of time is taken to transfer the sensed data to the sink with optimal utilization of network resources. Also, overall energy consumption of the network is reduced as only one node will participate in data transfer, while the rest nodes may be in sleep or inactive mode. There has been a lot of development proceeding in the field of data aggregation techniques. Cluster-based and tree-based data aggregation techniques are used very commonly. In [8], Krishnamchari et al. discussed energy saving and the delay trade-offs that are involved in data aggregation. In [9], Liao et al. set forth the concept of ant colony optimization algorithm by performing data aggregation in WSNs. Few other techniques of data aggregation include directed diffusion, in-network aggregation, fuzzy-based fusion, etc. [10–13].

2 Methodology

In this section, we are going to discuss the data fusion process which has been performed by us. In our work, we have used the discrete-event simulator NS2 (version 2.35) and the network performances have been evaluated.

2.1 Data Fusion

We have performed data fusion in several topologies. A simple topology with a single sink for fusion is shown in Fig. 2, and simulation settings are as shown in Table 1.

There are three transmitting nodes N0, N1, and N2, one intermediate node N3, and a sink node N4 which receives all the data packets that have been sent by the transmitting nodes. We assume that nodes N0, N1, and N2 send their fixed size data packets of 512 bytes at the rate of 600 Kbps. N0 sends its data packets for duration of 5 s. After N0 has completed sending its packets, N1 starts transmitting its data packets for another 5 s and similarly N2 sends its data packet for 5 more seconds. After each of the transmitting nodes has completed its allotted time for sending their respective

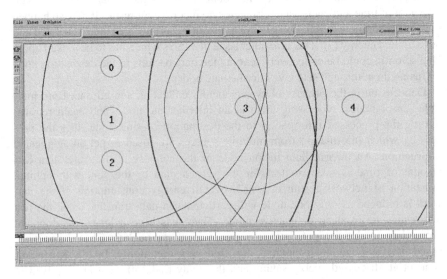

Fig. 2 Single sink scenario

Table 1 General simulation settings

Network dimension	500×500
Simulation duration	16 s
Number of nodes	5
Traffic type	CBR
Maximum queue length	200
Routing protocol	AODV
Channel propagation model	Two ray ground
Initial energy (J)	90

data packets, N3 which is the intermediate node performs the fusion. N3 receives the total number of packets sent by N0, N1, and N2, respectively. N3 now performs arithmetic fusion. Here, first, we have calculated the total number of data packets that can be sent by a node for the given amount of time it runs. For example, in our scenario, each transmitting node runs for 5 s and the packet size and transmitting rate is kept same, i.e., 512 bytes at the rate of 600 Kbps. Consider N0 to be transmitting its data packets for the first 5 s. For the given packet characteristics, in 1 s, the maximum amount of data packets that can be transmitted is 1172 (approx.). Therefore, for 5 s, the maximum number of packets that can be sent will be 5 * 1172 numbers of packets, which turns out to be 5860 number of packets. Now, the maximum number of packets that can be sent by each transmitting node has been calculated in a similar way, as shown in Eq. 1.

$$\text{Max. Packets received} = \text{Simulation Time} * \text{Bytes sent in 1 s} \qquad (1)$$

Hence, the maximum packets that can be received by N3 have been limited to the sum of maximum number of packets which are sent by N0, N1, and N2 as shown in Eq. 2.

$$\text{Max. Packets received by fusion node} = \text{Max. Packet sent by node0}$$
$$+ \text{Max. Packet sent by node1}$$
$$+ \cdots + \text{Max. Packet sent by nodeM} \quad (2)$$

The sum of data packets in output files f0, f1, and f2 are calculated, respectively. These output files store the amount of data packets that has been sent by the transmitting nodes N0, N1, and N2. We also calculate the amount of data packets that have been transferred between the fusion node N3 and the sink N4. It is found that these two values are approximately same, which means that the total amount of data packets sent to the fusion node and the total amount of data packets received by the sink node is same, thus justifying that data fusion has taken place.

To perform data fusion and sensor selection, two topologies have been considered. The first topology consists of 38 nodes with a centralized base station (labeled 0) having a mesh-like structure as shown in Fig. 3. The second topology consists of 21

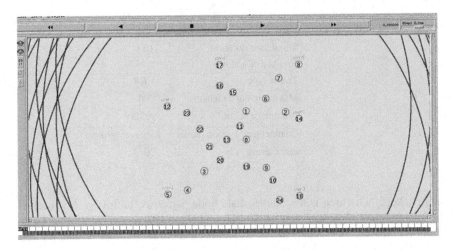

Fig. 3 Mesh topology

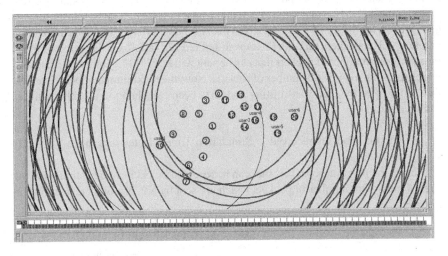

Fig. 4 Tree topology

nodes with a base station on the top (labeled 0) having a tree-like structure as shown in Fig. 4.

Out of 38 nodes in the first topology and 21 nodes in the second topology, only few nodes (active nodes) will be selected to participate in data fusion process and rest will be in sleep state. The active sensors are chosen in such a way that it can improve the overall performance of the network in terms of energy, throughput, packet delivery function, and delay. The path optimization procedure is further explained in Sect. 2.2.

2.2 Path Optimization

A routing protocol is responsible for finding a path between source and destination in a network. The intermediate or the relay nodes forward the data traffic through that path. In this routing algorithm, every user "U" tries to maintain a route which does not diverge much from the base station, thereby reducing the delay in the network. The minimum angle-based routing is mostly used in sensor networks where large numbers of node are deployed. When the user U1 tries to create a path with the base station, it initially draws a reference line to the destination node. Then, the intermediate node in its neighborhood that makes the minimum angle with the reference line is chosen as a relay. By following the same procedure for the rest of the nodes, the base station is finally reached. Hence, a wireless node is carefully chosen, based on minimum angle, from the neighbor nodes of the user to forward the data packets. As shown in Fig. 5, the packets from user U1 reach the BS through the route U1–node 3–node 6–BS [14, 15].

For selecting the sensors out of large number of sensors node in the network, an optimized path has been chosen. The optimized path and the sensor selection problem are solved by combining both minimum distance and minimum angle algorithms. We call it Minimum Distance/Minimum Angle (MD/MA) and Minimum Angle/Minimum Distance (MA/MD). In MD/MA, the first "H" hops will be taken by minimum distance and the rest of the hops will be taken by minimum angle [14]. Similarly, in MA/MD, the first "H" hops will be taken by minimum angle and the rest of the hops will be taken by minimum distance. "H" can consider values such as 1, 2, 3, 4, 5, etc. By considering different values of "H", the program has been run to select the most optimized routes. In order to optimize the network with all the users as a whole, each source–destination pair is optimized too. The optimized topologies for both 38 and 21 nodes have been shown in Figs. 6 and 7.

After the paths have been found by MA/MD and MD/MA algorithm, data fusion is performed in each topology. The topology with 38 nodes has been optimized to 25

Fig. 5 Routing by minimum angle [14]

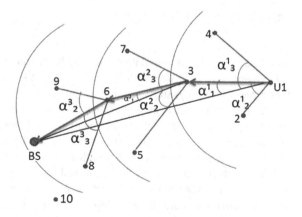

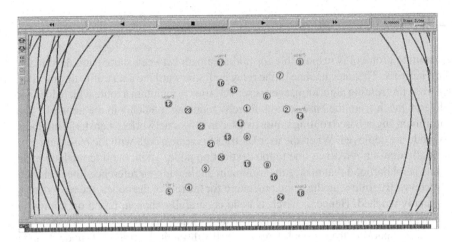

Fig. 6 Optimized path for mesh topology

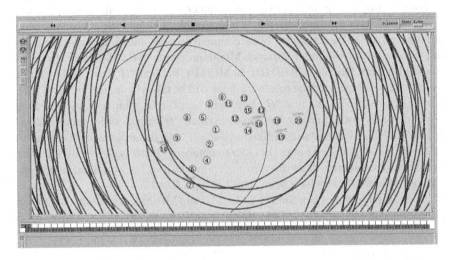

Fig. 7 Optimized path for tree topology

nodes; each user sends their data packets by fusing the data from the previous node. For example, in Fig. 8, node (N) 10 starts sending its data for 3 s. After 3 s are over, N9 (successor of N10) starts sending its data packet for 6 s. This is because it will fuse the data which has been sent to it by N10 for the last 3 s and it will also send its own data packets for another 3 s. In this manner, all the nodes in the topology will send their data packets by fusing the received packet from their preceding nodes.

For the mesh topology, MD/MA with $H = 1$ provides the optimized route and MA/MD with $H = 5$ provides the optimized route. For the tree topology, MD/MA with $H = 1$ and MA/MD with $H = 3$, respectively, provides the optimized route. The optimized routes are shown in Figs. 9, 10, and 11.

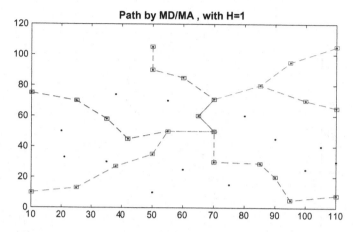

Fig. 8 Path by using MD/MA, with $H = 1$ for mesh topology

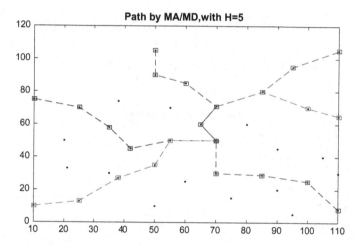

Fig. 9 Path by using MA/MD, with $H = 5$ for mesh topology [14]

3 Performance Evaluation

3.1 Performance Metrics

For performance evaluation, four network topologies have been considered which are optimized by MA/MD and MD/MA algorithm. The particulars are listed in Table 2.

In this paper, we have used NS (version 2.35) [16] and AWK Script [17] for evaluating the performance metrics. For performing data fusion, we have varied the number of nodes in our topology, whereas other network characteristics such as wireless channel, antenna gain, transmitting power, routing protocol, and ground propagation

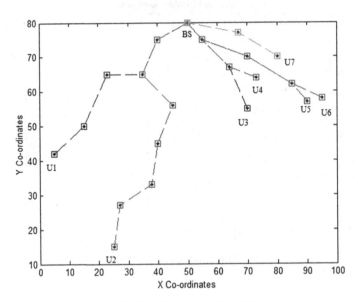

Fig. 10 Path by using MA/MD, with $H = 3$ for tree topology [14]

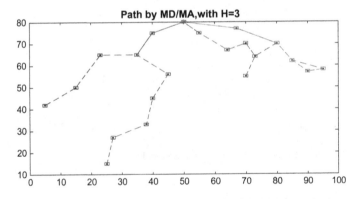

Fig. 11 Path by using MD/MA, with $H = 3$ for tree topology

model have been kept unchanged. MAC 802.15.4 has been used as wireless Ethernet for data link layer, UDP has been used for transport layer characteristics, and traffic type used for application layer is CBR.

Performance metrics are the parameters that determine the quality of a network. In this paper, we have considered the following performance metrics:

Throughput—Throughput is calculated as the total number of data packets that are transferred from one source to another in a given period of time.

$$\text{Throughput} = (\text{total packet received}/1000 * 512)/1024 \qquad (3)$$

Energy—The total transmission energy of the network is calculated as

$$\text{Energy} = (\text{number of nodes} * 10000) - \text{total energy consumed} \qquad (4)$$

Delay—Delay specifies the time it takes for a data packet to travel across the network.

$$\text{Delay} = \text{receiving time} - \text{sending time of each packet} \qquad (5)$$

Packet Delivery Ratio—Packet delivery ratio gives us the ratio of number of packets successfully received to that sent by a sensor node (Tables 3 and 4).

$$\text{Packet Delivery Ratio} = (\text{No. of packets received}/\text{No. of packets sent}) * 100 \quad (6)$$

Table 2 Network scenario particulars

Network topology	Number of nodes	Simulation area	Algorithm
Mesh topology	38	110×120	MD/MA ($H = 1$), MA/MD ($H = 5$)
Tree topology	21	110×120	MD/MA ($H = 3$), MA/MD ($H = 3$)

Table 3 Performance of mesh topology

Mesh topology	Throughput (Kbps)		Delay (ms)		Energy (J)		Packet delivery ratio (%)	
With fusion	MD/MA $T = 1$	MA/MD $T = 5$	MD/MA $T = 1$	MA/MD $T = 5$	MD/MA $T = 1$	MA/MD $T = 5$	MD/MA $T = 1$	MA/MD $T = 5$
	0.526	0.831	0.00376	0.0302	92.06	92.14	92.90	99
Without fusion	0.171	0.1445	0.00623	0.148	90.76	90.77	76.05	85.40

Table 4 Performance of tree topology

Tree topology	Throughput (Kbps)		Delay (ms)		Energy (J)		Packet delivery ratio (%)	
With fusion	MD/MA $T = 3$	MA/MD $T = 3$	MD/MA $T = 3$	MA/MD $T = 3$	MD/MA $T = 3$	MA/MD $T = 3$	MD/MA $T = 3$	MA/MD $T = 3$
	0.3039	0.3881	0.00409	0.00394	91.8	91.59	71.3	91.59
Without fusion	0.1168	.1364	0.00416	0.00417	90.6	90.66	61.2	87.7

4   Results and Discussion

From the simulation of all the four network topologies, we found that

1. The throughput is more when MA is applied. Also, the delay and energy con-
    sumption of the network are increased when MD is applied.
2. The throughput on performing data fusion for each of the topologies is found
    to be higher when MA is applied compared with the same topology without
    performing data fusion.
3. The delay and the energy consumption are more when data fusion is not per-
    formed for each of the topologies when compared with the same topology with
    data fusion being performed.
4. For both MD/MA and MA/MD, we found that the throughput of the entire net-
    work is more when data fusion is performed. Also, we can see that the delay of the
    network is less when we perform data fusion, whereas the delay is comparatively
    more when data packets are transferred without performing data fusion.
5. Since the number of packets that are sent from one node to another is more when
    we perform data fusion, the energy consumed by the network is slightly higher.
    But this can be justified by the fact that since the packet delivery ratio is very
    high when we perform data fusion, this slight increase in energy can neglected.
6. The packet delivery ratio is very high when we perform data fusion as compared
    to the network without performing data fusion.

The performance of network in terms of throughput, delay, packet delivery ratio,
and energy consumption with and without performing data fusion by applying
MD/MA and MA/MD is shown in the following graphs (Figs. 12, 13, 14, 15, 16, 17,
18 and 19).

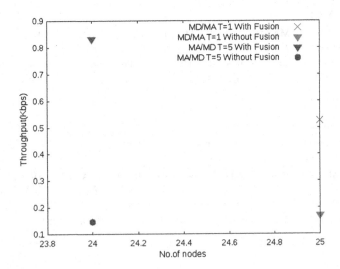

Fig. 12 Throughput of mesh topology

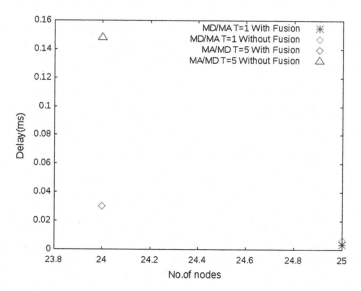

Fig. 13 Delay of mesh topology

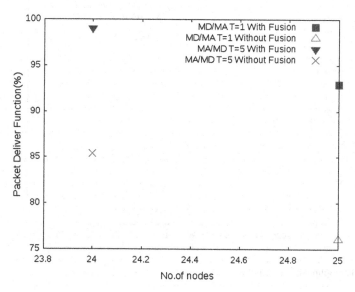

Fig. 14 Packet delivery ratio of mesh topology

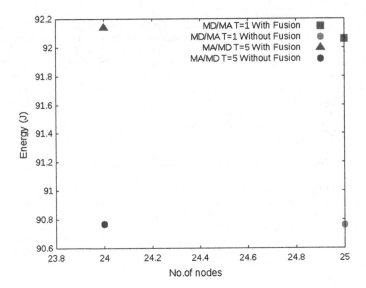

Fig. 15 Energy of mesh topology

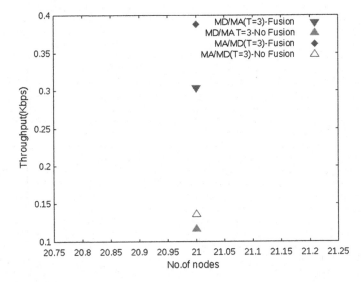

Fig. 16 Throughput of tree topology

5   Conclusion

In our paper, we have performed a simple data fusion method and we compared four important performance evaluation parameters of the network—throughput, delay, packet delivery ratio, and energy. It has been found that when data packets are

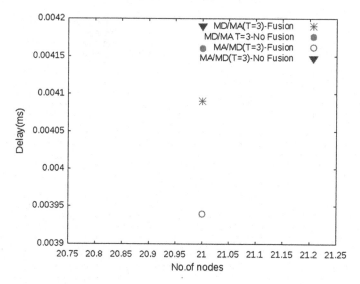

Fig. 17 Delay of tree topology

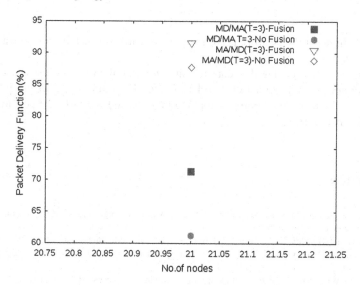

Fig. 18 Packet delivery ratio of tree topology

transmitted from source to destination by the following data fusion, the entire network throughput is more. Also, the delay in packet transmission is comparatively less when data fusion is performed. We also found that the number of packets received to the number of packets sent, also known as the packet delivery ratio, is very high when data fusion is performed. Although the energy taken for transmitting the data packets

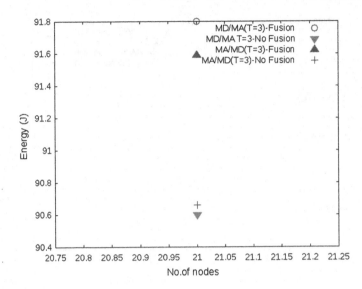

Fig. 19 Energy of tree topology

is slightly high for the network performing data fusion, this can be compensated by the high packet delivery ratio.

Thus, we can conclude that data fusion can be applied to wireless sensor networks for optimal utilization of network resources. For future work, we can perform data fusion for distributed network topology by considering different performance evaluation parameters.

References

1. Wireless Sensor Network-Wikipedia. https://en.wikipedia.org/wiki/Wireless_sensor_network/. Accessed 30 Jan 2018
2. Smart Dust, Berkley, U.C.: http://www-bsac.eecs.berkeley.edu/smartdust. Accessed 3 Feb 2018
3. Ahmed, A., Magdy, B.A.: In Resource-Aware Data fusion Algorithms for Wireless Sensor Networks, pp. 3–31. Springer, New York (2012)
4. TinyOS-Wikipedia. https://en.wikipedia.org/wiki/TinyOS. Accessed 3 Feb 2018
5. Alam, F., Mehmood, R., Katib, I., Albogami, N., Albeshri, A.: Data fusion and IoT for smart ubiquitous environments: a survey. IEEE Access 9533–9554 (2017)
6. Cheng, C.T., Leung, H.: A delay-aware network structure for wireless sensor networks with in-network data fusion. IEEE Sens. J. 1622–1630 (2013)
7. Patela, N., Vasavab, H., Jaliyab, U.: A survey on data fusion techniques and WSN simulators. IJIERE 96–102 (2015)
8. Krishnamchari, B., Estrin, D., Wicker, S.: Modeling data-centric routing in wireless sensor networks. In: Proceedings of IEEE Infocom, New York, IEEE Computer Society, pp. 1–19 (2002)

9. Liao, W.H., Kao, Y.C., Fan, C.M.: Data aggregation in wireless sensor networks using ant Colony algorithm. In: Proceedings of the First Revised Journal of Network and Computer Applications, pp. 387–401 (2008)
10. Sangolgi, N.B., Khaja, S., Zakir, A.: Energy aware data aggregation technique in WSN. IJSRP 1–7 (2013)
11. Farshid, B., Du, W., Ngai, E., Fu, X., Liu, J.: Cloud-assisted data fusion and sensor selection for internet-of-things. IEEE Internet Things J. 257–268 (2016)
12. Izadi, D., Abawajy, J.H., Ghanavati, S., Herewa, T.: A data fusion method in wireless sensor networks. Sensors (Basel) 1–4 (2015)
13. Zheng, Z., Lai, C., Chao, H.: A green data transmission mechanism for wireless multimedia sensor networks using information fusion. In: Proceedings of IEEE Wireless Communications, pp. 14–17 (2014)
14. Chakraborty, I., Hussain, A.: A simple joint routing and scheduling algorithm for a multi-hop wireless network. In: Proceedings of the International Conference on Computer Systems and Industrial Informatics, pp. 1–5 (2012)
15. Chakraborty, I., Sarmah, U.: A simple routing algorithm for multi-hop wireless network. In: Proceedings of the 9th International Conference on Intelligent System and Control, pp. 1–5 (2015)
16. The ns manual (formerly NS Notes & Documentation). https://www.isi.edu/nsnam/ns/doc/. Accessed 28 Jan 2018
17. GAWK.: An Effective AWK Programming. https://www.cse.iitb.ac.in/~br/courses/cs699-autu mn2013/refs/gawkmanual.pdf. Accessed 5 Feb 2018

An Overview of Hadoop MapReduce, Spark, and Scalable Graph Processing Architecture

Pooja P. Talan, Kartik U. Sharma, Pratiksha P. Nawade
and Karishma P. Talan

Abstract In today's technology era, Big Data has become a buzzword. Various frameworks are available in order to process this Big Data. Both Hadoop and Spark are open source framework to process Big Data. Hadoop provides batch processing while Spark supports both batch as well as stream processing, i.e., it is a hybrid processing framework. Both frameworks have their own advantages and drawback. The contribution of this paper is to provide a comparative analysis of Hadoop MapReduce and Apache Spark. In this paper, we also propose a scalable graph processing architecture that could be used to overcome traditional limitations of Hadoop framework.

Keywords Big Data · Hadoop MapReduce · Apache Spark · Graph processing

1 Introduction

Data is getting generated on a huge scale at every second. In today's era digitalization, cloud computing, IoT, and Big Data are taking over the world. Huge amount of data is generated by various sources like social media, business market, weather forecasting, blogs, Sensor, etc. Every minute more than 400 h of video content were uploaded to YouTube as of July 2015 and as of March 2017, everyday 1 billion hours of contents are consumed [1].

Till the first quarter of 2017, Facebook had more than 1.94 billion global monthly active users, including over close to 1.74 billion mobile monthly active users [1]. The use of Internet in the world has shown a tremendous increase in various regions. Internet user in the world by region is shown in Fig. 1. Talan and Sharma [2] have indicated various characteristics of Big Data as Volume, Velocity, Variety, Variability, Veracity, Visualization, Value, etc.

P. P. Talan (✉) · K. U. Sharma · P. P. Nawade
Prof Ram Meghe College of Engineering & Management, Badnera-Amravati, India
e-mail: pooja.talan@rediffmail.com

K. P. Talan
KPIT Technologies Ltd, Thane, Mumbai, India

© Springer Nature Singapore Pte Ltd. 2019
J. Kalita et al. (eds.), *Recent Developments in Machine Learning and Data Analytics*,
Advances in Intelligent Systems and Computing 740,
https://doi.org/10.1007/978-981-13-1280-9_3

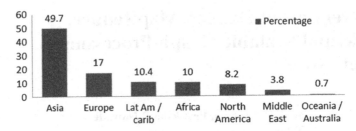

Fig. 1 Internet user in world by region—June 30, 2017

2 Literature Analysis

2.1 Hadoop Analysis

Barrachina and O'Driscoll [3] have provided a solution for proper categorizations and processing of thousands of customer queries at the technical support call centers, the proposed solution highlights the way to analyze all queries and identify the query pattern in order to improve the service and to identify the trend of customer needs. It gives a solution to storage space for thousands of customer queries. Authors have also examined five algorithms, namely k-means, fuzzy k-means, k-means with canopy clustering, Latent Dirichlet Allocation (LDA), and Dirichlet allocation. The approach proposed by Barrachina and O'Driscoll [3] have used Hadoop programming model and Mahout Big Data Analytics library and evaluates the results on VMware technical support data set, but does not provide practical implementation of the proposed approach. The current clustering output has limited relevance. Shahabinejad et al. [4] have presented binary locally repairable codes (LRC) with less complexity, less mean time to data loss, minimum size, and high reliability in order to provide data recovery instead of using the traditional method of data replication. The proposed code can be generalized in order to support more application.

Jacha et al. [5] have focused on storage and processing of data generated from the urban water distribution system. The performance of Hadoop and MySQL databases are evaluated and the comparative study proves the benefits of Hadoop over MySQL databases, further experiments with other NoSql databases can be conducted. Jain et al. [6] have mainly focused on the application of Hadoop to crime-related data so that some trend information can be obtained which will help to take proper action, as a future work, the proposed analysis can be implemented on fully distributed cluster mode. Jain et al. [6] have also provided experimental results for total number of crimes in year, total number of crimes occurring in each state, total number of crime by type, and total number of crime against women.

Jun et al. [7] have proposed HQ-Tree, a spatial data index based on Hadoop. HQ-Tree uses PR QuadTree to provide efficiency in parallel processing. Jun et al. [7] have proposed an approach called "Copy Write" in order to support the index update. It curtails the cost of network transferring and I/O operations, by supporting point query and range query, the proposed approach improves the performance of spatial query. In shared MapReduce environment, efficient scheduling is very important for the overall performance of system. Traditional scheduling algorithm supported by Hadoop shows poor response time under different workloads. Yao et al. [8] have proposed new Hadoop scheduler LsPS to reduce average job response time; experimental result shows that the proposed LsPS scheduler is effective, robust, and improves performance under different workloads

Traditional MapReduce technique suffers from a drawback that it does not support iteration, which reduces performance. Liu et al. [9] have proposed Meta-MapReduce algorithm to improve scalability and to reduce computational complexity and to reduce error rates. Liu et al. [10] have proposed a novel system for monitoring and analyzing large-scale network traffic data on commodity hardware, the algorithm can be further improved for better performance, and the experimental results indicate that the system can easily process large traffic data with high performance and less cost per day. There are different Hadoop MapReduce applications available in the market such as WordSort, TeraSort, and word count.

2.2 Spark Literature Analysis

Industries are using Hadoop on large scale. But the main point of concern is the speed at which the huge data is to be processed in terms of waiting time. Spark was basically introduced by Apache software foundation to speed-up the Hadoop computations; however, it is not a modified version of Hadoop.

Spark Architecture:

Spark is a general purpose, open source cluster computing, which supports various higher level tools such as Spark SQL for SQL, structured data processing, GraphX for graph processing, and Spark Streaming and MLib for machine learning. Spark architecture is shown in Fig. 2. It is based on two main concepts, RDD (Resilient Distributed Data sets) and DAG (Directed Acyclic Graph).

Spark supports two types of data sets, i.e.,

- Parallelized collection—It is based on existing scala collection
- Hadoop data sets—Created from files stored in HDFS

There are two operations supported by RDD, i.e.,

- Transformation—Create new data set from the input.
- Action—Return value after executing calculations on data sets.

Different jobs running on Apache Spark platform may show variable performance based on various parameters. Wang et al. [11] have presented a prediction model in order to predict the performance of job with high accuracy using iterative and non-iterative algorithms. Wang et al. [11] have also proposed a framework, which is evaluated with four real-world applications and shows more accurate performance. Huang et al. [12] have presented a model for analysis of large-scale Remote Sensing

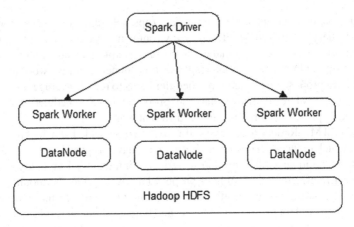

Fig. 2 Spark architecture

data using Spark on a Hadoop YARN platform and proposes strip-oriented parallel programming model, which assures high performance of Spark-based algorithm. Experimental evaluation shows that Spark-based parallel algorithms provide greater efficiency. Optimization of global cache for Apache Spark and estimation of global-scale soil moisture can be considered as future work. Apache Spark platform provides scalability, fault tolerance, performance, and productivity.

Yan et al. [13] have focused on the issue of scalability of Apache Spark platform to seismic data analytics. Some algorithms, for example, Fourier Transform, Hilbert Transform, and Jacobi stencil computations are used as experiments. The present SAC templates need more improvement and for the creation of advanced seismic data analytics models, high-level machine learning algorithm can be used. Harnie et al. [14] have discussed the Chemogenomics project within Janssen Pharmaceutica for discovering drugs. The original pipeline method uses set of programs which are not feasible to change in order to work on multi-node cluster. Harnie et al. [14] have also compared original pipeline with S-CHEMO which is the implementation of same pipeline with Apache Spark.

Ramirez-Gallego et al. [15] have mainly focused on the process of discretization of large data sets and presents multi-interval discretization method, which is based on entropy minimization with a completely distributed version of MDLP discretizer. The experimental result shows improved classification accuracy and time, the presented work can also contribute to MLib library. Mushtaq et al. [16] have presented a framework, which will implement an in-memory distributed version of the GATK pipeline. The proposed framework reduces the execution time and also supports dynamic load balancing algorithm, Mushtaq et al. [16] have also compared the proposed Spark implementation and Halvade on same machine, experimental result shows faster performance in Hadoop MapReduce framework.

Maarala et al. [17] have addressed the issue of low-latency analysis with real-time data, and it provides study of distributed and parallel computing, query and ingestion methods, storage, and evaluates tools for periodical and real-time analysis of heterogeneous data; it also introduces Big Data cloud platform with ingestion, analysis, storage, and data query APIs. For the Internet of Things applications, scalable real-time analysis and decision-making and integration of other technology with a proposed platform can be considered as a future work.

RDMA-based design has been proposed by Lu et al. [18] for data shuffle of Spark over InfiniBand and analyzes performance improvement for Spark over high-performance interconnects on modern HPC clusters. The proposed design by Lu et al. [18] can be integrated with updated Spark releases and shows performance improvement for GroupBy as compared to default Spark running with IP over Infini-Band (IPoIB) FDR. The experimental result shows that Spark can get benefit of high-performance interconnects and RDMA and plugin-based approach provides performance and productivity.

3 Proposed Architecture

Big Data is getting very popular in the business world. Both Hadoop and Spark are Big Data frameworks which provide tools for Big Data-related work. Talan and Sharma [2] have proposed an approach for graph data processing using Hadoop MapReduce. Hadoop performs multiple iterations on same data in most of the processing algorithms (e.g., PageRank). MapReduce basically performs three operations, reading data from disks, perform some iteration, sending result to HDFS, and again repeat this cycle for next iteration, which results in increase of latency and reduces speed of graph processing. To handle this issue, there are some tools supported by Hadoop for efficient graph processing, but they fail to support complex multi-stage algorithm.

This paper proposes architecture for scalable graph processing using Spark. In-build graph support and various graph computation libraries give better performance than that of traditional Hadoop MapReduce approach. The architecture involves three stages namely, loading the graph, Compute/Iterate, and Storing the graph. The first stage loads the graph (e.g., Subgraph data 1, Subgraph data 2, etc.), which will be processed by various workers (e.g., worker 0, worker 1, etc.). The second stage performs computation based on graph format and given condition and also performs various iterations on subgraph (e.g., Subgraph data part 1, Subgraph data part 2, etc.). The last stage stores the output graph as shown in Fig. 3.

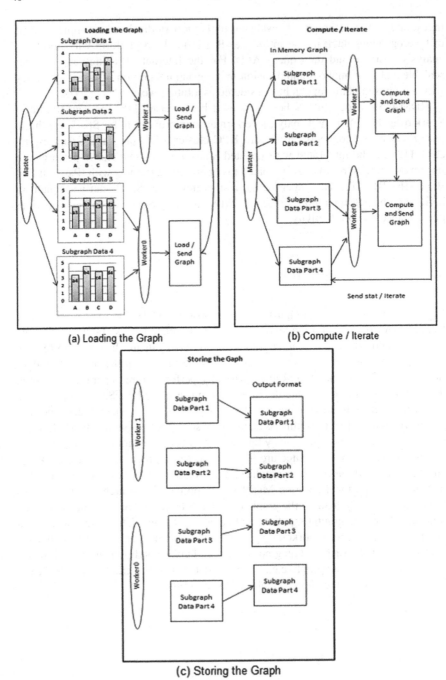

Fig. 3 Scalable graph processing architecture

Table 1 Source code and hands-on exercise

Source	Content	Link
Cloudera	How-to: do scalable graph analytics with Apache Spark	https://blog.cloudera.com/blog/2016/10/how-to-do-scalable-graph-analytics-with-apache-spark/
Ampcamp	GraphX hands-on exercise	http://ampcamp.berkeley.edu/big-data-mini-course/graph-analytics-with-graphx.html
Databricks	Graph analysis with GraphX tutorial	http://docs.databricks.com/spark/latest/graph-analysis/graph-analysis-graphx-tutorial.html
Edureka	Spark GraphX tutorial—graph analytics In Apache Spark	https://www.edureka.co/blog/spark-graphx/

Table 2 Graph data sets

Source	Content	Link
Yahoo	Graph and social data	http://webscope.sandbox.yahoo.com/catalog.php?datatype=g
Sanford University	Stanford large network data set collection	http://snap.stanford.edu/data/index.html
Laboratory for web algorithmics	Data sets	http://law.di.unimi.it/datasets.php

4 Online Resources

Table 1 lists some useful online resources for source code and hands-on exercise and Table 2 provides a list of graph data sets.

5 Conclusion

Graph processing and analysis is applicable to a lot of cases which basically deals with defining relationship between nodes and edges. The importance of graph is growing from language modeling to social media. The proposed architecture provides an efficient way to process graph data as Spark can handle various types of requirement such as graph processing, interactive processing, stream processing, batch processing, iterative processing, etc. The proposed scalable graph processing approach works in different stages which reduces overall processing complexity. Also proposed architecture efficiently supports multi-stage algorithm and also overcomes limitations of Hadoop MapReduce.

References

1. Statista [Online]. http://www.statista.com/. Accessed 5 Feb 2018
2. Talan, P., Sharma, K.: An overview and an approach for graph data processing using Hadoop MapReduce. In: 2nd International Conference on Computing Methodologies and Communication (2018) (submitted)
3. Barrachina, A.D., O'Driscoll, A.: A big data methodology for categorising technical support requests using Hadoop and Mahout. J. Big Data **1**, 1 (2014)
4. Shahabinejad, M., Khabbazian, M., Ardakani, M.: An efficient binary locally repairable code for Hadoop distributed file system. IEEE Commun. Lett. **18**(8), 1287–1290 (2014)
5. Jacha, T., Magieraa, E., Froelich, W.: Application of HADOOP to store and process big data gathered from an urban water distribution system. In: 13th Computer Control for Water Industry Conference, CCWI, pp. 1375–1380 (2015)
6. Jain, A., Bhatnagar, V.: Crime data analysis using pig with Hadoop. In: International Conference on Information Security & Privacy, pp. 571–578 (2016)
7. Jun, F., Zhixian, T., Mian, W., Liming, X.: HQ-Tree: a distributed spatial index based on Hadoop. China Commun., 128–141 (2014)
8. Yao, Y., Tai, J., Sheng, B., Mi, N.: LsPS: a job size-based scheduler for efficient task assignments in Hadoop. IEEE Transac. Cloud Comput. **3**(4), 411–424 (2015)
9. Liu, X., Wang, X., Matwin, S., Japkowicz, N.: Meta-mapreduce for scalable data mining. J. Big Data, 1–23 (2015)
10. Liu, J., Liu, F., Ansari, Nirwan: Monitoring and analyzing big traffic data of a large-scale cellular network with hadoop. IEEE Netw., 32–39 (2014)
11. Wang, K., Khan, M.M.H.: Performance prediction for apache spark platform. In: IEEE 17th International Conference on High Performance Computing and Communications, IEEE 7th International Symposium on Cyberspace Safety and Security, and IEEE 12th International Conference on Embedded Software and Systems, pp. 166–173 (2015)
12. Huang, W., Meng, L., Zhang, D., Zhang, W.: In-memory parallel processing of massive remotely sensed data using an apache spark on hadoop yarn model. IEEE J. Sel. Top. Appl. Earth Obs. Remot. Sens., 1–17 (2016)
13. Yan, Y., Huang, L., Yi, L.: Is apache spark scalable to seismic data analytics and computation. In: IEEE International Conference on Big Data, pp. 2036–2045 (2015)
14. Harnie, D., Vapirev, A.E., Wegner, J.K., Gedich, A., Steijaert, M., Wuyts, R., De Meuter, W.: Scaling machine learning for target prediction in drug discovery using apache spark. In: 15th IEEE/ACM International Symposium on Cluster, Cloud and Grid Computing, pp. 871–879 (2015)
15. Ramirez-Gallego, S., Garcia, S., Mourino-Talin, H., Martinez-Rego, D., Bolon-Canedo, V., Alonso-Betanzos, A., Benitez, J., Herrera, F.: Distributed entropy minimization discretizer for big data analysis under apache spark. In: IEEE Trustcom/BigDataSE/ISPA, pp. 33–40 (2015)
16. Mushtaq, H., Al-Ars, Z.: Cluster-based apache spark implementation of the GATK DNA analysis pipeline. In: IEEE International Conference on Bioinformatics and Biomedicine, pp. 1471–1477 (2015)
17. Maarala, A.I., Rautiainen, M., Salmi, M., Pirttikangas, S., Riekki, J.: Low latency analytics for streaming traffic data with apache spark. In: IEEE International Conference on Big Data, pp. 2855–2858 (2015)
18. Lu, X., Md. Wasi-ur-Rahman, Islam, N., Shankar, D., Panda, D. K.: Accelerating spark with RDMA for big data processing early experiences. In: IEEE 22nd Annual Symposium on High-Performance Interconnects, pp. 9–16 (2014)

Analyzing Social Trend Towards Girl Child in India: A Machine Intelligence-Based Approach

Sirshendu Hore and Tanmay Bhattacharya

Abstract In India, feticide or sex-selective abortion is the main reason for the sharp drop in the Child Sex Ratio (CSR). To remove this type of sick mindset from the society and to empower women by proper education, the Government of India has launched a social consciousness program under the banner "Beti Bachao Beti Padhao (BBBP)" when translated stand as "save girl child, educate her", in the year 2015. It has been observed that, nowadays, researchers are using the power of Natural Language Processing (NLP) to analyze the opinion or suggestions posted on social media. In the proposed work, the opinion expressed by various people on Twitter for a certain period of time has been taken up for analysis. Machine intelligence has been used to classify the opinion. In the current work, instead of binary classification, a multi-class sentiment classification has been introduced. To achieve the objective of the proposed work, four popular classifiers of machine intelligence domain have been used. Accuracy and Kappa statistics has been employed to measure the performance of the proposed model.

Keywords Natural language processing · Sentiment analysis
Sentiment polarity categorization · Social media · Twitter

1 Introduction

In Asia, the female sex ratio is declining on a frightening level and India is at the top of this decreasing ratio. The 2011 census of India suggests that, the sex ratio of children aged between 0 and 6 were 945 girls per 1000 boys in 1961. Now the ratio has dropped from 945 to 918 girls per 1000 boys. Although we are living in

S. Hore (✉)
Department of CSE, HETC, Hooghly, India
e-mail: shirshendu.hore@hetc.ac.in

T. Bhattacharya
Department of IT, TISL, Kolkata, India
e-mail: drtb@gmail.com

© Springer Nature Singapore Pte Ltd. 2019
J. Kalita et al. (eds.), *Recent Developments in Machine Learning and Data Analytics*,
Advances in Intelligent Systems and Computing 740,
https://doi.org/10.1007/978-981-13-1280-9_4

the twenty-first century, still some conservative families do not celebrate the birth of a girl child. It has been observed that this type of disgusting mindset prevails in every sector of our society from rural to elite class of the society. It is a sad and sobering reality that a newborn boy is still considered as a sign of fortune, whereas a newborn girl is considered a hardship, because the myth is, a boy can carry forward the legacy of the family; he can relieve the economic burden, also can bring the hefty amount in the form of dowry [1]. Many of our girl infants are either killed instantly after their birth or even in some cases, fetus is destroyed in the mother's womb. In reality, a girl child is a blessing for the society. They shine everywhere, from the classroom to the professional field. Girls, balance the sex ratio and moreover, only a woman can give birth to a child. Although to spread this message, government and different state bodies enforced strict laws, set up counseling centers across the country, but still, little progress has been made in this area. To bring changes in the mindset of people, the "Ministry of Women and Child Development", the "Ministry of Health and Family Welfare", the "Ministry of Human Resource Development", and the Government of India together flagged a country-wise consciousness campaign program from 09/10/17 to 17/10/2017. The purpose of the drive was to generate awareness amongst the people so that they welcome a newborn baby girl and empower girls through proper education [2]. Nowadays, people express their views or opinions in the form of a blog, post, or using tweets on social media platforms (like "Twitter", "Facebook", "Instagram"). Researchers and developers extract useful information from those media. A "Feeling" has a direct influence on individual's opinion, which is a combination of thought, attitude, and judgment. Opinion mining [3–5], also known as view analysis, judges people's views based on some specific issues. In the proposed work, tweets of various people about the Beti Bachao, Beti Padhao (BBBP) campaign have been collected and analyzed meticulously, using various APIs [6] offered by Twitter to build a successful model.

2 Literature Review

Data mining is the latest buzzword; there is a rising demand for data mining in the researcher's domain. Nowadays, many researchers are applying this technique extensively for the purpose of opinion analysis. Opinion analysis or sentiment classifications are purely a decision-making process. It has been observed that accuracy becomes a critical factor for all this decision-making process. The level of accuracy depends on the strength of the classification algorithm [7, 8]. To measure the sentiments of people, developers or researchers' need to have a set of positive words and a set of negative words along with the intensity of the words. Fortunately, Hu and Liu [9] summarized a group of such words based on client assessments. This group of words contains 4783 negative words and 2006 positive words. It also contains some words which are not part of the English dictionary, but, popularly used in a social media. Malhar et al. [10] successfully analyzes Twitter data using ANN, SVM, and Naïve Bayes classifiers. In their work, authors also investigated some factor that influences the overall mining process. Predicting the price of a particular piece of product has an immense impact on the stock market. Bing et al. [11] in their work attempts to predict the future price of some company products. The data set that has been used in their study, is a collection of tweets, which they collected from Twitter and analyzed those data to predict the price of an item. In politics, judging the impression of people and categorize those pieces of information plays a crucial role. Mahmood et al. [12] in their submission tried to predict the election result of Pakistan, which was held in the year 2013. Silva et al. [13] successfully applied random forests (ensemble learner) to build an automatic classification model to classify person's sentiments using Twitter data. Jani et al. [14] analyzed people's sentiment towards political leaders and parties in India using k-nearest neighbor, random forests, Bayes Net, and Naive Bayes classifier. In their proposed work, authors used both single classifier and ensemble classifiers to find which approach is more suitable for the said purpose. To categorize people's sentiment, three class labels, namely positive, negative, and neutral has been used. The findings of the authors work using single classifier have been tabulated. In a supervised machine learning domain, SVM has been used by many researchers for building their classifier model [15]. In classification domain, a lot of research work has been done using artificial neural network [16–22].

3 Proposed Work

Proposed work is divided into three parts. Figure 1 shows the block diagram of our proposed method.

(a) *Online collection of relevant data*: In the proposed work, tweets related to a special BBBP-campaign period has been collected from Twitter. On Twitter, people express their opinions by sending tweets which are a small text message having a word limit of 280 (except Korean, Chinese, and Japanese where word limit is 140). On Twitter, a user can like tweets, reply to the tweets, and follow a Twitter user/s or forward tweets. Forwarding a tweet is known as a retweet. Consequently, the process of tweets, reply, follow, and retweets creates a long list of opinion chain amongst the tweeter handlers.

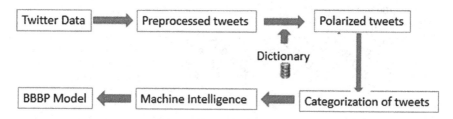

Fig. 1 Schematic description of proposed BBBP model

(b) *Preprocessing of collected data*: The sentiment expressed by the Twitter user, got some words, letters, and symbols which are not relevant or has no relevance in the English dictionary. Also, there is some grammar, stop words, and special character, which is also not suitable for opinion mining. Thus, some preprocessing need to be carried out. Finally, all the tweets are converted to lower case to avoid case sensitivity.

(c) *Analysis and classification of data*: In this part, preprocessed data are being analyzed. The power of machine intelligence has been applied to classify the opinions based on collected tweets. To classify tweets into various labels, "SentiWordNet 3.0.0" dictionary has been used. At the very beginning, the text parts of preprocessed tweets (which are in lowercase) are divided into words. The polarity of each word is checked and compared with polarity dictionary (having a large collection of positive and negative words). Then, each polarity is aggregated to find the total score. Eight different labels have been created, namely awful, poor, bad, neutral, satisfactory, good, very good, and great based on the score. Four established classifiers (Naïve Bayes, MLP_FFN, support vector machine, and random forests) have been engaged to construct the classification model. Accuracy and Kappa statistics have been applied to quantify the merit of the classifier. McNemar's test has been applied to judge the performance of the best classifier for the said work. Finally, the finding of the proposed work is compared with Jani and Katkar's work [14]. The work proposed also analyzes the frequency of tweets, number of retweets, and some active tweeter users for the said movement.

4 Results and Discussion

In the proposed work, nearly 5000 opinions of various people on Twitter in the context of BBBP-campaign has been collected. R language (version 3.41) has been used for the purpose of opinion collection and analysis [23]. Figure 2 shows the

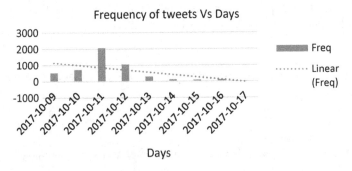

Fig. 2 Frequency of tweets on BBBP campaigning period

Fig. 3 Frequency of retweets

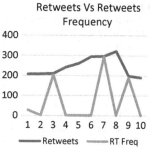

Retweets Vs Retweets Frequency

Fig. 4 Top five Twitter users

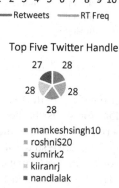

Top Five Twitter Handle

frequency of tweets on those special campaigning period. Figure 3 depicts that the people's having a tendency in retweeting rather than expressing his/her own opinion. Figure 4 displays the top five Twitter handlers with the number of tweets sent.

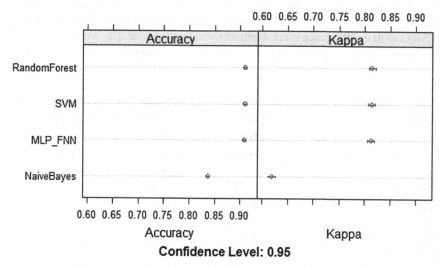

Fig. 5 Dot plot shows the classification results of different classifier without stopwords

Table 1 Evolution of classifiers performance based on Accuracy and Kappa statistics

Classifier	Accuracy	Kappa
SVM	0.9152542	0.8240644
RF	0.9152542	0.8245024
MLP_FFN	0.9152542	0.8240644
Naïve Bayes	0.8438819	0.6339234

Table 2 A comparative study of proposed finding with Jani et al. works (without stop words)

	Jain et al.	Proposed work
Classifier	Accuracy	Accuracy
Random forests	0.6566810	0.9152542
Naive Bayes	0.6031590	0.8438819

In Fig. 5, classification result has been plotted using a dot plot which shows random forests is the best classifier for this experimental work. Table 1 depicts the performance of classifiers based on Accuracy and Kappa statistics. Table 2 shows a comparative study of proposed finding with Jani and Katkar's work [14]. Table 3 exhibits the performance results of the random forests using the McNemar's test for all positive words and Table 4 illustrate the performance results of the random forests using the McNemar's test for all negative and neutral words

5 Conclusion and Future Work

In this study, a large volume of opinions has been used to obtain more accurate outcome to access the mindset of common people to improve the Child Sex Ratio (CSR) and to uplift women empowerment through education. The analysis revealed that many of those opinions are neither positive nor negative, even some of the

Table 3 Performance results of the random forests using McNemar's test for all positive words

Parameter	Class satisfactory	Class good	Class very good	Class great
Sensitivity	0.98850	0.00000	1.00000	1.00000
Specificity	0.9154	1.00000	1.00000	1.00000
PosPred value	0.66730	0	1.00000	1.00000
NegPred value	0.99790	0.92776	1.00000	1.00000
Prevalence	0.1466	0.07224	0.38640	0.00210
Detection rate	0.1446	0.00000	0.38640	0.00210
Detection prevalence	0.2171	0.00000	0.38640	0.00210
Balanced accuracy	0.95190	0.50000	1.00000	1.00000

Table 4 Performance results of the random forests using McNemar's test for all negative and neutral words

Parameter	Class awful	Class bad	Class poor	Class neutral
Sensitivity	1.00000	0.00000	0.97297	1.00000
Specificity	1.00000	1.0000	0.98440	0.99180
PosPred value	1.00000	0	0.66667	0.99640
NegPred value	1.00000	0.98486	0.99912	1.00000
Prevalence	0.00336	0.01512	0.03108	0.69090
Detection rate	0.00336	0.00000	0.03024	0.69090
Detection prevalence	0.00336	0.00000	0.04536	0.69340
Balanced accuracy	1.00000	0.50000	0.97868	0.99590

opinions are out of context. Four established classifiers and eight labels are used for an in-depth analysis. To evaluate the classifier performance, some advanced statistical methods have been used. The experimental results show that Random forests classifier performs better than the other classifiers. The predictive accuracy of random forests for all class labels has been evaluated by McNemar's test. Last, a comparative study has been made with another research work in the similar domain. The outcome suggests that the model is quite successful for classifying sentiments of people. In future, some nature-inspired searching techniques can be used to enhance the performance of the classifier. Opinions of people can also be collected from other social media sites such as Facebook or Instagram.

References

1. Clark, S.: Son preference and sex composition of children: evidence from India. Demography **37**(1), 95–108 (2000)
2. https://en.wikipedia.org/wiki/Beti_Bachao,_Beti_Padhao_Yojana, 2018
3. Liu, B.: Sentiment Analysis and Subjectivity. Handbook of Natural Language Processing, 2nd edn. Taylor and Francis Group, Boca Raton (2010)
4. Pak, A., Paroubek, P.: Twitter as a corpus for sentiment analysis and opinion mining. In: Proceedings of the Seventh Conference on International Language Resources and Evaluation. European Languages Resources Association, Valletta, Malta, pp. 1320–1326 (2010)
5. Gokulakrishnan, B., Priyanthan, P., Ragavan, T., Prasath, N., Perera, A.: Opinion mining and sentiment analysis on a Twitter data stream. In: Advances in ICT for Emerging Regions (ICTer), pp. 182–188. IEEE, New York (2012)
6. Twitter (2014) Twitter.aps. https://dev.twitter.com/start
7. Han, J., Kamber, M.: Data Mining Concepts and Techniques, 2nd edn, pp. 285–378. Morgan and Kaufmann (2005)
8. Agrawal, R., Imielinski, T., Swami, A.: Database mining: a performance perspective. IEEE Trans. Knowl. Data Eng. **5**(6), 914–925 (1993)

9. Hu, M., Liu, B.: Mining and summarizing customer reviews. In: Proceedings of the Tenth ACM SIGKDD International Conference on Knowledge Discovery and Data Mining, pp. 168–177. ACM, New York (2004). DOI-https://doi.org/10.1145/1014052.1014073

10. Malhar, A., Guddeti, R.M.R.: Influence factor based opinion mining of Twitter data using supervised learning. In: Sixth International Conference on Communication Systems and Networks (COMSNETS), pp. 1–8. IEEE, New York (2014)

11. Bing, L.I., Chan, K.C., Ou, C.: Public sentiment analysis in Twitter data for prediction of a company's stock price movements. In: 11th International Conference on e-Business Engineering (ICEBE), pp. 232–239. IEEE, New York (2014)

12. Mahmood, T., Iqbal, T., Amin, F., et.al.: Mining Twitter big data to predict 2013 Pakistan election winner. In: 16th International Multi Topic Conference (INMIC), pp. 49–54. IEEE, New York (2013)

13. Da Silva, NdiaFF, Hruschkaa, E.R., Hruschka, E.R.: Tweet sentiment analysis with classifier ensembles. Decis. Support Syst. **66**, 170–179 (2014)

14. Jani, P.A., Katkar, D.V.: Sentiments analysis of twitter data using data mining. In: International Conference on Information Processing (ICIP), pp. 807–810. IEEE, New York (2015)

15. Cristianini, N., Shawe-Taylor, J.: An Introduction to Support Vector Machines and Other Kernel-Based Learning Methods. Cambridge University Press, Cambridge (2000)

16. Hore, S., Chatterjee, S., Dey, N., et.al: In: Panigrahi B., Hoda M., Sharma V., Goel S. (eds.) Nature Inspired Computing. Advances in Intelligent Systems and Computing, vol. 652, pp. 109–115. Springer, Singapore (2015)

17. Bhattacharjee, A., Roy, S., Paul, S., et.al.: Classification approach for breast cancer detection using back propagation neural network: a study, biomedical image analysis and mining techniques for Improved Health Outcomes. In: IGI Global (2015)

18. Hore, S., Chatterjee, S., Dey, N., et.al.: Neural-based prediction of structural failure of multistoried RC buildings. Struct. Eng. Mech. **58**(3), 459–473. https://doi.org/10.12989/sem.2016.58.3.4592016

19. Dey, N., Ashour, A.S., Chakraborty, S., Samanta, S.: Healthy and unhealthy rat hippocampus cells classification: a neural-based automated system for Alzheimer disease classification. J. Adv. Microsc. Res. **11**(1), 1–10 2016

20. Hore, S., Bhattacharya, T., Dey, N., et al.: Areal-time dactylology based feature extraction for selective image encryption and artificial neural network. In: Image Feature Detectors and Descriptors Foundation and Applications. Studies in Computational Intelligence, vol. 630, pp. 203–226, Springer, Berlin (2016) https://doi.org/10.1007/978-3-319-28854-3

21. Chatterjee, S., Hore, S., Dey, N., et al.: Optimized forest type classification: a machine intelligence approach. In: Satapathy, S., Mandal, J., Udgata, S., Bhateja, V. (eds.) Information Systems Design and Intelligent Applications. Advances in Intelligent Systems and Computing, vol. 435, pp. 237–236. Springer, Berlin (2016)

22. Li, Z., Shi, K., Dey, N., et al.: Rule-based back propagation neural networks for various precision rough set presented KANSEI knowledge prediction: a case study on shoe product form features extraction. Neural Comput. Appl. **28**(3):613–630 (2017)

23. R Core Team (2017). R: a language and environment for statistical computing. R Foundation for Statistical Computing, Vienna, Austria. URL: https://www.R-project.org

Improvement in Boosting Method by Using RUSTBoost Technique for Class Imbalanced Data

Ashutosh Kumar, Roshan Bharti, Deepak Gupta and Anish Kumar Saha

Abstract Class imbalance problem is common in many fields, and it occurs due to imbalanced dataset. A dataset is considered as imbalanced when number of examples in one class is more or less compared to another class. Data mining algorithms may generate suboptimal classification models when trained with imbalanced datasets. Several techniques have been proposed to solve the class imbalance problem. One of them includes boosting which is combined with resampling technique. It has gained popularity to solve class imbalance problem, for example, Random Undersampling Boosting (RUSBoost) and Synthetic Minority Oversampling Boosting Technique (SMOTEBoost). RUSBoost method uses random undersampling technique as resampling technique. One of the disadvantages of random undersampling may include loss of important data which is overcome by redundancy-driven modified Tomek-link based undersampling. A new hybrid undersampling algorithm is proposed in which we use redundancy-driven modified Tomek-link based undersampling as resampling technique along with boosting for learning from imbalanced training data. Experiments are performed for various datasets which are related to various application domains. The results are compared with decision tree, Support Vector Machine (SVM), logistic regression, and K-Nearest Neighbor (KNN) to check the performance of proposed method.

Keywords RUSBoost · SMOTEBoost · Class imbalance problem
Boosting technique · SVM · Decision tree

A. Kumar · R. Bharti · D. Gupta (✉) · A. K. Saha
Computer Science and Engineering, National Institute of Technology, Yupia, Arunachal Pradesh, India
e-mail: deepakjnu85@gmail.com

A. Kumar
e-mail: ashutoshsingh227@gmail.com

R. Bharti
e-mail: giet12cse077@gmail.com

A. K. Saha
e-mail: anishkumarsaha@gmail.com

© Springer Nature Singapore Pte Ltd. 2019
J. Kalita et al. (eds.), *Recent Developments in Machine Learning and Data Analytics*,
Advances in Intelligent Systems and Computing 740,
https://doi.org/10.1007/978-981-13-1280-9_5

51

1   Introduction

Class imbalance [1] is a problem that occurs when the number of instances in one class is different compared to another class. The data preprocessing method includes two techniques, i.e., random oversampling and random undersampling. Random oversampling is applied in minority class, whereas random undersampling is applied in majority class until both the classes become equal. Major drawback of random oversampling is that the sampled training dataset is more as compared to the original training dataset, which is called as overfitting problem and also training time of the dataset is more as compared to original dataset in oversampling. Random undersampling method removes instances from majority class randomly until both the classes become equal. Since the number of instances in sampled training dataset is less compared to original training dataset, the training time of the classifier is also decreased. One of the major drawbacks of random undersampling is that some important instances may be deleted when we are removing instances from the majority class. SMOTEBoost [2] is a combination of Adaptive Boosting (AdaBoost) [3] and Synthetic Minority Oversampling TEchnique (SMOTE) [4]. SMOTE is an intelligent oversampling method which is specially designed for the treatment of imbalanced data problem. SMOTE is introduced to create new minority class instances by selecting some of the minority class instances which are nearest neighbor of each other. SMOTE method takes the difference between an instance of minority class and nearest neighbors, and then their difference is multiplied by a random number from 0 to 1 and the result is added to sample space under consideration. Since SMOTE method does not add the same instance from minority class. So, overfitting problem is obviated. Here, the data preprocessing (oversampling of minority class) is done by producing synthetic examples and the oversampled training dataset is given to AdaBoost to make a model. AdaBoost is most commonly used practical boosting algorithms that use entire dataset for training of the weak learners but during the training it gives more attention to incorrectly classified pattern. The importance of instances is calculated by a weight, which is initially same for all instances. Weights of misclassified patterns are increased, whereas the weight of actual classified pattern is decreased after each step or iteration. Boosting improves the accuracy of model by giving attention to examples, and at the same time SMOTE improves the performance of the model by increasing number of minority class instances. The only drawback associated with this method is that SOMTE is complex and time-consuming oversampling technique compared to other resampling methods. RUSBoost [5] is another technique which is a combination of random undersampling and AdaBoost. Random undersampling is applied to majority class examples to achieve desired balance ratio between minority and majority class, and then the balanced dataset is given to AdaBoost to build a model. The only drawback associated with RUSBoost is that the random undersampling technique removes the instances randomly from majority class which may discard instances not useful for classification. The performance is increased in RUSBoost compared to SMOTEBoost but still there is a loss of important data which is required by the classifier. A pair of instances is said to be

Tomek-link pair [6] if they are nearest neighbor of each other and belong to different classes. In this method, undersampling is performed by removing all Tomek-link pair from the training datasets.

Several techniques have been introduced to solve the class imbalanced problem. Galar et al. [7] provide some methods to manage class imbalance problem. They are classified into some categories based on how they deal with class imbalance problem. The first practical and most widely used boosting algorithm was AdaBoost algorithm. Hoang et al. [8] proposed a recognition-based approach, an alternative solution of class imbalance problem in which the classifier is trained with the examples of minority class in the absence of majority class examples. However, recognition-based approach is not applicable for classification algorithms such as decision tree, Naive Bayes, and associative classifier. Dennis L. Wilson proposed Edited Nearest Neighbor (ENN) [9], an undersampling method in which undersampling on majority class is performed by removing instances whose class label does not match with majority of its k-nearest neighbors, i.e., undersampling is performed on majority class by removing outliers. Another undersampling method known as Condensed Nearest Neighbor (CNN) was proposed by P. Hart et al. [10]. The main idea behind this method is to select a subset of original training dataset. Since CNN uses undersampling which requires many scans over training dataset; therefore, it is slow compared to other methods. SMOTEBoost and RUSBoost were also introduced to solve class imbalance problem. Readers can apply boosting techniques on [11–14] to improve the classification performance in terms of accuracy, G-mean, and CPU utilization.

Hence, in this paper, we are combining both RUSBoost and redundancy-driven modified Tomek-link based undersampling method to overcome class imbalance problem, and the algorithm is named as RUSTBoost (redundancy-driven modified Tomek-link based random undersampling boosting technique). Here, undersampling is done by redundancy-driven modified Tomek-link pair and boosting is done by AdaBoost algorithm with different learners, i.e., SVM, decision tree, KNN, and logistic regression.

Here, Sect. 2 represents the work related to RUSBoost algorithm and redundancy-driven modified Tomek-link undersampling method in detail; Sect. 3 describes the proposed RUSTBoost method in detail. In Sect. 4, we have compared the proposed RUSTBoost method with RUSBoost and SMOTEBoost in terms of accuracy and G-mean. Section 5 concludes the paper along with future work.

2 Background and Related Work

In this section, we have discussed RUSBoost algorithm and redundancy-driven modified Tomek-link based undersampling method.

2.1 RUSBoost Algorithm

This section describes the algorithm for RUSBoost proposed by Seiffert et al. [5].

Let A be the set of instances $(P_1, Q_1), \ldots, (P_n, Q_n)$ with minority class $Q^t \in B$, $|B| = 2$, T is the number of classifiers in the boosting, r where $1 \leq r \leq R$ and h_r is the iteration for weak learner hypothesis.

Output from the procedure is $H(P) = \arg\max(Q \in B) \sum_{r=1}^{R} h_r(P, Q) * \log\left(\frac{1}{\beta_r}\right)$

Procedure

WeakLearn ():
Begin

1. Initialize $D_1(i) = \frac{1}{n} \forall i$
2. Repeat for $r = 1, 2, \ldots, R$.
   (a) Temporary training dataset A'_r is created by applying random under-sampling with weight distribution D'_r on original training dataset.
   (b) Call WeakLearn, by providing A'_r and their weights D'_r.
   (c) Hypothesis is obtained as $h_r : X \times B \rightarrow [0, 1]$.
   (d) Pseudo-loss is calculated (for A and D_r) as

$$\epsilon_r = \sum_{(i,Q):Q_i \neq Q} D_r(i)(1 - h_r(P_i, Q_i) + h_r(P_i, Q))$$

   (e) Weight update parameter is calculated by using $\beta_r = \frac{\epsilon_r}{1-\epsilon_r}$.
   (f) Update D_r using $D_{r+1}(i) = D_r(i)\beta_r^{\frac{1}{2}(1+h_r(P_i, Q_i)-h_r(P_i, Q:Q \neq Q_i))}$
   (g) Normalize D_{r+1}: Let $Z_r = \sum_i D_{r+1}(i)$ then $D_{r+1}(i) = \frac{D_{r+1}(i)}{Z_r}$

2.2 Redundancy-Driven Modified Tomek-Link Based Undersampling

To reduce significance loss of information from training data during undersampling, Devi et al. [15] have proposed a Tomek-link and redundancy-based undersampling method where eliminations are done for outliers and noisy (Tomek-linked) patterns with greater redundancy and lower contribution factor, while undersampling was performed. The redundancy-driven modified Tomek-link boosting method works in the following three steps:

Step-1: Elimination of outliers from majority class,
Step-2: Detection of redundant pairs, and
Step-3: Elimination of redundant patterns.

In the first step, outliers are eliminated from majority class where outliers are determined based on given Tomek-link pair and a threshold value (r). Let A be a dataset with m number of patterns and n number of attributes in each pattern. Let each pattern is characterized as $(\vec{P}_a, Q_b)$ where $\vec{P}_a$ is the input vector for ath pattern, denoted as $\vec{P}_a = (P_{a1}, P_{a2}, \ldots, P_{an})$, $1 \leq a \leq m$. Let $Q_b = (Q_1, Q_2, \ldots, Q_c)$ is the designated class label out of c classes. Let us assume the following representation: $A_{m \times n}$ as imbalanced dataset, with m patterns and n attributes, $A_{\min}$ as set of minority class patterns, i.e., $A_{\min} = \{P_i | i = 1, 2, \ldots, f\}$, where f is the total count of minority class instances, A_{maj} as set of majority class patterns, i.e., $A_{\text{maj}} = \{P_j | j = 1, 2, \ldots, g\}$, where g is the number of majority class patterns, TL is the set of Tomek-linked patterns, indices$_{P_j}$ is the total minority patterns to which majority pattern P_j has associated as Tomek-linked pair, TL' is revised subset of TL after removal of P_j patterns determined as outliers, and A'_{maj} is the set of majority class patterns after removing outliers. Input to the procedure is $A_{\min}$, A_{maj} and threshold (r) to determine majority patterns P_j.

Then, the procedure to eliminate outliers from majority class is proposed as given below.

Begin

Step-1: Entire dataset $A_{m \times n}$ is divided into two subsets say $A_{\min}$ and A_{maj}.

Step-2: For each $P_i \in A_{\min}$, determine its k-nearest neighbor (By using KNN algorithm) from P_j such that $P_j \in A'_{\text{maj}}$ by using Euclidean distance.

Step-3: $TL = \{(P_i, P_j) | P_i \in A_{\min}, P_j \in A_{\text{maj}}\}$ is the nearest neighbor pair of (P_i, P_j).

Step-4: For each $P_j \in A_{\text{maj}}$, determine indices$_{P_j}$.

Step-5: If indices$_{P_j} > r$, then P_j is treated as outlier in the minority region and the output is stored in subset OUTER as OUTER $= \{P_j | \forall P_j, \text{indices}_{P_j} > r\}$.

Step-6: TL is modified to TL' as $TL' = TL = \{\text{OUTER}\}$.

Step-7: Subset A_{maj} is updated to A'_{maj} as $A'_{\text{maj}} = A_{\text{maj}} - \{\text{OUTER}\}$

Patterns creating the set OUTER are named as outliers and it must be removed. TL' contains remaining Tomek-link paired patterns.

Further, in step-2, redundant pairs are detected from the remaining Tomek-link paired patterns. Majority class patterns are detected by together considering redundancy and noise factors which is occupying the proximity of the decision boundary and high similarity with other majority patterns. Let A'_{maj} represent a new set of majority class patterns and RD represent a set of most redundant majority pair patterns for each P_n where $P_n \in A'_{\text{maj}}$ and TL_RD represents the intersection of TL' and RD.

Input to the procedure is A'_{maj}, TL' and k, where k is the number of redundant patterns to create RD. Set $k = 1$.

Now, the procedure to detect redundant pairs from the remaining Tomek-link paired patterns is given as follows:

Begin

Step 1: Redundant pair is determined from A'_{maj} by using the ED, CB, and CS $\forall P_n \in A'_{\text{maj}}$ and stored in the subset RD as

$$RD = \left\{ (P_n, P_t) | \forall P_n, P_t \in A'_{\text{maj}} \text{ and sim}(P_n, P_t) = \text{maximum} \right\}$$

Step 2: A revised subset TL_{RD} is created by the intersection of TL' and RD as

$$TL\_RD = \left\{ P_n \cap A'_{\text{maj}} | P_n \in TL', P_n \in RD \right\}$$

Redundant majority patterns which are stored in set TL_RD are considered for next step.

Here, after detecting the redundant patterns, it is eliminated at the last step, i.e., step-3. Now, to eliminate a redundant majority pattern from a majority redundant pair (P_u, P_v), where $P_u, P_v \in TL\_RD$ and a contribution factor Contri$_P$ is employed, which is determined as follows: Contri$_P = \frac{1}{N} \times \left\{ \sum_{i=1}^{m} \left(\sum_{k=1}^{n} \ln f(P_{ik}|C_n) \right) \right\}$

where N indicates the total number of patterns and $\ln f(P_{ik}|C_n)$ denotes the log-likelihood function. Hence, majority patterns reclined to the set TL_RD are calculated in terms of contribution factors and redundancy. Eliminations are done for patterns with lowest contribution factor and greater redundancy. Redundant eliminated patterns are stored in set TL_RD' by the following step:

$$TL\_RD' = \max\left(\frac{\text{sim}(P_u, P_v)}{\text{Contri}_{P_u}}, \frac{\text{sim}(P_u, P_v)}{\text{Contri}_{P_v}} \right); P_u, P_v \in TL\_RD$$

If two majority patterns are redundant, then the value with lesser contribution factor will be removed. Now, the refined dataset A_R is obtained by the following steps.

i.   Update A'_{maj} to A''_{maj} as $A''_{\text{maj}} = A'_{\text{maj}} - TL\_RD'$.
ii. $A_R = A_{\text{min}} \cup A''_{\text{maj}}$

Here, for preprocessing, the similarity measures considered are Euclidian Distance (ED), City Block distance (CB), and Cosine Similarity (CS). Now, the formula to calculate ED, CB, and CS is as follows:

$$ED = \sqrt{\left\{ \left((P_{i1} - P_{j1})^2 \right) + \left((P_{i2} - P_{j2})^2 \right) + \cdots + \left((P_{in} - P_{jn})^2 \right) \right\}}$$

$$CB = \left| P_{i1} - P_{j1} \right| + \left| P_{i2} - P_{j2} \right| + \cdots + \left| P_{in} - P_{jn} \right|$$

$$CS = \text{sim}\,(P_i, P_j) = \cos\,(P_i, P_j) = \frac{\vec{P_i}.\vec{P_j}}{|\vec{P_i}| \times |\vec{P_j}|}$$

3 Proposed Redundancy-Driven Modified Tomek-Link Based Undersampling (RUSTBoost) Method

In this section, to overcome the drawback of RUSBoost which used random under-sampling as resampling technique and AdaBoost as boosting technique, we have proposed a new hybrid sampling or boosting model which uses redundancy-driven modified Tomek-link based undersampling as undersampling technique along with AdaBoost as boosting technique to learn for imbalanced training data and named as RUSTBoost. It is designed to increase the performance and effectiveness in terms of accuracy and Geometric mean (G-mean).

The working of RUSTBoost method is shown as block diagram in Fig. 1. Here, original dataset is divided into two classes for preprocessing: the first one is majority class patterns and other one is minority class patterns. For majority class patterns, preprocessing is performed for undersampling using redundancy-driven modified Tomek-link undersampling algorithm as discussed above in related work.

After the preprocessing using Tomek-link is over, new training dataset is formed by combining the obtained refined majority class patterns and minority class patterns. New training dataset and its weight distribution both are now given to the boosting algorithm to be trained. Here, boosting is done by SVM, KNN, logistic regression, and decision tree. SVM is mainly used for regression and data classification which

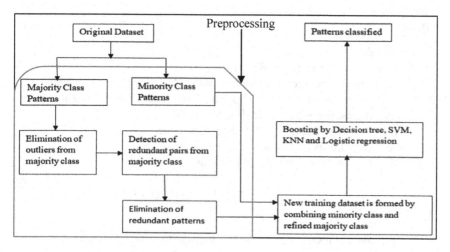

Fig. 1 Block diagram of RUSTBoost

became commonly used classifier due to its theoretical concepts. The objective is to find a linear separating hyperplane and the equation is given as

$$wf(x) + b = 0$$

where $f(x)$ is a nonlinear function for higher dimension space, w is the weight vector, and b is the bias. In case of KNN, input is the K-nearest training samples and output is the class membership for classification and property value of object for regression, where regression is the process of estimating the relationship among variables. Logistic regression associates in analyzing the link between categorical-dependent variables and a variable quantity that predicts the likelihood of incidence of an occasion by fitting knowledge to a supply curve. Binary logistic regression and multinomial logistic regression are two models of logistic regression. Decision tree in boosting method is a binary tree like structure which is used for the logical representation of decisions and decision-making. The tree consists of root node, leaf node, and internal nodes. As per the condition given in the node, we will decide whether we will move to left or right.

Now, we will elaborate the RUSTBoost algorithm in detail.

RUSTBoost is mainly performed on the refined dataset A_B but original dataset A is also required.

Input to the procedure is as follows: Set A of instances $(P_1, Q_1), \ldots, (P_n, Q_n)$ with minority class $Q^t \in B, |B| = 2$, T is the number of classifiers in the boosting, r where $1 \leq r \leq R$, and h_r is the iteration for weak learner hypothesis.

Output from the procedure is $H(P) = \arg\max(Q \in B) \sum_{r=1}^{R} h_r(P, Q) * \log\left(\frac{1}{\beta_r}\right)$

The procedure for RUSTBoost algorithm is given below:

WeakLearn():
Begin

1. Initialize $D_1(i) = \frac{1}{n} \forall i$
2. Repeat for $r = 1, 2, \ldots, R$
   (a) Create a temporary training dataset A_r' by applying redundancy-driven modified Tomek-link based undersampling on original training dataset with weight distribution D_r'.
   (b) Call WeakLearn, by providing A_r' and their weights D_r'.
   (c) Hypothesis is obtained as $h_r : X \times B \to [0, 1]$.
   (d) Pseudo-loss is calculated (for A and D_r) as

$$\epsilon_r = \sum_{(i,Q):Q_i \neq Q} D_r(i)(1 - h_r(P_i, Q_i) + h_r(P_i, Q))$$

   (e) Weight update parameter is calculated by using $\beta_r = \frac{\epsilon_r}{1 - \epsilon_r}$.

(f) Update D_r using $D_{r+1}(i) = D_r(i)\beta_r^{\frac{1}{2}(1+h_r(P_i,Q_i)-h_r(P_i,Q:Q\neq Q_i))}$

(g) Normalize D_{r+1}: Let $Z_r = \sum_i D_{r+1}(i)$, then $D_{r+1}(i) = \frac{D_{r+1}(i)}{Z_r}$

Here, in the first step, the weights of each pattern are initialized to $\frac{1}{n}$. In the second step, R weak hypotheses are iteratively trained. The undersampling algorithm is used to remove the majority class patterns in step 2a. As a result, A'_r will have a new weight distribution D'_r. A weak learner creates the weak hypothesis h_r in step 2c. Each hypothesis of a classifier is called as a weak learner. The final hypothesis $H(P)$ is returned as a weighted vote of the R weak hypotheses.

4 Experimental Results

We conducted experiments using 16 datasets downloaded from various application domains like the University of California, Irvine (UCI), and NASA [16, 17]. The detailed instruction about the downloaded datasets which is used to analyze the experimental result is given in Table 1.

We are using decision tree, SVM, KNN, and logistic regression as a base learner, and different evaluation metrics like accuracy and G-mean are considered. Our experiment shows that the proposed method (RUSTBoost) performs significantly better than RUSBoost and SMOTEBoost on benchmark datasets using F-measure and overall accuracy. After applying the algorithm, accuracy is being calculated for a total of 16 datasets and for all of the methods, i.e., SMOTEBoost, RUSBoost, and RUSTBoost, the difference in accuracy and G-mean is shown in Tables 2 and 3. Now, the graph for all the classifier, i.e., KNN, LR, DT, and SVM for accuracy and G-mean, is shown in Figs. 2, 3, 4, 5, 6, 7, 8, and 9.

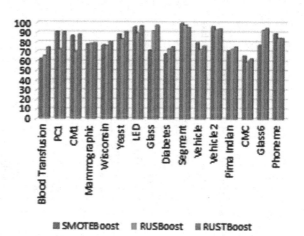

Fig. 2 Accuracy of SMOTEBoost, RUSBoost, and RUSTBoost for decision tree

Table 1 Detailed information about datasets

S.No	Dataset	Minority class	Majority class	Instances	Attributes	No. of minority	No. of majority	Imbalance ratio (IR)*
1	Blood transfusion	1	0	748	5	178	570	3.2022
2	PC1	1	0	1109	22	77	1032	13.4025
3	CM1	1	0	498	22	49	449	9.1632
4	Mammographic	1	0	1035	6	445	590	1.3258
5	Wisconsin	1	0	683	11	106	590	5.5660
6	yeast	1	0	1005	9	99	906	9.1515
7	Led	1	0	798	8	37	761	20.5675
8	Glass	1	0	897	10	60	837	13.95
9	Diabetes	1	0	768	9	268	506	1.8880
10	Segment	1	0	2308	20	329	1979	6.01519
11	Vehicle	1	0	846	19	217	846	3.8986
12	Vehicle2	1	0	846	19	199	846	4.2512
13	Pima Indian	1	0	768	9	268	500	1.8656
14	CMC	1	0	850	10	208	642	3.0865
15	Glass6	1	0	214	10	29	185	6.3793
16	Phoneme	1	0	5404	6	1586	3818	2.4073

Table 2 Performance comparison of SMOTEBoost, RUSBoost, and RUSTBoost in terms of accuracy

	Datasets	SMOTEBoost (KNN, LR, DT, SVM) (KNN, LR, DT, SVM)	RUSBoost (KNN, LR, DT, SVM) (KNN, LR, DT, SVM)	RUSTBoost (KNN, LR, DT, SVM) (KNN, LR, DT, SVM)
1	Blood transfusion	57.27,64.79,62.34,63.55	64.74,71.41,65.55,77.14	69.15,76.87,74.21,76.88
2	PC1	86.92,80.31,90.77,84.73	78.76,84.32,72.42,91.55	90.96,91.47,90.62,90.96
3	CM1	82.13,86.19,86.77,81.77	75.35,79.74,69.71,87.57	84.36,88.18,88.18,89.97
4	Mammographic	73.78,78.58,77.79,81.77	77.33,79.54,78.41,77.84	77.16,78.05,79.06,77.55
5	Wisconsin	77.61,78.5,77.05,78.77	78.2,79.21,76.34,77.06	80.2,80.8,80.34,77.65
6	Yeast	85.88,78.23,88.36,86.44	87.06,87.72,83.48,89.82	91.15,91.13,91.04,90.55
7	LED	95.12,93.86,96.05,95.56	88.99,91.21,89.75,91.21	96.77,96.65,96.89,97.26
8	Glass	97.75,89.32,71.51,86.61	80.46,89.41,91.97,89.7	93.97,94.75,97.65,93.33
9	Diabetes	68.21,73.81,68.02,74.1	74.1,75.4,73.05,74.21	70.18,75.65,75.14,75.14
10	Segment	98.57,77.18,99.65,99.48	97.18,88.39,97.79,97.79	99,88.39,95.19,99.61
11	Vehicle	72.58,72.87,79.44,76.83	72.47,78.83,72.47,76.14	73.4,79.8,75.9,77.56
12	Vehicle2	93.36,96.92,96.35,96.93	89.35,93.37,93.37,93.37	93.49,96.22,94.09,94.68
13	Pima Indian	69.41,75.53,71.36,75.53	73.17,75.66,73.18,75.8	70.18,75.65,75.14,75.14
14	CMC	62.63,57.59,66,57.59	54.35,61.52,60.01,64.11	59.22,65.67,62.9,67.48
15	Glass6	94.49,96.35,77.27,96.19	90.51,92.5,93.01,93.41	92.22,94.2,95,92.5
16	Phoneme	88.71,72.87,89.6,73.83	86.05,74.87,84.84,76.92	85.14,73.93,84.48,76.28

Table 3 Performance comparison of SMOTEBoost, RUSBoost, and RUSTBoost in terms of G-mean

	Datasets	SMOTEBoost (KNN, LR, DT, SVM) (KNN, LR, DT, SVM)	RUSBoost (KNN, LR, DT, SVM) (KNN, LR, DT, SVM)	RUSTBoost (KNN, LR, DT, SVM) (KNN, LR, DT, SVM)
1	Blood transfusion	0.23,0.24,0,0.33	0.11,0,0,0	0.24,0.32,0.24,0.34
2	PC1	0.49,0.13,0.5,0	0.59,0,0.79,0.33	0.5,0.23,0.49,0.42
3	CM1	0.42,0.01,0.44,0.44	0.43,0.72,0.58,0	0.44,0.45,0.44,0
4	Mammographic	0.56,0.62,0.54,0.64	0.6,0.6,0.9,0.1	0.67,0.65,0.63,0.67
5	Wisconsin	0.29,0.5,0,0.41	0.57,0.57,0.79,0.57	0.49,0.42,0.29,0.42
6	Yeast	0.72,0.41,0.64,0.6	0.72,0.49,0.93,0.82	0.59,0.51,0.71,0.76
7	LED	0.5,0.5,0.5,0.5	0.07,0,0.5,0	0.13,0.01,0,0.06
8	Glass	0.69,0.98,0.82,0	0.61,0.79,1,0	0.79,0.13,0.81,0
9	Diabetes	0.55,0.63,0.71,0.72	0.67,0.74,0.96,0.72	0.7,0.7,0.73,0.68
10	Segment	0.95,0.66,0.96,0.97	0.95,0.98,0.97,0.97	0.96,0.01,0.97,0.97
11	Vehicle	0.75,0.97,0.73,0.96	0.69,0.73,0.73,0.69	0.67,0.72,0.68,0.68
12	Vehicle2	0.91,0.93,0.97,0.96	0.89,0.97,0.77,0.94	0.89,0.95,0.94,0.93
13	Pima Indian	0.68,0.71,0.68,0.75	0.65,0.72,0.69,0.94	0.7,0.7,0.73,0.68
14	CMC	0.56,0.58,0.34,0.64	0.49,0.53,0.57,0.54	0.49,0.57,0.57,0.62
15	Glass6	0.72,0.95,0.79,0.69	0.65,0.69,0.79,0.69	0.72,0.64,0.82,0.69
16	Phoneme	0.54,0.71,0.84,0.71	0.84,0.7,0.8,0	0.83,0.72,0.8,0.72

Fig. 3 Accuracy of SMOTEBoost, RUSBoost, and RUSTBoost for KNN

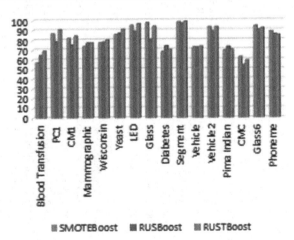

The performance can be measured by parameters like accuracy and G-mean which can be calculated as

$$\text{Accuracy} = \frac{TP + TN}{TP + TN + FP + FN} \text{ and G - mean} = \sqrt{\frac{TP}{TP + FN} \times \frac{TN}{TN + FP}}$$

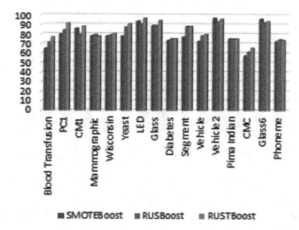

Fig. 4 Accuracy of SMOTEBoost, RUSBoost, and RUSTBoost for logistic regression

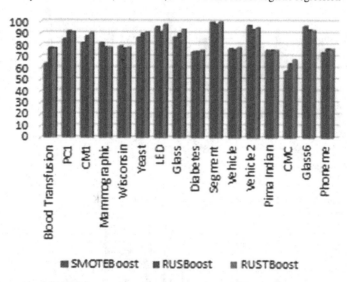

Fig. 5 Accuracy of SMOTEBoost, RUSBoost, and RUSTBoost for SVM

Here, True Positive (TP), True Negative (TN), False Positive (FP), and False Negative (FN) can be defined by the following confusion matrix which contains info regarding actual and foreseen classifications done by a classifier.

		Predicted	
		Positive	Negative
Actual	Positive	TP	FN
	Negative	FP	TN

A. Kumar et al.

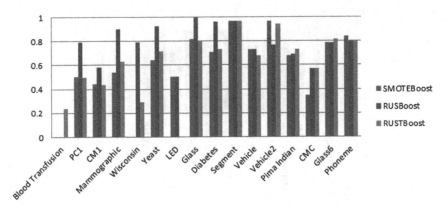

Fig. 6 G-mean of SMOTEBoost, RUSBoost, and RUSTBoost for decision tree

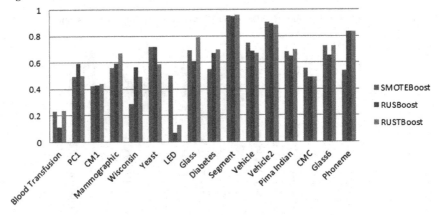

Fig. 7 G-mean of SMOTEBoost, RUSBoost, and RUSTBoost for KNN

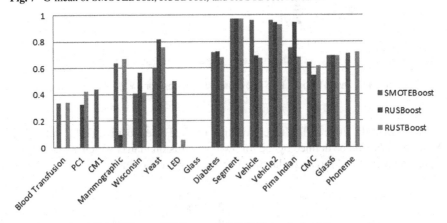

Fig. 8 G-mean of SMOTEBoost, RUSBoost, and RUSTBoost for logistic regression

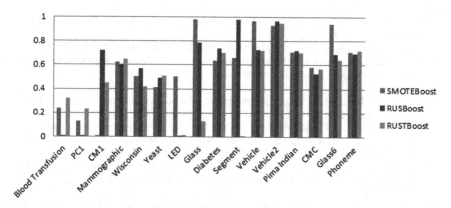

Fig. 9 G-mean of SMOTEBoost, RUSBoost, and RUSTBoost for SVM

5 Conclusion and Future Work

In this paper, we have proposed RUSTBoost method which performs significantly better than RUSBoost and SMOTEBoost to overcome the loss of information associated with RUSBoost. It is simple and faster approach similar to RUSBoost for learning from training datasets where class distribution is unequal. It is also well suitable for noisy imbalanced datasets. In future, we may further improve the accuracy and G-mean of RUSTBoost by using other learner and boosting methods. Since the proposed method is only for binary class, one can also design the same for multiclass problem.

References

1. Nyguyen, G.H., Bouzerdoum, A., Phung, S.L.: Learning pattern classification tasks with imbalanced data sets. University of Wollongong, Australia (2009)
2. Chawla, N.V., et al.: SMOTE: synthetic minority over-sampling technique. J. Artif. Intell. Res. **16**, 321–327 (2002)
3. Yoav, F., Schapire, R.E.: Experiments with a new boosting algorithm, machine learning. In: Proceedings of the Thirteenth International Conference (1996)
4. Chawla, N.V., Lazarevic, L., Hall, O., Bowyer, K.W.: SMOTEBoost: improving prediction of the minority class in boosting. Proc. Knowl. Discov. Databases **2838**, 107–119 (2003)
5. Seiffert, C., Khoshgoftaar, T., Van Hulse, J., Napolitano, A.: Rusboost: A hybrid approach to alleviating class imbalance. IEEE Trans. Syst. Man Cybern. A Syst. Humans **40**(1), 185–197 (2010)
6. Tomek, Ivan: The modifications of CNN. IEEE Trans. Syst. Man Cybern. **6**, 769–772 (1976)
7. Galar, M., Fernandez, A., Barrenechea, E., Bustince, H., Herrera, F.: A review on ensembles for the class imbalance problem: bagging-, boosting-, and hybrid-based approaches. IEEE Trans. Syst. Man Cybern. C Appl. Rev. **42**(4), 463–484 (2011)
8. Nguyen, G.H., Bouzerdoum, A., Phung, S.: Learning pattern classification tasks with imbalanced data sets. In: Yin, P. (ed.) Pattern recognition, pp. 193–208. In-Tech, Crotia (2009)

9. Wilson, D.L.: Asymptotic properties of nearest neighbor rules using edited data. IEEE Trans. Syst. Man Cybern. **3**, 408–421 (1972)
10. Hart, P.: The condensed nearest neighbor rule. IEEE Trans. Inf. Theory **14**(3), 515 (2016)
11. Balasundaram, S., Gupta, D.: On optimization based extreme learning machine in primal for regression and classification by functional iterative method. Int. J. Mach. Learn. Cybernet. **7**(5), 707–728 (2016)
12. Tanveer, M., Khan, M.A., Ho, S.S.: Robust energy-based least squares twin support vector machines. Appl. Intell. **45**(1), 174–186 (2016). https://doi.org/10.1007/s10489-015-0751-1
13. Gupta, D., Borah, P., Prasad, M.: A fuzzy based Lagrangian twin parametric-margin support vector machine (FLTPMSVM). In: 2017 IEEE Symposium Series on Computational Intelligence (SSCI), Honolulu, HI, pp. 1–7 (2017) https://doi.org/10.1109/ssci.2017.8280964
14. Balasundaram, S., Gupta, D., Prasad, S.C.: A new approach for training Lagrangian twin support vector machine via unconstrained convex minimization. Appl. Intell. **46**(1), 124–134 (2017). Springer
15. Devi, D., Biswasb, S.K., Purkayasthac, B.: Redundancy-driven modified Tomek-link based undersampling: a solution to class imbalance. Pattern Recogn. Lett. **93**, 3–12 (2016)
16. http://www.ics.uci.edu/~mlearn/~MLRepository.html
17. Metrics data program: http://mdp.ivv.nasa.gov

A Survey on Medical Diagnosis of Diabetes Using Machine Learning Techniques

Ambika Choudhury and Deepak Gupta

Abstract While designing medical diagnosis software, disease prediction is said to be one of the captious tasks. The techniques of machine learning have been successfully employed in assorted applications including medical diagnosis. By developing classifier system, machine learning algorithm may immensely help to solve the health-related issues which can assist the physicians to predict and diagnose diseases at an early stage. We can ameliorate the speed, performance, reliability, and accuracy of diagnosing on the current system for a specific disease by using the machine learning classification algorithms. This paper mainly targets the review of diabetes disease detection using the techniques of machine learning. Further, PIMA Indian Diabetic dataset is employed in machine learning techniques like artificial neural networks, decision tree, random forest, naïve Bayes, k-nearest neighbors, support vector machines, and logistic regression and discussed the results with their pros and cons.

Keywords Machine learning · Diabetes · Support Vector Machines
Decision Tree · Logistic Regression · k-Nearest Neighbors

1 Introduction

Machine learning is one of the most essential domains in the field of research with the goal of predicting and conducting a systematic review. According to Official World Health Organization data, India has the highest number of diabetes [1]. The total number of diabetics in India in the year 2000 is positioned as 31.7 million and by the year 2030, it is presumed to ascend to 79.4 million [2]. The disease, diabetes,

A. Choudhury · D. Gupta (✉)
Computer Science and Engineering Department, National Institute of Technology, Yupia,
Arunachal Pradesh, India
e-mail: deepakjnu85@gmail.com

A. Choudhury
e-mail: ambikachoudhury121@gmail.com

© Springer Nature Singapore Pte Ltd. 2019
J. Kalita et al. (eds.), *Recent Developments in Machine Learning and Data Analytics*,
Advances in Intelligent Systems and Computing 740,
https://doi.org/10.1007/978-981-13-1280-9_6

is chronic and has become one of the leading lifestyle ailments, characterized by prolonged elevated blood sugar levels. Failure of organs like liver, heart, kidneys, stomach, etc. is caused in the long run due to the effect of diabetes.

Classification [3] of diabetes mellitus can be described as follows:

i.   Type I diabetes: The diabetic condition that depends on insulin occurs mainly in children and adolescents because of the genetic disorders.
ii. Type II diabetes: Generally occurs in adults during the age of 40 years discernible by high blood sugar level.
iii. Pregnancy diabetes: The diabetes that occur during the pregnancy period.
iv. Diabetic retinopathy: This type of disorder leads to eye blindness.
v.   Diabetic neuropathy: Nerve disorder is the cause of this type of diabetes.

Factors Responsible for Diabetes:

i.   Combination of genetic susceptibility and environmental factor can cause diabetes.
ii. Overweight may lead to cause diabetes in the long run.
iii. If a parent or sibling has diabetes, then the risk is supposed to be increased.
iv. Aging increases risk of diabetes.
v.   More than 140/90 mm of Hg is linked to an increased risk of diabetes.
vi. Low levels of high-density lipoprotein (HDL) are also the cause of occurring the risk.

Complications Arise Due to Diabetes:

The complications progress moderately. Possible complications those are included for arising:

i.   Cardiovascular disease: Diabetes vividly increases the risk of various cardiovascular problems;
ii.   Damage in the nerves (Neuropathy);
iii. Damage in the kidneys (Nephropathy);
iv.   Damage in eyes (Retinopathy);
v.   Damage in foot: Deficient blood flow to the feet increases the risk;
vi.   Acute skin condition: Bacterial and fungal infections may happen;
vii. Impairment of hearing: The problems of hearing are common;
viii. Alzheimer's disease: Increases the chance of Alzheimer's disease.

2   Techniques Used for Diabetes Detection

The data samples are partitioned into the target class in classification technique and the same is predicted for each and every data point. For instance, we may classify a patient as "higher risk" or "lower risk" on the premise of their diseases using the data classification methods. Some of habitually used techniques are discussed in the following.

2.1 Support Vector Machine (SVM)

Support vector machine [4] is a supervised learning technique and represents the dataset as points in n-dimensional space, n being the number of features. The purpose of SVM is to establish a hyperplane which partitions the datasets in different sorts and the hyperplane should be at utmost margin from the various sorts. For robustness, the hyperplane is needed to be chosen in such a way that it is having high margin and maximizes the distances between the nearest data point of either class.
Advantages:

- SVM provides better accuracy and can easily handle complex nonlinear data points.
- It removes the overfitting nature of the samples.

Disadvantages:

- It is difficult to use in large datasets.
- Execution is comparative slow.

2.2 k-Nearest Neighbors (k-NN)

In this classification technique, the anonymous data points are discovered using the familiar data points which are known as nearest neighbors. k-Nearest neighbors (k-NN) [5] is conceptually simple and is also called as lazy learning, where "k" is the nearest neighbor. In k-NN algorithm, the aim is to vigorously recognize k samples in the training dataset which are identical to a new sample.
Advantages:

- It is easy to implement.
- Training is done in a faster manner

Disadvantages:

- Time becomes prohibitive for finding the nearest neighbor in the training data which is of huge size, thus making it slow.
- It requires large storage space.
- The transparency of knowledge representation is very poor.

2.3 Decision Tree

In the decision tree technique [6], on the premise of parameters, a tree- or graph-like shape is constructed and it contains a predefined target variable. Traversing from root to leaf is done for the decision to take, and the traversing is done till the criteria are met.

Advantages:

- Domain knowledge is not required for the decision tree construction.
- Inexactness of complex decision is minimized which results to assign exact values to the outcome of various actions.
- Easy interpretations and it can handle both numerical and categorical data.

Disadvantages:

- One output attribute is restricted for decision tree.
- Decision tree is an unstable classifier.
- Categorical output is generated.

2.4 Random Forest

Random forest [7] is a classifier that constitutes a few decision trees and considered as one of the dimensionality reduction methods. It is one of the ensemble methods for classification, regression, and other terms. It can be used to rank the importance of variables.
Advantages:

- Random forest improves classification accuracy.
- It works well with the dataset of large number of input variables.

Disadvantages:

- Random Forest is fast to train but once trained, it becomes slow to create predictions.
- It is slow to evaluate.
- Interpretation is very hard.

2.5 Naïve Bayes Approach

Based on the Bayes' theorem, naïve Bayes [8] is a supervised learning technique. To classify the text documents, it is one of the most successful known algorithms for learning because of its better outcomes in multi-class problems and rules of independence.
Advantages:

- Simple and easy to implement.
- More accuracy in result due to higher value of probability

Disadvantages:

- Strong assumption on the shape of data distribution.
- Loss of accuracy.

2.6 Artificial Neural Network

The artificial neural network [9] is aroused by the neural network of human being, and it is a combination of three layers, i.e., input layer, hidden layer, and output layer, which is also called as MLP (Multilayer Perceptron). The hidden layer is similar to neuron, and each hidden layer consists of probabilistic behavior.
Advantages:

- Ability to learn and model nonlinear and complex relationships.
- Ability to generalize the model and predict the unseen data.
- Resistant to partial damage.

Disadvantages:

- Optimizing the network can be challenging because of the number of parameters to be set in.
- For large neural networks, it requires high processing time.

2.7 Logistic Regression

Logistic regression is also known as logit model [10] for dichotomic output variables and was comprised for classification prediction of diseases. It is a statistical method for analyzing. Here, one or more than one independent variables ascertain the consequences.
Advantages:

- Easy to implement and very efficient to train.
- Can handle nonlinear effect and interaction effect.

Disadvantages:

- Cannot predict continuous outcomes.
- Vulnerable to overconfidence, i.e., the models can appear to have more predictive power than they actually do, resulting in overfitting.
- Requires quite a large sample size to achieve stable results.

3 Literature Review

S. No.	Author	Methodology	Central idea	Merits and demerits
1	Alade et al. [11]	• Artificial neural network • Backpropagation method • Bayesian regulation algorithm	In this paper, a four-layer artificial neural network is designed. Backpropagation method and Bayesian regulation (BR) algorithms are used to train and avoid overfitting the dataset. The training of data is done in such a way that it forms a single and accurate output displaying in the regression graphs	*Merits* Easy accessibility due to the web-based application Diagnosis can be communicated without having to be around the patient *Demerits* The system is in online platform so it will not work where there will be loss of connectivity
2	Alic et al. [12]	• Artificial neural network • Bayesian networks	Here, the application of two classification techniques, i.e., artificial neural network and naïve Bayes, are compared for classification and artificial neural network gives more accurate result than the Bayesian networks	*Merits* This paper compared the mean accuracy and is an improved version of 20 main papers *Demerits* Assuming the independence among the observed nodes, the naïve Bayes gives less accurate results
3	Khalil et al. [13]	• SVM • K-means algorithm • Fuzzy-C means algorithm • Probabilistic neural network (PNN)	Prediction of depression operation by comparing, developing, and applying, machine learning techniques is consolidated by this paper	*Merits* SVM gives better accuracy result than the other techniques *Demerits* More accuracy can be measured Optimization is needed
4	Carrera et al. [14]	• SVM	A computer-assisted diagnosis is proposed in this paper on the basis of digital processing of retinal images to detect diabetic retinopathy. Classification of the grade of non-proliferative diabetic retinopathy at any retinal image is the main goal of this study	*Merits* Maximum sensitivity of 95% and a predictive capacity of 94% are attained Robustness has also been evaluated *Demerits* Both accuracy and sensibility are needed to be improved for the application of texture analysis

S. No.	Author	Methodology	Central idea	Merits and demerits
5	Lee et al. [15]	• Naïve Bayes • Logistic regression • 10-fold cross-validation method	The aim of this study is to analyze the relation between hypertriglyceridemic waist (HW) phenotype and diabetes type II and to assess the predictive power of different phenotypes containing combinations of individual anthropometric measurements and triglyceride (TG) levels. WC or TG levels are weaker than the combination of HW phenotype and type II diabetes	*Merits* Can be easily used to identify the best phenotype or predictor of type II diabetes in different countries *Demerits* According to gender, the actual WC and TG values combined in predictive power are not the best method to predict type II diabetes
6	Huang et al. [16]	• SVM • Decision tree • Entropy methods	Support vector machines and entropy methods were applied in this paper to evaluate three different datasets, including diabetic retinopathy Debrecen, vertebral column, and mammographic mass. The most critical attributes in the dataset were used to construct decision tree and comparing the classification accuracy of test data with the others. It was found that the mammographic mass dataset and diabetic retinopathy Debrecen dataset had better accuracy on the proposed method	*Merits* To solve the problem layer by layer, analysis of the training dataset step by step using the combination of support vector machine and deep learning is performed *Demerits* Improvisation of accuracy is needed in the future as the classification accuracies are not comparable to the result of the training datasets
7	Lee et al. [17]	• Naïve Bayes • Logistic regression	The main aim of this paper is to predict the fasting plasma glucose (FPG) status which has been used to diagnose diabetes disease. Both the machine learning algorithms, i.e., Naïve Bayes and logistic regression, are compared here. As an outcome, the naïve Bayes shows better result than logistic regression	*Merits* The prediction performances in females are comparatively higher than the males. The number of parameters that are measured is one of the main reasons for the difference between the previous study and this particular study *Demerits* This study cannot establish a cause–effect relationship. For higher FPG status, the models of this study may cause incorrect diagnosis

4   Numerical Results

In this paper, we have applied five supervised machine learning techniques termed as support vector machine, k-nearest neighbors, decision tree, naïve Bayes approach, and logistic regression for the classification of diabetes disease samples. We have performed these algorithms on PIMA Indian Diabetic dataset [18] in which the total number of samples is 768 and the total number of attributes is 9. Last column represents a class label where positive class represents person is diabetic and negative class represents person is not diabetic. In our experiments, 250 samples are chosen as training data and remaining 518 samples for test data. Here, true positive (TP) represents number of samples with the absence of diabetes predicted as the absence of diabetes, false positive (FP) represents number of samples with the presence of diabetes predicted as the absence of diabetes, true negative (TN) represents number of samples with the presence of diabetes predicted as the presence of diabetes, and false negative (FN) represents number of samples with the absence of diabetes predicted as the presence of diabetes. Here, we have used the following quality measures to check the performance of machine learning techniques:

- Accuracy = (TP + TN)/(TP + TN + FP + FN)
- Recall (Sensitivity or true positive rate) = TP/(TP + FN)
- Specificity (True negative rate) = TN/(TN + FP)
- Precision = TP/(TP + FP)
- Negative predicted value (NPV) = TN/(TN + FN)
- False positive rate (FP rate) = FP/(FP + TN)
- Rate of misclassification (RMC) = (FP + FN)/(TP + TN + FP + FN)
- F_1-measure = 2 * (precision * recall)/(precision + recall)
- G-mean = sqrt(precision*recall)

The confusion matrix of prediction results for support vector machine, k-nearest neighbors, decision tree, naïve Bayes approach, and logistic regression are tabulated in Tables 1, 2, 3, 4, and 5. One can observe from these tables that SVM is having the highest number of true positive (absence of diabetes predicted as the absence of diabetes) and naïve Bayes is having the highest number of true negatives (presence of diabetes predicted as the presence of diabetes). Further, SVM is having the lowest number of false negative and naïve Bayes is having the lowest number of false positive. We have also drawn the classification results of these methods in Fig. 1.

We have computed the value of TP, FP, TN and FN for SVM, k-NN, Decision Tree, Naïve Bayes and Logistic Regression and depicted in Table 6. Further, we have computed quality measures termed as accuracy, recall, true positive rate, precision, negative predicted value, false positive rate, rate of misclassification, F_1-measure, and G-mean based on predicted result by using SVM, k-NN, decision tree, naïve Bayes, and logistic regression and depicted in Table 7 and also shown in Fig. 2. Here, one can conclude from this Table 7 that logistic regression has performed better among all five algorithms, whereas decision tree is having the lowest accuracy. In terms of F_1-measure, SVM performed better than other compared methods.

Table 1 Confusion matrix using support vector machine (SVM)

Support vector machine (SVM)		Predicted class		
		Absence	Presence	Actual total
Actual class	Absence	310 (77.5%)	36 (30.5%)	346
	Presence	90 (22.5%)	82 (69.5%)	172
	Total predicted	400	118	518

Table 2 Confusion matrix using k-nearest neighbors (k-NN)

k-nearest neighbors (k-NN)		Predicted class		
		Absence	Presence	Actual total
Actual class	Absence	281 (81.5%)	65 (37.6%)	346
	Presence	64 (18.5%)	108 (62.4%)	172
	Total predicted	345	173	518

Table 3 Confusion matrix using decision tree

Decision tree		Predicted class		
		Absence	Presence	Actual total
Actual class	Absence	234 (80.7%)	112 (49.1%)	346
	Presence	56 (19.3%)	116 (50.9%)	172
	Total predicted	290	228	518

Table 4 Confusion matrix using naïve Bayes

Naive Bayes		Predicted class		
		Absence	Presence	Actual total
Actual class	Absence	272 (85.3%)	74 (37.2%)	346
	Presence	47 (14.7%)	125 (62.8%)	172
	Total predicted	319	199	518

Table 5 Confusion matrix using logistic regression

Logistic regression		Predicted class		
		Absence	Presence	Actual total
Actual class	Absence	308 (79.8%)	38 (28.8%)	346
	Presence	78 (20.2%)	94 (71.2%)	172
	Total Predicted	386	132	518

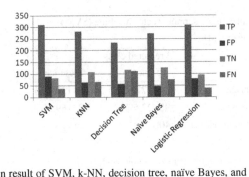

Fig. 1 Classification result of SVM, k-NN, decision tree, naïve Bayes, and logistic regression in terms of TP, FP, TN, and FN

Table 6 Values of TP, FP, TN, and FN for SVM, k-NN, decision tree, naïve Bayes, and logistic regression

	SVM	k-NN	Decision tree	Naïve Bayes	Logistic regression
TP	310	281	234	272	308
FP	90	64	56	47	78
TN	82	108	116	125	94
FN	36	65	112	74	38

Table 7 Classification performance measure indices of SVM, k-NN, decision tree, naïve Bayes, and logistic regression

Measures/ Methods	SVM	k-NN	Decision tree	Naïve Bayes	Logistic regression
Accuracy	0.7568	0.751	0.6757	0.7664	**0.7761**
Recall	0.8960	0.8121	0.6763	0.7861	0.8902
TP rate	0.4767	0.6279	0.6744	0.7267	0.5465
Precision	0.7750	0.8145	0.8069	0.8527	0.7979
NPV	0.6949	0.6243	0.5088	0.6281	0.7121
FP rate	0.5233	0.3721	0.3256	0.2733	0.4535
RME	0.2432	0.2490	0.3243	0.2336	0.2239
F_1-measure	0.8311	0.8133	0.7359	0.8180	0.8415
G-mean	0.6536	0.7141	0.6754	0.7559	0.6975

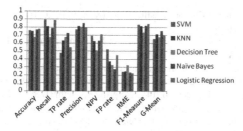

Fig. 2 Classification performance measurements of SVM, k-NN, decision tree, naïve Bayes, and logistic regression

5 Conclusion and Future Scope

At an early period of diagnosing the diabetic disease, machine learning techniques can help the physicians to diagnose and cure diabetic diseases. This paper presents a comprehensive comparative study on various machine learning algorithms for PIMA Indian Diabetic dataset. The comparative study is based on the parameters such as accuracy, recall, specificity, precision, negative predicted value (NPV), false positive rate (FP rate), rate of misclassification (RMC), F_1-measure, and G-mean. Increase in classification accuracy helps to improvise the machine learning models and yields better results. The performance analysis is analyzed in terms of accuracy rate among all the classification methods such as decision tree, logistic regression, k-nearest neighbors, naïve Bayes, and SVM. It is found that logistic regression gives the most accurate results to classify the diabetic and nondiabetic samples. Future work can be done in such a manner that type I and type II diabetes can be made possible to identify in a single classifier.

References

1. Mohan, V., et al.: Epidemiology of type 2 diabetes: Indian scenario. Indian J. Med. Res. **125**(3), 217 (2007)
2. Kaveeshwar, S.A., Cornwall, J.: The current state of diabetes mellitus in India. Australas. Med. J. **7**(1), 45 (2014)
3. Sumangali, K., Geetika, B. S. R., Ambarkar, H.: A classifier based approach for early detection of diabetes mellitus. In: 2016 International Conference on Control, Instrumentation, Communication and Computational Technologies (ICCICCT). IEEE (2016)
4. Cortes, C., Vapnik, V.: Support-vector networks. Mach. Learn. **20**(3), 273–297 (1995)
5. Friedman, J.H., Baskett, F., Shustek, L.J.: An algorithm for finding nearest neighbors. IEEE Trans. Comput. **100**(10), 1000–1006 (1975)
6. Argentiero, P., Chin, R., Beaudet, P.: An automated approach to the design of decision tree classifiers. IEEE Trans. Pattern Anal. Mach. Intell. **1**, 51–57 (1982)
7. Breiman, L.: Random forests. Mach. Learn. **45**(1), 5–32 (2001)
8. Kotsiantis, S.B., Zaharakis, I., Pintelas, P.: Supervised machine learning: a review of classification techniques. Emerg. Artif. Intell. Appl. Comput. Eng. **160**, 3–24 (2007)

9. Zhang, G.P.: Neural networks for classification: a survey. IEEE Trans. Syst. Man Cybern. C Appl. Rev. **30**(4), 451–462 (2000)
10. Hosmer Jr., D.W., Lemeshow, S., Sturdivant, R.X.: Applied Logistic Regression, vol. 398. Wiley, New York (2013)
11. Alade, O.M., Sowunmi, O.Y., Misra, S., Maskeliūnas, R., Damaševičius, R.: A neural network based expert system for the diagnosis of diabetes mellitus. In: International Conference on Information Technology Science, pp. 14–22. Springer, Cham (Dec 2017)
12. Alić, B., Gurbeta, L., Badnjević, A.: Machine learning techniques for classification of diabetes and cardiovascular diseases. In: 2017 6th Mediterranean Conference on Embedded Computing (MECO), pp. 1–4. IEEE (June 2017)
13. Khalil, R.M., Al-Jumaily, A.: Machine learning based prediction of depression among type 2 diabetic patients. In: 2017 12th International Conference on Intelligent Systems and Knowledge Engineering (ISKE), pp. 1–5. Nanjing (2017)
14. Carrera, E.V., González, A., Carrera, R.: Automated detection of diabetic retinopathy using SVM. In: 2017 IEEE XXIV International Conference on Electronics, Electrical Engineering and Computing (INTERCON). IEEE (2017)
15. Lee, B.J., Kim, J.Y.: Identification of type 2 diabetes risk factors using phenotypes consisting of anthropometry and triglycerides based on machine learning. IEEE J. Biomed. Health Inform. **20**(1), 39–46 (2016)
16. Huang, Y.-P., Nashrullah, M.: SVM-based decision tree for medical knowledge representation. In: 2016 International Conference on Fuzzy Theory and Its Applications (iFuzzy). IEEE (2016)
17. Lee, B.J., et al.: Prediction of fasting plasma glucose status using anthropometric measures for diagnosing type 2 diabetes. IEEE J. Biomed. Health Inform. **18**(2), 555–561 (2014)
18. Smith, J.W., Everhart, J.E., Dickson, W.C., Knowler, W.C., Johannes, R.S.: Using the ADAP learning algorithm to forecast the onset of diabetes mellitus. In: Proceedings of the Symposium on Computer Applications and Medical Care, pp. 261–265. IEEE Computer Society Press (1988)

How Effective Is the Moth-Flame Optimization in Diabetes Data Classification

Santosh Kumar Majhi

Abstract Diabetes is one of the most common disease in both male and females worldwide and is a major cause of deaths. Many statistical techniques and cognitive science approaches like machine intelligence are used to detect the type of diabetes in a patient for getting correct accuracy. This paper uses a metaheuristic algorithm named Moth-Flame Optimization for diabetes data classification. The MFO is used to update the weights of the feed foreword neural network. The performance of the proposed algorithm is evaluated by experimenting Wisconsin Hospital dataset. A comparative analysis has been performed with respect to the recently published works.

Keywords Classification · Diabetes data · MFO · Feed forward neural network

1 Introduction

Diabetes is one of the most dangerous abnormalities in human beings. It has the notorious distinction of being the fifth leading cause for deaths in the past decade. With passage of time, more and more people are getting affected by this disease. It usually occurs to people having higher blood sugar levels. The reason behind the disease is lesser insulin production in the body. It is reported that over 382 million people all over the world suffer from diabetes. Diabetes can be classified into two categories, namely infectious and constitutional. Diabetes Mellitus (DM) is a constitutional disease caused by the consumption of processed foods and lack of exercises. The insufficient secretion of insulin leads to diabetes. This, in turn, starts abnormality in the body and hampers the metabolism. The treatment involves glucose level control in the body. Advances in machine learning techniques have been used for prognosis, diagnosis, and screening of diabetes [1]. The neural networks are vital part

S. K. Majhi (✉)
Department of Computer Science and Engineering, Veer Surendra Sai
University of Technology, Burla 768018, Odisha, India
e-mail: smajhi_cse@vssut.ac.in

© Springer Nature Singapore Pte Ltd. 2019
J. Kalita et al. (eds.), *Recent Developments in Machine Learning and Data Analytics*,
Advances in Intelligent Systems and Computing 740,
https://doi.org/10.1007/978-981-13-1280-9_7

of the medical decision support system. In literature, many works have been reported regarding the use of neural network in medical systems for data classification such as neural-fuzzy classifiers [2], fuzzy neural network, fuzzy probabilistic neural networks [3], and fuzzy min–max neural network [4]. Neural network is highly regarded amongst decision techniques to classify medical data, as it provides rapid processing and maintains multifarious relation between the input parameters and the target values in the given dataset [5, 6]. Keeping in view the key features of neural networks, the feed foreword neural network (FFANN) has been investigated for designing a decision support system for diabetes diseases in this paper. To further enhance the FFANN for diabetes data classification, some nature-inspired algorithms such as Moth-Flame Optimization (MFO) have been used to improve accuracy in the system. Here, the motivation is to use the advantages of optimization algorithms in neural network weight adjustment to produce correct classification results. Moreover, the hybrid learning algorithm—FFANN has the capability to learn incrementally, predict output, and classify correctly.

2 Materials and Methods

2.1 Overview of FFANN

In our work, we considered FFANN for classification due to its proven applications in many engineering domains. In the present work, we have considered three layers, namely input layer, output layer, and one hidden layer for the specification of the feed forward network. In diabetes dataset, eight different parameters are used to classify the type of diabetes. In our work, we have considered all the eight parameters for diabetes data classification. Here, nine input nodes correspond to each of the attributes from the dataset, four nodes in the hidden layer, and one node in the output layer specifies for the design of the neural network (Refer Fig. 1). In Fig. 1, the inputs from $x1$ to $x8$ are multiplied by some weights and fed into each node of hidden layer. All the inputs are summed and fed to the activation function to get the output of each node. The same process is repeated for the output layer. Sometimes, we require biases in each node of hidden layer and output layer. These biases are added to the results of the activation function of that node and become the output of that node. From the output of each node of output layer, we can understand the classes to which the corresponding input belongs. The weights are adjusted by the MFO algorithm to minimize error.

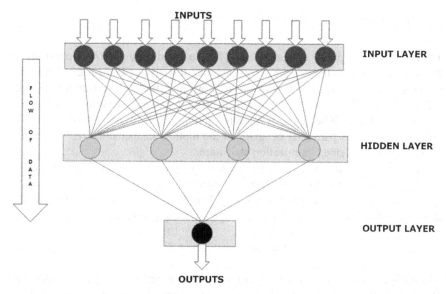

Fig. 1 Feed foreword network model

2.2 Dataset Information

The Pima Indian diabetes dataset is used here. This dataset is originally taken from National Institute of Diabetes and Digestive and Kidney Diseases. The objective is to predict the diabetes based on diagnostic measurements whether a patient has diabetes. The dataset samples are taken from the population living near Phoenix, Arizona, USA. This diabetes check was conducted only on female patients during the times of their pregnancy. It contains nine parameters: eight input parameters and one output parameter.

Attributes:

1. Pregnancies: Number of times the patient was pregnant
2. Glucose: Plasma glucose concentration for 2 h in an oral glucose tolerance test
3. Insulin: 2-h serum insulin (mu U/mL)
4. Blood Pressure: Diastolic blood pressure (mm Hg)
5. Skin Thickness: Triceps skin fold thickness (mm)
6. BMI: Body Mass Index (weight in kg/(height in m)^2)
7. Diabetes Pedigree Function: Diabetes pedigree function
8. Age: Age (in years)
9. Outcome: Class variable (class 0 or class 1)

2.3 Moth-Flame Optimization Algorithm

The Moth-Flame Optimization algorithm [7] is inspired from the travel in a straight line at night, moths reference the moon such that it is always to their left. Since the moon is far away, so referencing it helps to move in a straight line. But they are not able to distinguish between the moon and artificial lights such as flames. So if a moth is nearer to a flame, it becomes its reference point. This leads to the moth spiraling into the flame. This particular behavior of the moth is mathematically modeled to devise the Moth-Flame Optimization algorithm.

Since, moths go towards the flame in a spiral manner, so a spiral function is used to move a moth towards a single flame. In every iteration, one flame is assigned to each moth so that they move towards it only. This is done so as to avoid getting stuck at local optima. Here, a logarithmic spiral given in Eq. (1) is used.

$$D \times ebt \times \cos(2\pi t) + F. \tag{1}$$

where b is a constant, t is a random number present in range $[-1, 1]$, F is flame, and D is distance between moth M and flame F. The distance is calculated by Eq. (2).

$$D = |F - M|. \tag{2}$$

Due to these equations, the next position of the moth under consideration is always with respect to its corresponding flame. "t" defines the closeness of the next position of the moth to the flame. If $t = 1$, the moth is farthest from its flame and if $t = -1$, the moth is closest to its flame. Thus, the moth flies around the flame and of course, not necessarily towards the flame. Here, there is exploration as well as exploitation. To make exploitation increase with every iteration, t can be any random number within the range values $[r, 1]$, where r changes linearly from -1 to -2 with multiple iterations.

The F matrix always has the n best solutions. The OF matrix and OM matrix are calculated after each iteration. The F matrix is sorted according to OF matrix. If a better solution is found by a moth at any iteration (known from OM), then that position replaces the last row of the F matrix. The moths are assigned flames in a chronological order. The last moth takes the last flame (worst) and first moth takes the first flame (best). This ensures that there is no trapping in local optima.

To increase exploitation in the later iterations, Eq. (3) has been used to minimize the number of flames.

$$\text{Number of flames} = n - (\text{iteration number}) * (n - 1)/\text{Total Iterations} \tag{3}$$

Due to this, in the last iterations, the positions of moths are updated only with respect to the best few possible flames.

The time complexity of this algorithm is given in Eqs. (4)–(6).

$$O(t \times (O(\text{sorting}) + O(\text{Position update}))) \tag{4}$$
$$= O(t \times (n \times \log(n)) + n \times d) \tag{5}$$
$$= O(t \times n \times \log(n) + t \times n \times d) \tag{6}$$

where t is the iteration count and d is the dimensions count.

2.4 Proposed Algorithm

The dataset has eight attributes. So, in the feedforward artificial neural network, eight nodes are used in the input layer and only one node is used in the output layer. By trial-and-error method, it has been found that only one hidden layer with four nodes is sufficient to produce satisfactory results. Since some values in the dataset are missing, so these need to be handled. There are different techniques to do so. Some of them are using mean of all the values for that attribute or using the mean of the values of that attribute that belongs to the same class or deleting that tuple, etc. Here, the whole tuple is deleted. After deletion, the dataset is updated and this dataset is used. Once the network is set up, the weights and biases need to be balanced. To do this, search agents are formed. Each search agent is an array of 45 floating point values in range $[-1, 1]$. The range is set arbitrarily. Here, 150 search agents are used. The determination of 45 search agents is determined as follows:

- There are eight input nodes to be connected to four hidden nodes. So, the total number of weights needed to connect them is $8 \times 4 = 32$.
- There is only one output node which is connected to the four hidden nodes. So, the number of weight needed is $4 \times 1 = 4$.
- Each of the nodes in hidden and output layer has a bias value. Thus, there are $4 + 1 = 5$ bias values.
- Thus, each search agent has $32 + 4 + 5 = 41$ values.

Initially, the 150 search agents are initialized with random values in range $[-1, 1]$. The values of each search agent are assigned as weights and biases of the feed forward artificial neural network. The neural network is then run to get the output. The error for each search agent is calculated by comparing the target with the output. This error is the objective function and needs to be minimized by changing the values of the search agents. To optimize the values of the search agents, optimization algorithms are used which are described in the next sections.

After optimization, the fittest search agent (the search agent with least error) is found out. The values of the best search agent are used as the weights and biases of the network and the network is tested.

Here, a total of 768 instances are used. The first 538 instances are used for training and next 230 instances are used for testing. In the end, a confusion matrix is created for training, testing, and total dataset.

The flowchart given in Fig. 2 shows the flow of the algorithm used.

Fig. 2 Flowchart of
Moth-Flame Optimization
algorithm

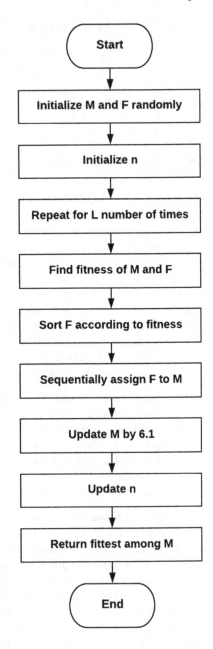

3 Experimental Result and Discussion

3.1 Evaluation

Evaluation uses some standard parameters [8] to evaluate the usefulness of the proposed model. We have used accuracy, sensitivity, error rate, prevalence, and specificity are calculated for model evaluation. The parameters are described below along with their formulae.

Error rate: The error rate of a classifier is the number of incorrectly classified dataset by the classifier expressed in terms of percentage.

Accuracy: The accuracy of the classifier is the percentage of dataset that is classified correctly by the classifier.

Specificity: Specificity is the percentage of correctly classified negative samples with respect to actual negative sample.

Prevalence: Prevalence is defined as the percentage of actual positive samples to total number of samples in the dataset.

Sensitivity: Sensitivity is also known as true positive rate. It is the percentage of correctly classified positive samples with respect to total number of actual positive samples.

Based on the above parameters, the proposed method is evaluated and values of each are presented in Table 1. The convergence curve of the MFO-FFANN for diabetes data classification is given in Fig. 3.

It is evident from Table 1 that the error rate is 22.8% and the accuracy is 77.9% for MFO. The proposed algorithm has been compared in terms of performance with other algorithms and works in Table 2.

4 Conclusion

In this paper, we have proposed a diabetes data classification technique named as MFO-tuned Feed foreword neural network. The experimental study has been done with the UCL machine learning repository Prima Indian dataset. The proposed technique has given an overall classification accuracy of 80.686%. This method is

Table 1 Performance parameters and values

S. No.	Evaluation parameters	Values MFO
1	Error rate	19.61
2	Accuracy	80.68
3	Specificity	75.25
4	Prevalence	52.08
5	Sensitivity	81.79

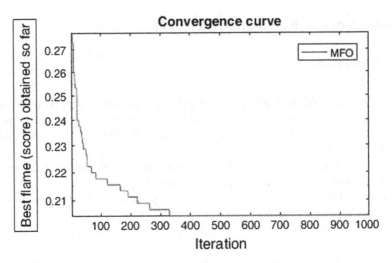

Fig. 3 Convergence curve

Table 2 Comparison with other works

First author [source]	Method	CCR %
Proposed method	**FFNN-MFO**	**80.686**
Luckka [1]	SIM	75.97
Orcku and Bal [9]	Real coded GA	77.60
Seera and Lim [4]	FMM-CART-RF	78.39
Coubey and Paul [10]	GA-MLP-NN	79.13
Coubey and Paul [11]	GA-NB	78.69
Polat et al. [12]	GDA-LSSVM	79.16
Khashei et al. [13]	LDA, QDA, KNN,SVM	80.00
Bozkurt et al. [14]	ANN, AIS	76.00

compared with the reported methods in literature. Experimental analysis conducted in the work has shown that the proposed method in the work has performed better than the existing approaches.

References

1. Luukka, P.: Feature selection using fuzzy entropy measures with similarity classifier. Expert Syst. Appl. **38**(4), 4600–4607 (2011)
2. Kahramanli, H., Allahverdi, N.: Extracting rules for classification problems: AIS based approach. Expert Syst. Appl. **36**(7), 10494–10502 (2009)
3. Sekar, B.D., Dong, M.C., Shi, J., Hu, X.Y.: Fused hierarchical neural networks for cardiovascular disease diagnosis. IEEE Sens. J. **12**(3), 644–650 (2012)

4. Seera, M., Lim, C.P.: A hybrid intelligent system for medical data classification. Expert Syst. Appl. **41**(5), 2239–2249 (2014)
5. Downs, J., Harrison, R.F., Kennedy, R.L., Cross, S.S.: Application of the fuzzy ARTMAP neural network model to medical pattern classification tasks. Artif. Intell. Med. **8**(4), 403–428 (1996)
6. Hayes-Roth, F., Waterman, D., Lenat, D.: Building expert systems (1984)
7. Mirjalili, S.: Moth-flame optimization algorithm: A novel nature-inspired heuristic paradigm. Knowl. Based Syst. **89**, 228–249 (2015)
8. Hashem, E.M., Mabrouk, M.S.: A study of support vector machine algorithm for liver disease diagnosis. Am. J. Intell. Syst. **4**(1), 9–14 (2014). p-ISSN: 2165-8978, e-ISSN: 2165-8994. https://doi.org/10.5923/j.ajis.20140401.02
9. Örkcü, H.H., Bal, H.: Comparing performances of backpropagation and genetic algorithms in the data classification. Expert Syst. Appl. **38**(4), 3703–3709 (2011)
10. Choubey, D.K., Sanchita, P.: GA_MLP NN: A hybrid intelligent system for diabetes disease diagnosis. Int. J. Intell. Syst. Appl. **8**(1), 49 (2016)
11. Choubey, D.K., Paul, S., Kumar, S., Kumar, S.: Classification of Pima Indian diabetes dataset using naive bayes with genetic algorithm as an attribute selection. Communication and Computing Systems, pp. 451–455. CRC Press, Boca Raton (2016)
12. Polat, K., Güneş, S., Arslan, A.: A cascade learning system for classification of diabetes disease: generalized discriminant analysis and least square support vector machine. Expert Syst. Appl. **34**(1), 482–487 (2008)
13. Khashei, M., Eftekhari, S., Parvizian, J.: Diagnosing diabetes type II using a soft intelligent binary classification model. Rev. Bioinform. Biometrics (2012)
14. Bozkurt, M.R., Yurtay, N., Yilmaz, Z., Sertkaya, C.: Comparison of different methods for determining diabetes. Turkish J. Electr. Eng. Comput. Sci. **22**(4), 1044–1055 (2014)

Evaluating Big Data Technologies for Statistical Homicide Dataset

Roland Askew, Sreenivas Sremath Tirumala and G. Anjan Babu

Abstract Data for police and crime analysis is becoming large and complex and increasing the difficulty for technological implementations particularly storage and retrieval mechanisms. Crimes are committed each day which demand for data engineering techniques that are flexible enough to handle complex formats and high volumes of data. NoSql technologies are typically used efficient implementations of data with processing of increasingly large volumes of cases and related data. This further helps in processing unstructured data, and providing rapid processing times to users. However, it is always a challenge to decide on the type of NoSql database for crime data. In this paper, we evaluated two NoSQL database technologies: Cassandra and MongoDB for storage and retrieval of homicide dataset. Initially, we developed a mechanism to store these datasets in Cassandra and MongoDB followed by a systematic implementation of test and evaluation criteria for read, write, and update scenarios. The experiment results were analyzed and compared and from the results, it can be concluded that Cassandra overthrows MongoDB in efficiency, reliability, and consistency.

Keywords Crime dataset · NoSQL · Cassandra · MongoDB · Big data

1 Introduction

Data storage and retrieval is complex and often involves efficient systems for storage and analysis particularly with complex data [1]. With the continuous raise of crime

R. Askew · S. S. Tirumala
Unitec Institute of Technology, Auckland, New Zealand
e-mail: stirumala@unitec.ac.nz

G. Anjan Babu (✉)
S V University, Tirupati, India
e-mail: gabsvu@gmail.com

J. Kalita et al. (eds.), *Recent Developments in Machine Learning and Data Analytics*,
Advances in Intelligent Systems and Computing 740,
https://doi.org/10.1007/978-981-13-1280-9_8

89

rate across the world, crime data is considered as big complex data which requires considerable amount of space for storage as well as time for analysis [2, 3]. Crimes are committed each day and require a database flexible enough to handle the various formats of agencies and needs for growing database size and data complexity [3]. For creation of statistical reports, fast retrieval and reliable storage is a high demand. The database is required to be always online and near real time to contain the newest data for relevant and correct statistical analysis. Due to the databases distribution across many states and regions, data is being collected in a high volume and with a fast increase in size. In terms of use, the read command is of the highest use and requires the fastest throughput time. Write is a secondary requirement, as the focus is on querying and analytic use (read/data retrieval). Updating records will be used less and therefore of a lower importance in terms of its speed. Deletion of records will rarely be used as the data within the analysis database should be correct. The data in use is anonymous (includes no names or clear identification of people) as the data is used for statistical reasons, and this data is not needed. The data managed is structured, while incoming data to the statistical database may be unstructured and semi-structured since the data is coming from different states and regions with their own formats and structures. With the structured data, the form of data can be assumed and/or guaranteed to provide faster and simple searches.

A NoSQL database will fulfill these requirements needed by the collected statistical homicide data. After a quick comparison of different NoSQL database types, we shortlisted Cassandra, Voldemort, MongoDB, and HBase. CAP theorem is considered as the fundamental principle for choosing big database technology. Following the CAP theorem, Cassandra and Voldemort covered the A-P combination while MongoDB and HBase covered C-P. The C-P combination favors consistency over availability, and this ensures the queried homicide data to be correct and most up to date or output an error. A-P will output the data even if it is not the most up to date; always providing data without an error response. With both benefiting from partition tolerance, the homicide data can be partitioned and can still operate unless a total network failure occurs. Upon further analysis and consideration of the shortlisted databases with various aspects, we found Cassandra and MongoDB to be more suitable for this research [4].

The purpose of this paper is to explain two ways for providing a partition tolerant and distributed NoSQL based solution for big homicide data. For simplicity, the scenario is limited to the statistical store and reporting system commonly needed in police information systems.

2   Literature Review

There are several works on the comparison of NoSQL databases [5–8]. Wide column stores join functionality from both key value and column-oriented databases; key values fast read/write and column-oriented flexible addition of columns. Data is stored column by column providing a faster retrieval from storage and a high flexibility due to being schema-free [9].

Cassandra records all operations to a commitlog (efficiently persisted to disk). Changes are made to Memcaches (in RAM), then persisted to SSTables on disk under configured conditions (either when a batch of Memcaches gets big enough, or after a recurring period). Search queries check Memcaches first, then SSTables if necessary.

Like other NoSQL databases, Cassandra does not benefit from the use of joins, but normalized data. Availability and a strong partition tolerance is Cassandra's focused area, although during implementation, can increase consistency with some changes to be eventually consistent [8]. Cassandras nodes allow distributed access and are easily restored upon failure. Apache Cassandra is open source and free [10].

Since Cassandra is alterable consistency, consistency can be tuned from eventually consistent to immediate depending on system changes. In the case of searching for a record, a search table needs to be created in order to efficiently search. With a high efficiency, reads, and inserts, Cassandra is a highly beneficial database. Cassandra is highly flexible and allows distribution. The level of availability and consistency is alterable to benefit the data use.

For the homicide case study, Cassandra has many advantages and features that are beneficial to the data use and storage. The homicide data requires fast retrieval, near real-time, and reliable storage to handle large volumes. Cassandra offers fast and efficient read and inserts, turnable consistency and availability and NoSQL ability to handle a high volume of data.

As document stores store semi-structured data without a schema, schema techniques such as normalization cannot be applied. Instead, relationships can only be represented through sub-documenting or using references. MongoDB groups documents into collections (equivalent to databases under SQL schemas). MongoDB has a strong partition tolerance and consistency. It is less focused in availability but still claims to be highly available. MongoDB is highly scalable and can map business objects (such as police cases and suspects) directly to internal documents. MongoDB storage of documents is ACID compliant, to specified levels. MongoDB is open source and free.

MongoDB offers ACID compliance, this ensures all users to see the same documents and prevents inconsistencies in updates [11]. MongoDB includes no query language, so queries have to be preprocessed at application layer, which adds to its complexity [12]. No Triggers or transactions are available and cannot represent relationships as per NoSQL.

MongoDB offers a good solution to the homicide case study. It provides a high availability and partition tolerance, offers flexibility to handle structured and unstructured data, easy addition of new kinds of data to the system, and offers good reliability and responsiveness for real-time analytics.

Flexible, documents can vary in structure. It is very easy to add new properties such as raw text from reports, or identikit images. MongoDB includes no query language, so queries have to be preprocessed at application layer, which adds to its complexity.

No Triggers or transactions are available and cannot represent relationships as per NoSQL.

MongoDB Master Slave replication means a longer stabilization time after a master node failure. Depending on MongoDB routing and configuration server instances, redundancy and disaster recovery measures are needed. Documents are limited to a maximum size of 16 KB.

3   Experiments Results and Discussion

Due to the security, privacy and other sensitive needs of police systems, either a separate application, or integration with existing applications will be used when requesting writes. This application can prefilter personally identifiable information, and acts as an entity recognized and trusted by the interfacing servers. A web-interface for reading data can be present for untrusted public clients.

Technical Details For implementation and testing of MongoDB, we used Windows 10 64-bit operating system on Intel i7-2630QM processor, 4 GB RAM with 2 GB Hard Drive Space allocated. We used MongoDB 3.4.2 along with Mongo Java Driver 3.4.2, ava JDK 1.8.

For implementation and testing of Cassandra, we have used Toshiba SatPro laptop i7-4700MQ Processor (8 Cores), 12 GB memory. The Operating system used was Ubuntu 16.04 Xenial LTS, Desktop. We used Apache Cassandra 3.0 development version with Datastax Cassandra Drivers for java, Custom framework written java, JUnit 4.6, Slf4j, log4J (logging subsystem, required by Cassandra development version).

Performance tests All read queries were performed against 100,000 inserted records. On average, read queries took around 22 s, or 4–5 records per ms. Except for insertion, all operations involved 100,000 records. Deletion is very slow. At default configuration, if MongoDB reaches the internal memory limit, it will hang until more memory is available. This can crash other applications and processes. All tests were restricted to a single node cluster. No multi-node tests were performed.

All tests were performed independently with cool down time between them. This was to prevent the node from fail over due to excessive write demand, and to prevent overly optimistic performance times due to the node already warmed and ready for action.

All tests were done with Consistency at the default of $LOCAL_ONE$. All tests captured timing data around the moments a query were executed. The times shown do not include any processing time by the application layer. Unless otherwise specified, all tests were performed with one thread.

The data architecture consists of master tables (directly written to), and search tables (directly read from). Master tables are semi-denormalized to represent individual people (perpetrators and victims) and whole cases. The people in a case is actually a group of relationships between people entities. Relationships are in the

master_case	
case_id	timeUUID
is_solved	boolean
when_created	bigint
evidence	set<varchar>
city	varchar
state	varchar
agency_name	varchar
agency_type	varchar
crime_type	varchar
people	set<tuple< bigint, varchar, bigint>>

PK(state, city, agency_type, case_id)

master_person	
person_id	bigint
ethnicity	varchar
gender	varchar
age	smallint

PK(ethnicity, gender, age, person_id)

Fig. 1 Data architecture of Cassandra

form of tuples (Perpetrator PersonID, relationship name, Victim PersonID). All creates and updates are performed over the master tables only. Note that dates and times are stored as UNIX epochs. The partition key will be the state, ordered by city, agency and finally case id (to be unique). This will distribute records evenly across states or regions. Note that any updates to case details will require specifying the state, city, and agency as well as the case id. Search tables hold data specific to each data analysis question. This includes aggregate data such as counts of people, where possible. The search table approach speeds up read operations, which now can consist of a single get everything from table X statement, filterable with a where clause. Further aggregation can be performed at the application layer if needed. In addition, the approach above allows more flexible data analysis querying beyond the original five questions (Fig. 1).

Read operations involve simple retrievals of the search table data. Write operations, however, are more complicated. Writing to all tables requires complex logic, retrieving and examining data, and many batch writes. Therefore the approach is to write to the master table data only, and also to a log table which lists recent master table operations. A separate process regularly checks for changes to the master table, and updates the query table accordingly. This approach adds a logical eventual consistency, as query tables do not show the most recent (master) data state, but combined with the appropriate consistency level, reduces likelihood of node failover due to excessive write demand.

Records are replicated across 3–5 nodes, which provides a form of backup to the data. Upon a node failover, the remaining nodes will shard. If the Master node cannot be contacted, a new master node within the local shard will be negotiated. Two access rights groups will be constructed, with one group or user having write access and used by police clients, and another group or user having no write access, for untrusted clients.

Software will be developed in Java, with Datastax Java drivers. Datastax Cassandra mapping libraries will be used as an ORM, to map application objects (representing the cases and people) to CQL queries and query execution. Spring will be used to make the software more manageable. Cassandra Unit and JUnit will support software testing.

The interface server layer will be constructed first, with the Client application constructed in parallel when the server layer reaches sufficient maturity. Client will be written in Java with Spring/FX. The interface server will run multiple threads per instance. At least one thread will be used for search table changes, and multiple threads used for inserts or updates with one thread per batch query.

All operations work in a similar manner. The Mongo Driver provides document handling in the form of Filter objects (comparison conditions on document properties) and Projection objects (generating a new document with a subset of properties from another document). Additionally, collections can be map-reduced before filtering to speed read queries and aggregation. Inserted documents will be converted to JSON/BSON format as required by the driver.

As explained previously, reads simply require finding matching documents as specified in the filter, then aggregation, grouping and projection through MapReduce. Writes will simply involve adding new Cases to the collection, deleting or updating cases which match a filter.

Software will be written in Java, with Spring framework to make the software maintainable. Datastax Java Drivers for Mongo will be used. Data will be received in JSON/XML, and converted for read/write operations. An ORM like Morphia may be used.

The Client will be developed in parallel, and as specified will be thin (no DB information will be stored locally). The language and libraries for the Client are still to be determined, but are more likely to involve Python, CSharp or Java with Spring/FX. Client and Server software will be threaded, with 2–10 threads per instance.

4   Evaluation

4.1   Operations

These mandatory operations must be performed for time collections and comparisons.

Query 1 Select all the unsolved murders or manslaughters from a particular city and state, from 1980 to 1990 which has the maximum victim count, handled by either the municipal or county police and are investigating a black male who is responsible for murdering victims younger than him.

Query 2 Select all solved crimes and weapons used in those crimes, involving people of different race and sex from the year 1980 to 1991 where the difference in age is greater than 2 grouped by agency type within state order by weapons.

Query 3 Select the average age of all the victims from all the states where victim count is greater than the average victim count from all the states such that the victim and perpetrator were acquaintances with each other (Fig. 2).

Query 4 Select the weapons most used to commit crimes per state.

Query 5 Select the count of all unsolved crimes per state group by month with year.

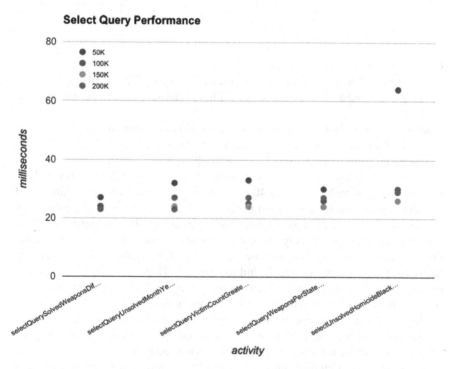

Fig. 2 Select Query performance

Table 1 Update operations timings for Cassandra

Type of update	Time (ms)	
	Avg.	Max.
CaseAddEvidence	2.5895	120
CaseRemoveEvidence	2.5529	68
CaseAddRelationship	2.8736	435
CaseRemoveRelationship	1.8642	74
CaseSolved	2.5178	131

Fig. 3 Read performance for MongoDB

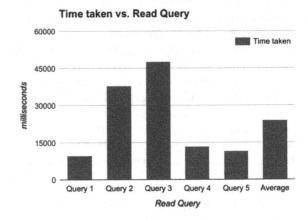

Insert Add full records to the system.

Update Status Change the status from unsolved to solved for a number of records.

Additional update operations along with operational timings is presented in Table 1.

Update Relationships Add or delete a relationship in existing cases.

Update evidence Add or delete evidence in existing cases.

Delete cases Remove a number of cases from the system (Fig. 3).

The unusually high value for selectUnsolvedHomicide at 50K records is an outlier. Queries tend to be less than 40 ms and average around 28 ms, regardless of data volume. Graphs for inserts and updates could not be produced due to the high volume of raw values. All updates were performed across 50 thousand records. The high maximums are most likely due to background processing and persisting of data. Overall updates are very fast. All insert tests used 50 thousand records, except the heavy load test, which inserted 20 million records. There is no significant difference between the average times. There appears to be no significant difference if the software uses a background thread for write operations.

Regular peaks were apparent in the raw test values, likely from another background process (or possibly writeouts to the SSTables). Read slowdown was more pronounced than write slowdown due to background activity.

Batching across multiple tables is possible but is slower, and cannot use conditional inserts (if not exists clause). Some tests would crash due to a write timeout, while starting—the Memcaches were unavailable to write to because they were too busy writing out to the SSTables. Because the tests crashed only when starting, it may be a tombstone issue—the whole Keyspace was deleted at the end of each test, so maybe Cassandra, under a periodic writeout, tried to persist the keyspace deletion, recreation, and inserts at once. This may indicate a risk with Cassandra node failovers.

There was a significant slowdown in running tests for a while after the remove relationships update test. This was likely due to tombstones. Some other insert tests were performed with 1 million records already inserted into the master tables, with no noticeable difference in average time.

5 Conclusion and Future Work

Cassandra outperformed MongoDB. Cassandra reads were faster than Mongo reads, and Cassandra writes were much faster than Mongo writes. Mongo is more flexible with structure changes due to having no schema, whereas Cassandra needs additional columns designed into the data architecture. Cassandra likely requires more volume than Mongo due to data duplication across multiple search tables. Mongo presents a more consistent and up to date view of data compared to Cassandra. The Mongo API is less consistent between versions compared to Cassandra.

Cassandra and Mongo match one another in regards to partition tolerance [13]. Cassandra has better availability due to less downtime and is more flexible in regards to multiple data centers than Mongo. Cassandra has better node rejoin time compared to Mongo. Mongo has better shard recovery time compared to Cassandra, when a node fail overs. Mongo shows more durability than Cassandra.

Overall, Cassandra is concluded to be better than MongoDB at this time, mainly due to higher availability, and faster reads and writes.

5.1 Future Work

Investigate a Graph or Graph Hybrid (Graph with Column or Document store) system. This could better represent the relationships between cases and people, and provide for improved analysis relating to victims and perpetrators. Evaluation criteria should be the same for effective comparisons.

Investigate unstructured data, such as raw text from reports and statements, in documents in document stores. Look at analysis techniques for this text, and techniques and methods supporting effective analysis and storage at the database level.

References

1. Tirumala, S.S., Narayanan, A.: Hierarchical data classification using deep neural networks. In: International Conference on Neural Information Processing, pp. 492–500. Springer (2015)
2. Jäckle, D., Stoffel, F., Mittelstädt, S., Keim, D.A., Reiterer, H.: Interpretation of dimensionally-reduced crime data: a study with untrained domain experts. In: 12th International Conference on Computer Vision, Imaging and Computer Graphics Theory and Applications (VISIGRAPP 2017), 2017, pp. 164–175
3. Smith, G.J., Bennett Moses, L., Chan, J.: The challenges of doing criminology in the big data era: towards a digital and data-driven approach. Br. J. Criminol. **57**(2), 259–274 (2017)
4. Marungo, F.: A primer on nosql databases for enterprise architects: the cap theorem and transparent data access with MongoDB and Cassandra. In: Proceedings of the 51st Hawaii International Conference on System Sciences (2018)
5. Gessert, F., Wingerath, W., Friedrich, S., Ritter, N.: NoSQL database systems: a survey and decision guidance. Comput. Sci.-Res. Dev. **32**(3–4), 353–365 (2017)
6. Siddiqa, A., Karim, A., Gani, A.: Big data storage technologies: a survey. Front. Inf. Technol. Electron. Eng. **18**(8), 1040–1070 (2017)
7. Rodrigues, R.A., Lima Filho, L.A., Gonçalves, G.S., Mialaret, L.F., da Cunha, A.M., Dias, L.A.V.: Integrating NoSQL, relational database, and the hadoop ecosystem in an interdisciplinary project involving big data and credit card transactions. In: Information Technology-New Generations, pp. 443–451. Springer (2018)
8. Branch, S., Branch, B., Boroujen, I.: A novel method for evaluation of NoSQL databases: a case study of cassandra and redis. J. Theor. Appl. Inf. Technol. **95**(6) (2017)
9. Haseeb, A., Pattun, G.: A review on nosql: applications and challenges. Int. J. Adv. Res. Comput. Sci. **8**(1) (2017)
10. Okman, L., Gal-Oz, N., Gonen, Y., Gudes, E., Abramov, J.: Security issues in nosql databases. In: 2011 IEEE 10th International Conference on Trust, Security and Privacy in Computing and Communications (TrustCom), pp. 541–547. IEEE (2011)
11. Katkar, M.: Performance analysis for NoSQl and SQL, vol. 2, pp. 12–17 (2015)
12. Truică, C.O., Boicea, A., Trifan, I.: Crud operations in mongodb. In: Proceedings of the 2013 international Conference on Advanced Computer Science and Electronics Information. Ed. Atlantis Press (2013)
13. Wodehouse, C.: Should you use mongodb or cassandra for your nosql database? https://www.upwork.com/hiring/development/mongodb-vs-cassandra/ (2016)

Journal Recommendation System Using Content-Based Filtering

Sonal Jain, Harshita Khangarot and Shivank Singh

Abstract Recommendation systems provide an approach to facilitate the user's desire. It is helpful in recommending the things from various domains. Researchers express their ideas and experience in an academic article for the research community. However, they have ample of options when they aspire to publish. At times, they end up with incorrect submission resulting in waste of time and effort of editor as well as himself. Journal selection has been a very tedious task for the novice authors. In this paper, Journal Recommendation System (JRS) is proposed, which will solve the problem of publication for many authors. Content-based filtering method is used for this purpose. The dataset used is prepared by the authors and distance algorithm is used for recommendation.

Keywords Journal recommendation system · Preprocessing
Single-value decomposition · Euclidean distance

1 Introduction

In the modern world, the data is increasing continuously, and there is a need to have an engine that can filter the information for predicting the preference according to the content that the user gives to a system. Various algorithms and approaches are already used to make recommendation system but people need personalized recommendation system that will fulfill their custom requirement. This research elaborates the people's preference according to their interest in which they want to expand the knowledge.

S. Jain (✉) · H. Khangarot · S. Singh
Department of Computer Science Engineering, Institute of Engineering & Technology,
JK Lakshmipat University, Jaipur, India
e-mail: sonaljain@jklu.edu.in

H. Khangarot
e-mail: harshitakhangarot@jklu.edu.in

S. Singh
e-mail: shivanksingh@jklu.edu.in

© Springer Nature Singapore Pte Ltd. 2019
J. Kalita et al. (eds.), *Recent Developments in Machine Learning and Data Analytics*,
Advances in Intelligent Systems and Computing 740,
https://doi.org/10.1007/978-981-13-1280-9_9

Some recommendation systems recommend research papers while some recommend articles, books, and useful chapters on behalf of their interesting research area or topic. It is a difficult challenge for recommendation system to provide a perfect match as per requirement of people. Various approaches and methods based on collaborative filtering, content-based filtering, and knowledge-based recommendation exist in the literature. After recommendation, decision support is implemented to process the data mining for noisy data and big data repository. After decision support, prediction and assumption make the decision in favor to get the fruitful results by recommendation. Recommendation system for faculty, students, industry, website owners, etc., can be designed using algorithms available in the literature. In this research, authors propose the system for researchers intending to publish. It has become a challenging task to publish a research paper in a suitable journal and also eliminating the risk of rejection by the authorized journal. This problem is affecting the inexperienced authors while showcasing their career path recommending the appropriate journal for research papers. For this purpose, content-based filtering technique is used. LDA is used for dimensionality reduction and semantic analysis has been performed to understand the meaning of the abstract and title of the paper to be published.

2 Literature Review

Various researchers have worked on recommendation systems.

Martin et al. [1] introduced Personal Health Recommendation System (PHRS) for patient-oriented decision-making system. For this purpose, the author used context-based approach, to put patients in control of their own health data. To evaluate the quality of the proposed approach he designed a controlled experiment in which an advanced HRS implementation is compared against a naive HRS implementation by means of a test collection (gold standard).

Manish et al. [2] discussed an online news recommender system for the popular social network, Facebook. The system fetches the news articles and filters them based on the community description to prepare the daily news digest. Crawling is used for collecting the data from specified sources. This novel approach consists of both content-based and collaborative filtering method. For the future work, the author recommends tagging articles in newsletters and incorporating them into the system.

Tanmoy et al. [3] had proposed a framework of Faceted Recommendation for Scientific Articles (FeRoSA) and also related to the given query by the user through four predefined facets. The author had used random walk-based framework and subnetwork consist of nodes as papers related to query. The system is evaluated using human judgment and performance improvement.

Punam et al. [4] discuss the recommendation system that uses the knowledge stored in the form of ontologies for creating semantic web. The generation of recommendation in the form of intuitionist fuzzy set is based on the trust network that exists between the peers. For this purpose, a case study on tourism recommendation system is used and validated by manually generated results.

Paul et al. [5] had looked upon YouTube recommendations brought up by deep neural network architecture learning. It provides practical insights into enormous user-facing impact. It requires special representations of categorical and continuous features which we transform with embeddings and quantile normalization, respectively.

Prateek et al. [6] had highlighted the performance of recommendation system on the MovieLens dataset. A scalable model using Apache Spark is built to perform recommendation system-based analysis. Baseline methods, user-based collaborative filtering, and content-based filtering algorithms are used by the author to measure the lowest mean square error, and the results are interpreted to find the best approach (collaborative).

Klesti et al. [7] had proposed content-based recommender system for a modular system for research papers written in Albanian. The author had designed a highly modular system that indexes and allows for searching of scientific articles.

The above literature shows that recommendation system is used in various domains for the proper utilization of data available over the network. For making these systems successful, methodologies are incorporated into it according to the requirement of the author. In article search, various researchers had contributed in searching through keywords, previous searches and filter-based approaches. The inexperienced researcher is able to make a good research paper, taking the help from various recommendation systems. After contributing their hard work in the form of research article, they must need to publish in a good and relevant journal. Apart from making a research article, publishing it in an appropriate journal is a challenging task, which authors have tried to solve (Fig. 1).

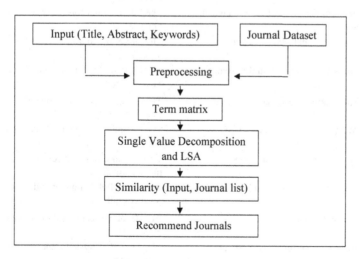

Fig. 1 Flowchart for JRS

3   Methodology

The purpose is to provide a journal recommendation system to the authors who are interested in publishing their articles or papers in the suitable journal. Dataset consists of journal records that have been prepared to validate the model. The journal dataset contains the title, keywords, aim, and scope related to it (Table 1). The journal dataset is prepared using web scraper designed with Python. Websites of some of the reputed journal publishers like Elsevier, Springer, IEEE, ACM, and Inderscience have been scrapped to prepare dataset. This system consists of six stages:

Stage 1: In the first stage, the input is taken from the user which consists of title, abstract, and keywords. Web portal (Fig. 2) is created for this purpose, from where the user can give the input of article. For summarizing the whole system, the results are evaluated by taking example of an abstract "Analysis of arrhythmia dataset using decision tree" and recommending the journal for its publication. The user gives the paper details and wants the system to recommend journals by clicking on the button at the end of the page.

Stage 2: Preprocessing is one of the major steps in preparing the raw data such that it can be used for further analysis. Removal of punctuations and stop words is performed at this stage as it improves the accuracy of term matrix.

Stage 3: Term matrix is made using the TfidfVectorizer imported from sklearn library [8]. It gives the score to the importance of words (terms) in the input data

Table 1   Dataset of journal details

S. No.	Title	Description
1	Computers in Biology and Medicine	Computers in Biology aid Medicine is a medium o…
2	International Journal of Mechanical Sciences	The International Journal of Mechanical Science…
3	Computational Materials Science	The aim of the journal is to publish papers th…
4	Journal of Computer and System Sciences	The Journal of Computer and System Sciences publ…
5	The Computer Journal	The Computer Journal is one of the longest estab…
6	ACM Computing Surveys CSUR	These comprehensive readable tutorials and sur…
7	Theoretical Computer Science	Theoretical Computer Science is mathematical an…
8	Journal of Computer and System Sciences	The Journal of Computer and System Sciences publ…
9	Acta Informatica	Acta Informatica provides international dissemi…
10	Journal of Systems and Software	The Journal of Systems and Software publishes p…

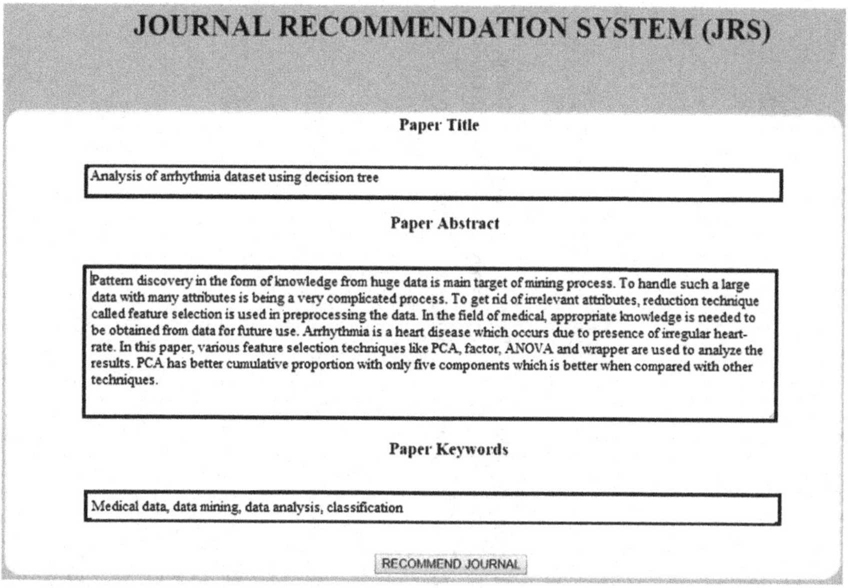

JOURNAL RECOMMENDATION SYSTEM (JRS)

Paper Title

Analysis of arrhythmia dataset using decision tree

Paper Abstract

Pattern discovery in the form of knowledge from huge data is main target of mining process. To handle such a large data with many attributes is being a very complicated process. To get rid of irrelevant attributes, reduction technique called feature selection is used in preprocessing the data. In the field of medical, appropriate knowledge is needed to be obtained from data for future use. Arrhythmia is a heart disease which occurs due to presence of irregular heart-rate. In this paper, various feature selection techniques like PCA, factor, ANOVA and wrapper are used to analyze the results. PCA has better cumulative proportion with only five components which is better when compared with other techniques.

Paper Keywords

Medical data, data mining, data analysis, classification

RECOMMEND JOURNAL

Fig. 2 Web portal of JRS for the user

depending on how often they appear in the list. The output is given in the form of sparse matrix as given in Table 2.

Stage 4: This stage includes converting the huge set of words and their frequency into a linear combination of words or components. As thousands of keywords are picked by the dataset, they are reduced to two components (Table 3) by calculating eigenvalues using Single-Value Decomposition (SVD). Other decomposition techniques produce the ill-conditioned matrix which gives unstable solutions and floating point overflow. Words and documents are mapped using LSA, where 3343 columns are reduced according to the total number of journals in the dataset (Table 4). Along with it, LSA (Latent Semantic Analysis) compares the words and document file including the meaning of it. It assumes that the choice of dimensionality in which all of the local word context relations simultaneously represented can be of great importance, and that reducing the dimensionality (the number parameters by which a word or passage is described) of the observed data from the number of initial contexts to a much smaller—but still large—number will often produce much better approximations to human cognitive relations [9] (Fig. 3).

Stage 5: The system has to recommend top journals. For this purpose, there is need to find the similarity between the input paper details given by the user and journal details in the dataset. For this purpose, Euclidean distance is measured giving the normalized squared distance between two vectors, where vectors consist of normalized journal description word frequencies.

Table 2 Sparse matrix calculating the score of terms

Title	Analysis	Analysis arrhythmia	Arrhythmia	Arrhythmia dataset	Technology	Technology feature	Theoretical	Tree	Tree technology	Using	Using decision
Analysis of Arrhythmia dataset using decision…	0.24254	0.242536	0.242536	0.242536	0.242536	0.242536	0	0.24254	0.242536	0.24254	0.24254

Table 3 Decomposition of variables into two components

	0.036220	0.036220	0.036220	0.036220	0.005094	0.018520	0.018520	0.013442	0.013442	0.013442	...	0.0025
Component_1												
Component_2	0.057971	0.057971	0.057971	0.057971	−0.003347	−0.003347	−0.013781	−0.013781	−0.013183	−0.013183	...	−0.0023

2 rows × 3343 columns

Table 4 Results after applying LSA

Component_1	Component_2	Title
0.305337	−0.952244	Analysis of Arrhythmia dataset using decision …
0.541670	−0.840591	Computers in Biology and Medicine
0.675908	−0.736936	International Journal of Mechanical Sciences
0.833817	−0.552040	Computational Materials Science
0.653568	0.756868	Journal of Computer and System Sciences
0.933551	−0.130470	The Computer Journal
0.696582	−0.717477	ACM Computing Surveys CSUR
0.998524	−0.054314	Theoretical Computer Science
0.567719	0.809065	Journal of Computer and System Sciences
0.992902	0.118937	Acta Informatica

```
([[0.    , 0.261, 0.429, ..., 0.22 , 0.091, 0.281],
  [0.261, 0.    , 0.17 , ..., 0.042, 0.171, 0.02 ],
  [0.429, 0.17 , 0.    , ..., 0.212, 0.339, 0.149],
  ...,
  [0.22 , 0.042, 0.212, ..., 0.    , 0.129, 0.062],
  [0.091, 0.171, 0.339, ..., 0.129, 0.    , 0.191],
  [0.281, 0.02 , 0.149, ..., 0.062, 0.191, 0.    ]])
```

Fig. 3 Similarity matrix of vectors

Stage 6: The purpose of the system is to recommend top journals to the user. The vectors which have the smallest distance from the input vector are selected and recommended, considering the similarity matrix.

4   Results

Manual evaluation has been done for cross-checking of the results. Sample input given by user from medical domain is showcased in Fig. 2. After applying LSA, the input title received the eigenvalue as 0.305337 (component_1), and thus journals having nearby eigenvalues are recommended as shown in Table 5.

Top 5 journals recommended to the user, according to their vector distances nearest to the input vector, are being plotted on scatter graph (Fig. 4). The graph depicts the id number of the journal. The system had predicted the journals which are near to the domain given by the user. Table 5 contains the recommended journals along with the eigenvalues of component_1.

Table 5 Recommended journals and its component values

Component_1	Journal Id	Journal name
0.319759	30	International Journal of Biomedical Data Mining
0.384592	33	International Journal of Data Warehousing and Mining (IJDWM)
0.390679	49	Organization Studies (OS)
0.403017	31	Translational Medicine
0.505765	48	Manufacturing & Service Operations Management (M&SOM)

Fig. 4 Scatter plot between the two components recommending top journals (Ids) to the user

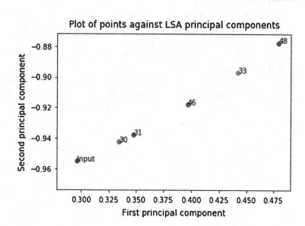

5 Conclusion and Future Work

The proposed system is fully dependent on input data given by the user and on journal dataset. It is based on content-based filtering technique. The results show that it will help the authors in finding the appropriate journals and fastening their submission process, and further enhance user's experience. Assigning the score to the importance of all the keywords is the vital part of the work. Moreover, workshops and seminars can also be recommended considering the various parameters required by the user using the similar approach. For the future work, other similarity measuring techniques can be applied to get closer results in other domains also, for building recommendation system.

References

1. Martin, W., Daniel, P.: Health recommender systems: concepts, requirements, technical basics and challenges. Int. J. Environ. Res. Public Health **11**, 2580–2607 (2014)
2. Manish, A., Maryam, K., ChengXiang, Z.: An Online News Recommender System for Social Networks. SIGIR-SSM, Boston (2009)

3. Tanmoy, C., Amrith, K., Mayank, S., Niloy, G., Pawan, G., Animesh, M.: FeRoSA: a faceted recommendation system for scientific articles. In: PAKDD: Advance in Knowledge Discovery and Data Mining. Lecture note in Computer Science, vol 9652, pp. 528–541. Springer, Cham (2016)
4. Punam, B., Harmeet, K., Sudeep, M.: Trust based recommender system for the semantic web. In: IJCAI-2007, Proceedings of the 20th International Joint Conference on Artificial Intelligence, pp. 2677–2682
5. Covington, P., Adams, J., Sargin, E.: Deep neural networks for youtube recommendations. In: RecSys, Proceedings of the 10th ACM Conference on Recommender Systems, pp. 191–198 (2016)
6. Prateek, S., Yash, S., Pranit, A.: Movie recommender system. Search Engine Architecture, Spring 2017, NYU Courant
7. Hoxha, K., Kika, A., Gani, E., Greca, S.: Towards a modular recommender system for research papers written in albanian. Int. J. Adv. Comput. Sci. Appl 5. https://doi.org/10.14569/IJACSA. 2014.050423 (2014).
8. SKLEARN Library documentation: http://scikit-learn.org/stable/documentation.html. Retrieved on 12 Jan 2018
9. Landauer, T.K., Foltz, P.W., Laham, D.: An introduction to latent semantic analysis. Introduction to latent semantic analysis. Discourse Process. **25**:259–284 (1998)

Recommending Top *N* Movies Using Content-Based Filtering and Collaborative Filtering with Hadoop and Hive Framework

Roshan Bharti and Deepak Gupta

Abstract Nowadays, the recommender system plays an important role in the real world by which we can recommend the most useful and perfect movies to the users from a large set of movies list and their ratings based on different users. Since the number of users and the movies are increasing day by day, computing the recommended movies list in a single node machine takes a very large time. Hence to reduce the computation time, we are using Hadoop framework to work in a distributed manner. Further, we have proposed a hybrid approach to recommend movies to the users by combining both the filtering techniques, i.e., user-based collaborative filtering and content-based filtering to overcome the problems of these techniques. In content-based filtering, we recommend items that are similar to the previous items which are highly rated by that user. Whereas in case of user-based collaborative filtering technique, we find out the most similar users with respect to the current user based on their cosine similarity and centered cosine similarity, and based on best similarity values, top *N* movies are recommended to the user by predicting the ratings of the movies. Further, to reduce the computation complexity, Hive database for Hadoop framework is used for developing SQL type scripts to perform MapReduce operations.

Keywords Content-based filtering · User-based collaborative filtering
Hadoop framework · Hive · Cosine similarity · Centered cosine similarity
Recommendation system · MapReduce operations

R. Bharti · D. Gupta (✉)
Computer Science and Engineering, National Institute of Technology, Yupia, Arunachal Pradesh, India
e-mail: deepakjnu85@gmail.com

R. Bharti
e-mail: giet12cse077@gmail.com

© Springer Nature Singapore Pte Ltd. 2019
J. Kalita et al. (eds.), *Recent Developments in Machine Learning and Data Analytics*,
Advances in Intelligent Systems and Computing 740,
https://doi.org/10.1007/978-981-13-1280-9_10

109

1 Introduction

A recommender system [1] is software that filters the list based on user's choice and interests and suggests the items to user that might be of interest to him. The use of recommender system is increased in recent years over Internet and is utilized in a variety of areas including movies, music, news, books, search queries, advertisements, etc. For advertisement recommendation, we can take an example of any e-commerce websites like Flipkart or eBay or Amazon. In these websites, whenever we are searching any product say jeans then different types of jeans are recommended on the homepage of these sites. So we can say that a recommender system produces a list of recommendations to the users based on his interests and recent searches.

One can describe the recommender systems in two categories such as content-based filtering [2, 3] and collaborative filtering [3–5]. Content-based filtering algorithms try to recommend items that are similar to previous items which are highly rated by that user. Let us suppose that the user X likes

Then, we can say that X is interested in either red or circle or triangle. Assume that there are few other items available, i.e.,

Now as per X's interest, red square and blue circle can be recommended to user X, i.e.,

In collaborative filtering, it finds the set of N other similar users based on similarity values with respect to the user X's rating to recommend top M movies. Here, we define the terms $\text{Sim}(X, Y)$ which represent similarity between users X and Y, and r_X represents rating vector of user X. There are three different methods defined to find the similarity between two users:

1. Jaccard similarity:

$$\text{Sim}(A, B) = |r_A \cap r_B| / |r_A \cup r_B|$$

Table 1 User's rating for different movies of different users

User\movies	HP1	HP2	HP3	HP4	HP5	HP6
A	4			5	1	
B	5	5	4			
C				2	4	5

2. Cosine similarity:

$$\text{Sim}(A, B) = \cos(r_A, r_B) = \frac{(\vec{r_A} \cdot \vec{r_B})}{|\vec{r_A}| \cdot |\vec{r_B}|}$$

3. Centered cosine similarity: Here, first normalize ratings by subtracting row mean from each element of the corresponding row and then compute $\cos(r_A, r_B)$. It is also called as "Pearson correlation".

Now, let us explain these techniques with an example. Suppose we have three users say A, B, and C and there are six movies named Heartstopper 1 (HP1), Heartstopper 2 (HP2), Heartstopper 3 (HP3), Heartstopper 4 (HP4), Heartstopper 5 (HP5), and Heartstopper 6 (HP6) as shown in Table 1.

In this example, we have considered the rating from 0 to 5. First, we find out the top similar users from this table. Suppose we are using Jaccard similarity then $\text{Sim}(A, B) = \frac{1}{5} = 0.20$ and $\text{Sim}(A, C) = \frac{2}{4} = 0.50$. Here, one can notice that $\text{Sim}(A, C) > Sim(A, B)$; therefore, user A is more similar to user C. But if we see the rating values in Table 1, then we can conclude that B is more similar to A as both A and B likes HP1 whereas A likes HP4 but dislikes HP5 and C dislikes HP4 but likes HP5. So, the problem with the Jaccard similarity is that it ignores rating values.

Further, one can use the cosine similarity as

$$\text{Sim}(A, B) = \cos([4, 0, 0, 5, 1, 0], [5, 5, 4, 0, 0, 0])$$

$$= \frac{4 * 5}{\sqrt{16 + 0 + 0 + 25 + 1 + 0)} * \sqrt{25 + 25 + 16 + 0 + 0 + 0}} = 0.38$$

Similarly, $\text{Sim}(A, C) = 0.32$.

Hence, user A and user B are more similar. So we recommend movies HP2 and HP3 to user A. But the problem with cosine similarity is that it has considered rating value as 0 for unrated movies but the users who have not rated any movies do not mean that the user dislikes the movie. That is why centered cosine similarity, i.e., Pearson correlation comes into the picture. So in this method, we have calculated the average rating of all the rated movies and subtract it from each movie. Here, the unrated movies are considered as average value, i.e., 0. After finding the average, we will calculate the cosine similarity and from top similar users, we have recommended movies to the users as in the previous case of cosine similarity.

In this work, we have used different techniques. Hadoop [6] is an Apache open-source framework which is written in Java to handle a large amount of data, i.e., big data in an efficient and effective manner. It works better in a cluster of N number of systems, i.e., an environment that provides distributed storage and computation. It has two main modules, i.e., HDFS [7] and MapReduce [8].

HDFS is a Hadoop distributed file system designed to run on commodity hardware. It is designed to store a very large amount of datasets reliably and to stream those datasets at high bandwidth.

MapReduce is a programming model for processing parallelizable problems across large datasets using a large number of computers, i.e., cluster. It has three phases: (1) map phase, (2) shuffle and sort phase, and (3) reduce phase.

HIVE [9] is data warehouse software that facilitates reading, writing, and managing large datasets residing in a distributed storage using SQL. It resides on top of Hadoop to summarize big data to make querying and analyzing easy. Hive can also be defined as a platform to develop SQL type scripts to do MapReduce operations.

PyHive is a collection of Python DB-API for HIVE. It is used to run the HIVE commands using Python program.

The content of the paper is organized as follows: Sect. 2 briefly describes the related work; Sect. 3 shows the environment setup for designing recommender system; Sect. 4 shows the proposed work in the form of flowchart along with the explanation; and Sect. 5 shows the experimental results and we draw the conclusion with some future work in Sect. 6.

2 Related Work

De Pessemier et al. [10] have proposed a content-based recommendation algorithm for analyzing Wikipedia articles using Hadoop framework. They also proposed MapReduce algorithms for keyword extraction and for generating content-based suggestions for the end user. Jure Leskovec et al. [11] have discussed the basic principles of content-based recommendation system which has to analyze the massive datasets for making relevant recommendations. Deshpande et al. [12] introduced an item-based collaborative filtering technique algorithm to recommend top N most similar items to the users.

Kadam et al. [13] are used big data analytics for recommendations using Hadoop framework. Guillermo et al. [14] have introduced movie recommendation systems for ephemeral group of users. Tharun et al. [15] have proposed item-based collaborative filtering technique to recommend movies to the users. But for a very large set of users and their reviews, it will take a huge amount of time, and also, it was very difficult to search for similar users for a newly entered user. Saravanan [16] used content-based recommender system to recommend movies to users using Hadoop MapReduce framework to handle very huge data. But the major problem was that it was only able to recommend only those genres of movies which user has already watched and rated.

Fig. 1 Large-scale
recommender system
architecture

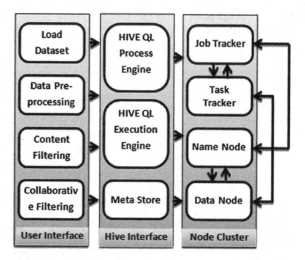

Here in this paper, we have proposed a hybrid approach by combining content-based filtering and collaborative filtering based on the type of user. For the new user, we are using content-based filtering, and for the old user, we are using collaborative filtering for the recommendations. Also, Hive is used for storing movies and users details in the database. Further, Hadoop framework is used as a platform for handling large size dataset so that the recommendations can process in a parallel manner based on no. of clusters.

3 Environment Setup

To conduct the experiment, we have considered five Linux flavor systems (i.e., Ubuntu 14.x) to run Hadoop and Hive with 4 GB RAM and 500 GB of HDD. We have downloaded Hadoop and Hive from its official website [6, 9] and further installed in the system for multinode clusters. To connect the Hive database, we have included a pyHive library for Python. Numpy library is required to create arrays of zero values components in our program and scipy library is used to calculate cosine similarity and Pearson correlation. In Fig. 1, we have shown the large-scale recommender system architecture.

4 Proposed Work

The main objective is to recommend relevant movies to users from a huge set of movies list and ratings provided by the different users. We have shown a detailed

Fig. 2 Flowchart of the
proposed hybrid approach

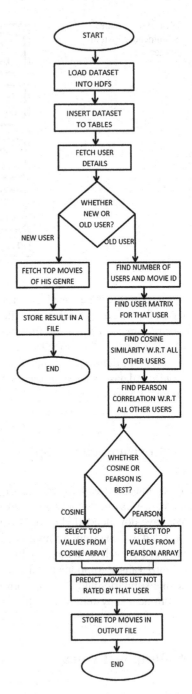

flowchart for our proposed algorithm in Fig. 2. In this algorithm, first, we are loading the dataset which is downloaded from movielens.com in the form of an external table which is stored into Hadoop distributed file system and then insert dataset into internal table of Hive named as movies for movies.csv and ratings for ratings.csv files. Thereafter, we are fetching the details of the particular user (say *X*) for which we have to recommend movies and stored in two arrays for further processing. Then, we will decide whether the user is a new user or old user depending on how many times user *X* has rated the movies. User *X* is a new user if user *X* has not rated till now or rated only one time; otherwise, the user will be old user. For the new user, we have applied content-based filtering and recommended top rated latest movies if user *X* has not rated any of the movies. But if the user has rated only one movie, in this case first, we check whether he has rated more than or less than the average rating if it is more then we have fetched highest rated movies similar to same genre movies. Otherwise, highest rated movies which are not belonging to the same genre are recommended to user *X*.

For an old user, we have applied collaborative filtering and calculated the number of users and movies available in the dataset. To find out the recommended movies for user *X*, first, we have stored the movie ID along with their ratings in a matrix form of all others users. Now, we have calculated both cosine and centered cosine similarity with respect to all other users and filtered only top *N* similarity values with their respective user id and has kept in two arrays of size *N*. In next step, we have calculated average values for both similarities and the best one is selected based on highest average value for further calculations. Now from top *N* similar users with their ratings, we have extracted the top *M* movie list which is not rated by user *X* and then calculated the rating for each movie by using weighted average formula which is defined as

$$r_{Xi} = \frac{\sum_{Y \in N} S_{XY} * r_{Yi}}{\sum_{Y \in N} S_{XY}}$$

where S_{XY} represents similarity between user *X* and *Y*; r_{Xi} and r_{Yi} represent rating of movie *i* for user *X* and *Y*, respectively; and *N* is the total number of users.

Further, based on their highest rating values, top *M* movies are recommended to user *X* and have stored the result in the following format as shown in Figs. 3 and 4.

Fig. 3 Result format for new user

User_ID:- uid		
Movie_ID:- [mid0,mid1,....]		
Movie_ID	**Title**	**Genre**
mid 0	"Movie Name1"	genre1|genre2
mid1	"Movie Name2"	genre3|genre4

Fig. 4 Result format for old
user

User_ID:- uid

Movie_ID:- [mid0,mid1,....]

Better Result For:- Cosine/Centered Cosine

Movie_ID	Title	Genre
mid0	"Movie Name1"	genre1|genre2
mid1	"Movie Name2"	genre3|genre4
---	---	---

5   Experimental Results

To check the effectiveness of our proposed method, we have considered Movielens
dataset [17] which is having around 27,278 movies list in the file named movies.csv
and 69,139 numbers of users with around 9,999,999 numbers of rows of their rating in
the file named ratings.csv which is downloaded from movielens.com. By creating the
Hadoop framework, we have written the program in Python programming language
and execute the program for Movielens dataset and having the output for new user
and old user in the following form as shown in Figs. 5 and 6. Figure 5 represents the
top 10 recommended movies list for new user i.e., UID 0 and Fig. 6 represents the
top 10 recommended movies list for old user i.e., UID 2001.

User_ID:- 0

Movie_ID:- [1956,2571,1196,1200,1036,1242,1387,527,1208,2427]

Movie_ID	Title	Genre
1956	"Ordinary People"	Drama
2571	"Matrix, The (1999)"	Action|Sci-Fi|Thriller
1196	"Star Wars: Episode V – The Empire Strikes Back (1980)"	Action|Adventure|Sci-Fi
1200	"Aliens (1986)"	Action|Adventure|Horror|Sci-Fi
1036	"Die Hard (1988)"	Action|Crime|Thriller
1242	"Glory (1989)"	Drama|War
1387	"Jaws (1975)"	Action|Horror
527	"Schindler's List (1993)"	Drama|War
1208	"Apocalypse Now (1979)"	Action|Drama|War
2427	"Thin Red Line, The (1998)"	Action|Drama|War

Fig. 5 Recommended list of top 10 movies for new user, i.e., User ID:- 0

User_ID:- 2001

Movie_ID:- [1682,2028,551,750,1213,4963,1784,2916,3897,1252]

Better Result For:- Cosine Similarity

Movie_ID	Title	Genre
1682	"Truman Show, The (1998)"	Comedy|Drama|Sci-Fi
2028	"Saving Private Ryan (1998)"	Action|Drama|War
551	"Nightmare Before Christmas, The (1993)"	Animation|Children|Fantasy|Musical
750	"Dr. Strangelove or: How I Learned to stop Worrying and Love the Bomb (1964)"	Comedy|War
1213	"Goodfellas (1990)"	Crime|Drama
4963	"Ocean's Eleven (2001)"	Crime|Thriller
1784	"As Good as It Gets (1997)"	Comedy|Drama|Romance
2916	"Total Recall (1990)"	Action|Adventure|Sci-Fi|Thriller
3897	"Almost Famous (2000)"	Drama
1252	"Chinatown (1974)"	Crime|Film-Noir|Mystery|Thriller

Fig. 6 Recommended list of top 10 movies for old user, i.e., User ID:- 2001

6 Conclusion and Future Work

In this paper, a recommendation system is proposed for recommending movies to the users in an easy, efficient, and reliable manner for the well-known dataset, i.e., Movielens dataset. Even the size of the dataset is huge in terms of large number of users and movies list, though we can recommend the movies in a very less time compared to standalone system by creating the Hadoop MapReduce framework. Since we have considered the offline data which is stored in Hive database, so one can consider the online dataset as a future work, i.e., the data will be updated with user visits, and based on their reviews, recommendations can be suggested in the real time.

References

1. Resnick, P., Varian, H.R.: Recommender systems. Commun. ACM **40**, 56–58 (1997)
2. Mooney, R.J., Roy, L.: Content-based book recommendation using learning for text categorization. In: Workshop Recommendation System: Algorithm and Evaluation (1999)
3. Ricci, Francesco, Rokach, Lior, Shapira, Bracha: Introduction to Recommender Systems Handbook, pp. 1–35. Recommender Systems Handbook, Springer (2011)
4. Breese, J.S., Heckerman, D., Kadie, C.: Empirical analysis of predictive algorithms for collaborative filtering. In: Proceedings of the Fourteenth conference on Uncertainty in artificial intelligence (1998)
5. Jafarkarimi, H., Sim, A.T.H., Saadatdoost, R.: A naïve recommendation model for large databases. Int. J. Inf. Educ. Technol., June (2012)
6. "Welcome to Apache Hadoop!". hadoop.apache.org. Retrieved 25 August 2016
7. "What is the Hadoop Distributed File System (HDFS)?". ibm.com. IBM. Retrieved 30 Oct 2014
8. "Google Research Publication: MapReduce". Retrieved 9 March 2016
9. "Apache Hive TM". Retrieved 9 March 2016

10. De Pessemier, T., Vanhecke, K., Dooms, S., Martens, L.: Content-based recommendation algorithms on the Hadoop map reduce framework. In: 7th International Conference on Web Information Systems and Technologies, pp. 237–240 (2011)
11. Leskovec, J., Rajaraman, A., Ullman, J.D.: Mining of Massive Datasets, pp. 322–331 (2014)
12. Deshpande, M., Karypis, G.: Item-based top-N recommendation algorithms. ACM Transac. Inf. Syst. 22(1), 143–177 (2004)
13. Kadam, S.D., Dilip M., Siddhesh A.V.: Big data analytics—recommendation system with Hadoop framework. In: International Conference on Inventive Computation Technologies (ICICT) August (2016)
14. Fernández, G., et al.: Let's go to the cinema! A movie recommender system for ephemeral groups of users. In: Proceedings of the 2014 Latin American Computing Conference, CLEI 2014, Institute of Electrical and Electronics Engineers Inc. (2014)
15. Ponnam, L.T.: Movie recommender system using item based collaborative filtering technique. In: International Conference on Emerging Trends in Engineering, Technology and Science (ICETETS). 24–26 Feb 2016, https://doi.org/10.1109/icetets.2016.7602983 (2016)
16. Saravanan, S.: Design of large-scale content-based recommender system using Hadoop map reduce framework. In: 8th International Conference on Contemporary Computing, IC3 2015, Institute of Electrical and Electronics Engineers Inc., pp. 302–307 (2015)
17. Movielens Dataset http://grouplens.org/datasets/movielens/latest/

WSD for Assamese Language

Pranjal Protim Borah, Gitimoni Talukdar and Arup Baruah

Abstract Word sense ambiguity comes about the use of lexemes associated with more than one sense. In this research work, an improvement has been proposed and evaluated for our previously developed Assamese Word-Sense Disambiguation (WSD) system where potential outcomes of using semantic features were evaluated up to a limited extent. As semantic relationship information has a good effect in most of the natural language processing (NLP) tasks, in this work, the system is developed based on supervised learning approach using Naïve Bayes classifier with syntactic as well as semantic features. The performance measure of the overall system has been improved up to 91.11% in terms of $F1$-measure as compared to 86% of the previously developed system by incorporating the Semantically Related Words (SRW) feature in our feature set.

Keywords Word sense ambiguity · Naïve Bayes classifier · Semantic feature
Corpus · Prior probability

1 Introduction

The existence of multiple senses for a single lexeme is one of the common characteristics of natural languages. This particular criterion of words creates ambiguity

P. P. Borah (✉)
Department of Design, Indian Institute of Technology Guwahati, Guwahati, India
e-mail: pranjalborah777@gmail.com

G. Talukdar
Department of Computer Science and Engineering, Royal Group of Institutions,
Guwahati, India
e-mail: talukdargitimoni@gmail.com

A. Baruah
Department of Computer Science and Engineering, Assam Don Bosco University,
Guwahati, India
e-mail: arup.baruah@gmail.com

© Springer Nature Singapore Pte Ltd. 2019
J. Kalita et al. (eds.), *Recent Developments in Machine Learning and Data Analytics*,
Advances in Intelligent Systems and Computing 740,
https://doi.org/10.1007/978-981-13-1280-9_11

by making the user to deduce more than one meaning for a distinct word. Most of the time it is easy for a human to detect the appropriate sense but coming to automated language processing systems, this disambiguation becomes a critical task. In computational linguistics, we can consider WSD to be a phenomenon whereby main work is to focus on determining the correct word sense provided the word is creating an ambiguity in the context.

Ambiguity is the quality of having more than one permissible interpretation. In computational linguistics, a sentence is said to be ambiguous if it can be understood in two or more possible ways [1]. Word sense ambiguity occurs when a word or a lexeme is associated with more than one meaning or sense. It is a long-standing problem in NLP, which has been discussed in reference to machine translation [2].

For example, in English language:

Sentence A1 He was mad about stars at the **age** of nine.
Sentence A2 About 20,000 years ago, the last ice **age** ended.

In the above sentences, the word 'age' is ambiguous. In Sentence A1, the word 'age' means 'how long something has existed' and in Sentence A2 the word 'age' refers to 'an era of history having some distinctive feature'.

For example, in Assamese language:

Sentence A3 মিটিং খনত মানুহে বাহ পাতি আছে
(meeting khanat mAnuhe bAh pAti Ase)
People have assembled together in the meeting.
Sentence A4 যোৱা কালিৰ ধুমুহা জাকে চৰাই বাহ বোৰ ভাঙি পেলালে
(juwA kAli dhumuhA jAke sorAi bAh bur vAngi pelAle)
Last night storm has broken the birds' nests.

In the above sentences, the word 'bAh' (বাহ) is ambiguous. In Sentence A3, the word 'bAh' (বাহ) means 'to gather together' or 'a drove' and in Sentence A4 the word 'bAh' (বাহ) refers to 'a structure in which animals lay eggs or give birth to their young ones' or 'a nest'.

WSD is one of the challenging tasks in the area of computational linguistics. In the later part of 1940s, WSD came to be recognized as a very important computational task, specially when the days of machine translation began to start [2]. In 1949, Weaver introduced WSD when he presented his popular memorandum based on machine translation [3]. Machine translation is the area in which the first attempt to perform WSD was carried out [3]. According to Weaver the contexts as well as statistical semantic studies play crucial parts in WSD [3].

WSD for Assamese language reported by Borah et al. [4] achieved 86% $F1$-measure by incorporating four features (Unigram Co-occurrences, POS of Target Word, POS of Next Word and Local Collocation) with Naïve Bayes classifier. Another Naïve Bayes Classifier-based WSD task reported by Sarmah and Sarma [5] obtained a result of accuracy 71%, which achieved 7% improvement in accuracy by adopting iterative learning mechanism. The work by Sarmah and Sarma [6] has used Decision Tree model for Assamese WSD task and reported an average F-measure of 0.611 for 10 Assamese ambiguous words.

2 Methodology

Assamese is a highly inflectional language [7]. Word sense disambiguation in Assamese is difficult due to its rich morphology. A subset of homonymous and polysemous words of Assamese language has been selected for this research work. In this project, the WSD system is designed based on supervised learning approach, which demands a large set of resources such as annotated corpora and lexical database in Assamese language.

The inputs to the Assamese WSD system are the training data (training corpus) and test data (test corpus). The system uses external knowledge source the Lexicon for feature extraction of both training and testing phase. The sense-tagged test corpus is the expected output of the system. In our previous work, 86% of $F1$-measure was obtained using the features Unigram Co-occurrence (UCO), Parts of Speech of Target word (POST), Parts of Speech of Next word (POSN) and Local Collocation (LC) [4]. A new feature Semantically Related Words (SRW) has been employed in addition to the above-mentioned features in this research work.

2.1 Classification Process

Naïve Bayes machine learning process, when collaborated with richer set of features, can help to obtain high accuracies in classification process [8]. Naïve Bayes approach is a famous and a common classification approach, and Gale, Church and Yarowsky were the first researchers to use Naïve Bayes technique for WSD task in their work named 'A Method for Disambiguating Word Senses in a Large Corpus' in the year 1992 [9].

Naïve Bayes classifier works on the assumption that all the features, which are used to classify the test case are class conditionally independent. If the feature vector is defined by $F = (f_1, f_2, ..., f_n)$ and k multiple senses for the ambiguous word is defined by $S = (S_1, S_2, ..., S_k)$, then in order to classify the true sense of the target ambiguous word (w), we have to find the sense S_i that maximizes the conditional probability represented as $P(w = S_i | F)$.

2.2 Features

In supervised word sense disambiguation where corpus plays an important role, the classification system becomes solely reliable on the information obtained from the training corpus. Syntactic features are extracted from POS information incorporated in training corpus, whereas semantic features can also be used based on the availability of lexical database. The five different features used in this Assamese WSD system are described below.

Unigram Co-occurrences (*UCO*). Co-occurrences are word pairs, which mostly have the tendency to occur in the same context (not necessarily in a particular order) with a variable number of intermediate words [10]. Assuming a window of size *m* (that is *m* number of previous words and *m* number of next words are available with respect to the target ambiguous word where *m* ranges from 1 to 5), a list of unigrams (most frequently appearing words) is maintained in the Lexicon for every sense of the ambiguous words.

POS of Target Word (*POST*). It may happen that all or some of the senses of a multi-semantic word appear in different parts of speech. The contribution of this feature depends on whether or not all or some of the possible senses of a multi-semantic word have different POS. This POS information is maintained in the Lexicon for every sense of each ambiguous word. The computation is done according to likelihood estimation.

POS of Next Word (*POSN*). The structure of Assamese language is such that it makes use of auxiliary verbs often to indicate the action with regard to a noun. Frequently, these auxiliary verbs appear in next positions to that noun in particular. This is why the next word's part of speech (POSN) information has a contribution in WSD. This information is obtained from the training corpus by likelihood estimation.

Local Collocation (*LC*). Collocation represents a group of words most often occurring together in a particular sequence, which notifies a distinct sense for the multi-semantic word appearing in a particular set of occurrences [11]. Collocation has a great contribution in WSD if unique collocations are possible for all or some of the possible senses of the multi-semantic words. A list of collocations is maintained in the Lexicon for every senses of each ambiguous word.

Semantically Related Words (*SWR*). Information about semantic relations has a quite good effect in most of the language processing tasks. Generally, for the task of WSD, this information is collected from the WordNet [12]. In our work, the required semantic information is provided by the Lexicon as Assamese WordNet is not readily available. This Lexicon contains synonyms of ambiguous words corresponding to every particular sense, and for each synonym, there is a separate entry in the Lexicon containing the information such as sense-id, POS, synonyms, unigrams and collocations. The semantic closeness of the target ambiguous word with its synonyms in a particular sense represents the closeness between the ambiguous word and that particular sense. Relation between the synonyms and the context is measured in terms of Unigram Co-occurrences. If the synonyms are also multi-semantic words, then the same process is repeated.

3 Results and Discussion

The process of WSD starts with the phase of discovering ambiguous words in the test corpus, and finally, performs the classification of these ambiguous words. Once the WSD task is completed, we need to focus on the performance measure of the system. The accuracy of the system can be measured in terms of the following:

Fig. 1 $F1$-measure
considering prior probability
with respect to change in
window size (m)

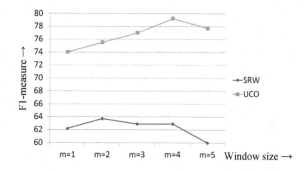

Precision (P). Precision is the ratio of relevant results returned to total number of results returned.

Recall (R). Recall is the ratio of relevant results returned to the number of results possible.

F1-measure. $F1$-measure is calculated as the harmonic mean of precision and recall.

$$F1\text{-measure} = 2 * (\text{Precision} * \text{Recall})/(\text{Precision} + \text{Recall})$$

The disambiguation process for the test corpus has been carried out several times by considering the features in different combinations along with varying window sizes (range 1–5) for the feature UCO. Moreover, for each combination of features, we have two set of results, respectively, by considering and avoiding the prior probabilities as shown in Table 1.

For the rows 2, 3, 5, 10, 12, 14 and 23 in Table 1, there is no change in $F1$-measures with respect to the change in window size (m) as these respective combinations of features neither have UCO nor SRW feature, which relies on window size (m).

As shown in Fig. 1, the line diagram of $F1$-measure (considering prior probability) with respect to the change in window size (m) clearly indicates that the maximum value of $F1$-measure obtained for the features UCO and SRW are, respectively, at $m = 4$ and $m = 2$.

As shown in Fig. 2, the line diagram of $F1$-measure (without considering prior probability) with respect to the change in window size (m) indicates that the maximum value of $F1$-measure obtained for the features UCO and SRW are, respectively, at $m = 3$, 4 and $m = 2$. However, the combination of UCO and SRW gives same accuracy values for $m=3$ and $m=4$ (Row 8 in Table 1).

The bar diagram in Fig. 3 representing the $F1$-measures for individual features UCO, POST, POSN, SRW and LC when $m=3$, indicates that contribution of UCO is highest as compared to other features. It is also found that 76% of times $F1$-measure without prior probability is greater than $F1$-measure with prior probability when window size $m = 3$.

Figure 4 represents $F1$-measures without considering prior probabilities with window size $m=3$ for a selected set of feature combinations. The combination of all the five features together gives an accuracy measure of 91.1%.

Table 1 F1-measure for different combination of features and window sizes of UCO

S. No.	Combination of features	F1-measure with prior probability					F1-measure without prior probability				
		$m = 1$	$m = 2$	$m = 3$	$m = 4$	$m = 5$	$m = 1$	$m = 2$	$m = 3$	$m = 4$	$m = 5$
1	UCO	74	75.5	77	79.2	77.7	79.2	82.2	82.9	82.9	81.4
2	POST	67.4	67.4	67.4	67.4	67.4	62.9	62.9	62.9	62.9	62.9
3	POSN	59.2	59.2	59.2	59.2	59.2	58.5	58.5	58.5	58.5	58.5
4	SRW	62.2	63.7	62.9	62.9	60	57	59.2	57.7	57.7	56.2
5	LC	66.6	66.6	66.6	66.6	66.6	68.8	68.8	68.8	68.8	68.8
6	UCO+POST	80.7	82.9	83.7	84.4	82.9	82.9	85.9	88.1	89.6	88.1
7	UCO+POSN	70.3	73.3	76	77	80	74.8	76.2	80	81.4	80
8	UCO+SRW	77	79.2	80.7	80.7	77.7	79.2	82.2	82.2	82.2	80
9	UCO+LC	82.2	84.4	85.9	86.6	85.9	82.9	85.9	88.1	88.8	88.1
10	POST+POSN	66.6	66.6	66.6	66.6	66.6	68.1	74	68.1	68.1	68.1
11	POST+SRW	72.5	74.8	74.8	74	71.8	71.8	74.8	73.3	73.3	71.8
12	POST+LC	73.3	73.3	73.3	73.3	73.3	74.8	74.8	74.8	74.8	74.8
13	POSN+SRW	60.7	62.9	62.9	62.9	62.2	61.4	62.9	62.9	62.9	62.2
14	POSN+LC	67.4	67.4	67.4	67.4	67.4	70.3	70.3	70.3	70.3	70.3
15	SRW+LC	74	75.5	75.5	74.8	73.3	75.5	77.7	76.2	75.5	74.8
16	UCO+POST+POSN	75.5	79.2	80.7	81.4	80.7	80	80.7	83.7	83.7	82.2
17	UCO+POST+SRW	82.9	85.1	85.9	85.1	82.9	82.9	85.9	87.4	88.8	86.6
18	UCO+POST+LC	86.6	88.1	88.8	89.6	88.8	85.1	88.8	90.3	91.1	90.3
19	UCO+POSN+SRW	74	77	87.5	87.5	76.2	77	78.5	80.7	80	78.5
20	UCO+POSN+LC	76.2	79.2	82.2	84.4	84.4	82.2	84.4	88.1	88.1	87.4

(continued)

Table 1 (continued)

S. No.	Combination of features	F1-measure with prior probability					F1-measure without prior probability				
		$m=1$	$m=2$	$m=3$	$m=4$	$m=5$	$m=1$	$m=2$	$m=3$	$m=4$	$m=5$
21	UCO+SRW+LC	82.9	85.9	87.4	86.6	85.9	82.9	85.9	86.6	86.6	85.9
22	POST+POSN+SRW	68.8	71.1	71.1	71.1	71.1	70.3	69.6	71.8	71.8	71.1
23	POST+POSN+LC	73.3	73.3	73.3	73.3	73.3	75.5	75.5	75.5	75.5	75.5
24	POSN+SRW+LC	69.6	71.8	72.5	73.3	73.3	74	75.5	75.5	75.5	74.8
25	UCO+POST+POSN+SRW	77.7	81.4	82.9	82.9	81.4	80.7	81.4	83.7	83.7	82.2
26	UCO+POST+POSN+LC	80.7	83.7	85.9	86.6	86.6	85.1	85.9	88.8	90.3	89.6
27	POST+POSN+SRW+LC	74.8	77	77	77	77	77	78.5	79.2	80	79.2
28	POSN+SWR+LC+UCO	80.7	83.7	85.9	85.9	85.1	83.7	85.9	88.1	87.4	86.6
29	UCO+POST+LC+SRW	86.6	88.1	88.8	88.8	88.1	85.1	88.8	89.6	90.3	89.6
30	ALL	82.9	85.9	88.1	88.1	87.4	88.1	88.8	91.1	91.1	90.3

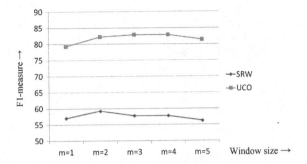

Fig. 2 $F1$-measure without prior probability with respect to change in window size (m)

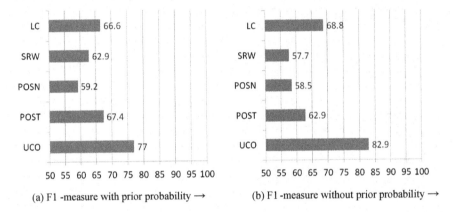

(a) F1 -measure with prior probability →   (b) F1 -measure without prior probability →

Fig. 3 Features with prior probability (**a**) and without prior probability (**b**)

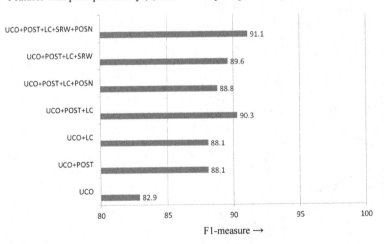

Fig. 4 Change in $F1$-measure with respect to addition of features

4 Conclusion

In this paper, we reported advancement in our Assamese WSD task by addition of a semantic feature to the Naïve Bayes classification process. A number of experiments have been carried out considering different combinations of features to disambiguate 135 multi-semantic words present in the test corpus (size 1300 words). Half of the experiments were carried out considering the use of prior probability and other half without considering the use of prior probability. For all these experiments, the obtained accuracy values ranges from 56.2 to 91.11%. Highest $F1$-measure of 91.11% (without considering the use of prior probability) was obtained using all the five features UCO, POST, POSN, SRW and LC with window size $m = 3$. However, considering the use of prior probability for the same set of features and window size, the $F1$-measure was obtained to be 88.1%. The performance measure of the overall system has been improved up to 91.11% in terms of $F1$-measure as compared to 86% of the previously developed system by incorporating the Semantically Related Words (SRW) feature in our existing feature set. The basic drawback of this system is the size of the training corpus and test corpus due to which the current system is suitable for a selected set of nouns, adjectives, verbs, pronouns and quantifiers with less effects of morphology. A large set of words of different parts of speech can be included for the task of word sense disambiguation with the use of WordNet and the morphology being excessively handled. The contribution of the features and effect of prior probability discussed in this paper will be helpful for future works in Assamese WSD.

References

1. Jurafsky, D.: Speech & Language Processing. Pearson Education, India (2000)
2. Kaplan, A.: An experimental study of ambiguity and context. Mech. Transl. 2(2), 39–46 (1955)
3. Weaver, W.: Translation. Mach. Transl. Lang. 14, 15–23 (1955)
4. Borah, P.P., Talukdar, G., Baruah, A.: Assamese word sense disambiguation using supervised learning. In: 2014 International Conference on Contemporary Computing and Informatics (IC3I). IEEE (2014)
5. Sarmah, J., Sarma, S.K.: Word sense disambiguation for Assamese. In: 2016 IEEE 6th International Conference on Advanced Computing (IACC). IEEE (2016)
6. Sarmah, J., Sarma, S.K.: Decision tree based supervised word sense disambiguation for Assamese. Int. J. Comput. Appl. 141(1) (2016)
7. Sharma, P., Sharma, U., Kalita, J.: Suffix stripping based NER in Assamese for location names. In: 2012 2nd National Conference on Computational Intelligence and Signal Processing (CISP). IEEE (2012)
8. Le, C.A., Shimazu, A.: High WSD accuracy using Naïve Bayesian classifier with rich features. In: Proceedings of the 18th Pacific Asia Conference on Language, Information and Computation (2004)
9. Gale, W.A., Church, K.W., Yarowsky, D.: A method for disambiguating word senses in a large corpus. Comput. Humanit. 26(5-6), 415–439 (1992)
10. Smadja, F.A.: Lexical co-occurrence: The missing link. Literary Linguist. Comput. 4(3), 163–168 (1989)

11. Yarowsky, D.: One sense per collocation. Pennsylvania University Philadelphia, Department of Computer and Information Science (1993)
12. Pedersen, T., Patwardhan, S., Michelizzi, J.: WordNet:: similarity: measuring the relatedness of concepts. Demonstration papers at HLT-NAACL 2004. Association for Computational Linguistics (2004)

Aptitude Question Paper Generator and Answer Verification System

Meghna Saikia, Saini Chakraborty, Suranjan Barman and Sarat Kr. Chettri

Abstract Aptitude test plays a vital role in assessing the ability of a person to perform various tasks and inculcates the ability of numerical reasoning, logical thinking, speed, accuracy and other such skills. Generating an effective aptitude question paper for aptitude test is a non-trivial task and manual generation of aptitude question paper is a conventional method. In this paper, a novel method is proposed, which automatically generates aptitude-based questions with certain keywords using randomization technique. The proposed system has the feature of generating multiple-related answers including the correct option for every generated question, and at the same time, it verifies the user's response in real time and generates score. It overcomes the major limitations of the existing automated system where question papers are generated by random selection of questions from question banks prepared by the examiner. The implementation of the proposed system has been shown along with the performance evaluation on the basis of repetitiveness of same questions.

Keywords Aptitude question paper generator · Randomization technique
Answer verifier · Android application

M. Saikia (✉) · S. Chakraborty · S. Barman · S. Kr. Chettri
Department of Computer Science & Engineering and Information Technology,
Assam Don Bosco University, Guwahati, India
e-mail: meg2008s@gmail.com

S. Chakraborty
e-mail: saini.c814@gmail.com

S. Barman
e-mail: suranjanb908@gmail.com

S. Kr. Chettri
e-mail: sarat.chettri@dbuniversity.ac.in

© Springer Nature Singapore Pte Ltd. 2019
J. Kalita et al. (eds.), *Recent Developments in Machine Learning and Data Analytics*,
Advances in Intelligent Systems and Computing 740,
https://doi.org/10.1007/978-981-13-1280-9_12

1 Introduction

Aptitude test assesses one's logical reasoning and thinking ability where one needs to answer multiple-choice questions. The importance of such test can be widely understood by the fact that they are used for various purposes from choosing career to placement purpose and so on. Traditionally, composing such question papers and verifying the answers are done manually using the writers' knowledge, experience and style, which is tedious and dependency on the intelligence of question paper setters and evaluators might raise the probability of error. With the advent of latest technologies in information technology, the automation of aptitude question paper generation and verifying the responses provided by the user is the need of the hour. In this context, the proposed Aptitude Question Paper Generator and Answer Verification System (AQPG and AVS) is conceived with the idea of an automated generation of aptitude-based questions and answers in a randomized manner. The main objective is to automatically generate a wide variety of multiple-choice-based aptitude questions with certain minimal inputs. In the proposed system, randomization technique is used to achieve the goal. Compilation of the questions and probable answers results in generating the aptitude question paper. Moreover, the system would verify the correct answer and generate score of the student in real time. An Android application is developed to facilitate the entire process of question paper generation and answer verification.

2 Background and Related Work

There are few related works existing in the literature, the proposed systems [1, 2] makes use of the Fuzzy Logic algorithm and aims at an unbiased selection of questions in a question paper reducing the manpower and time required for the same. In the model proposed by Gadge et al. [3], there is flexibility for the users to choose the subjects. They can also add, delete or update the questions and the admin has rights to provide complexity level and marks for each question.

In the work titled 'Automated Question Paper Generator System using Apriori Algorithm' [4], the main focus is to efficiently generate question paper using Apriori algorithm. But while encountering dense data due to a large number of long patterns, the Apriori algorithm's performance degrades. An improved algorithm of association rules along with the classical Apriori algorithm was also proposed to tackle the issue. Some more related works [5–7] exists in the literature where soft computing techniques [8] are used. Other systems [9–11] use shuffling algorithms as randomization technique [12] where the question paper is generated from a database of questions. These systems generally make use of a large set of questions, so that the questions are less repetitive, and where the performance evaluation based on repetition of the same question is done using various data mining techniques [13–15]. This results in the system requiring higher storage capacity, but with an increase in the size of the database of questions, the variation of questions in the final question paper increases.

All these existing systems require human staff to chalk out questions that may appear in a question paper. In some cases, algorithms have been developed for automated question paper generation, and evaluations of the performance of existing system are also done in terms of accuracy and speed. But to the best of our knowledge, automated generation of questions using a set of keywords and verification of correct answer given by the user does not exist in the literature. The common point of the existing systems is that they require a large database of questions along with their answers where a set of questions that are given as an output, are in a randomized or shuffled manner, taken from the database itself. Thus, there is a fixed amount of questions that can be taken from the database maintained by the examiner and as such it does not overcome the problem of biasness by the examiner, and literally cannot be termed as fully automated system.

3 Proposed System

In the proposed Aptitude Question Paper Generator and Answer Verification System (AQPG and AVS), there is a set of questions in each category. The basic structure of the question remains the same, but the key values/words, in specific locations of the questions changes. This, in turn, changes the answer to the question, although the way to solve the question remains unchanged. The system then calculates the answer according to the changed question and verifies it to the user's response. The advantage of such a system is that there is a wide variety of questions that can be generated using the given set of keywords.

In the proposed system, the use of a bulky database of questions is eliminated. Moreover, with the increase in the amount of keywords, the variation in the questions in turn increases. Thus, there is a least chance of repetition of same questions. In this manner, the system proves to be lightweight, easy to use, generate more variations of the non-repetitive questions and verifies user's response with the correct answer in real time.

In the proposed system (Fig. 1), two types of questions are taken into consideration; (a) Multiple choice and (b) Non-multiple choice. The question belongs to various categories like Blood Relations, Profit and Loss, Time and Work, etc. A template is designed for each type of question under every category. In each question, there are indexes, to be replaced by the set of keywords. With the random selection and insertion of different keywords both the question and answer changes.

For each type of question, keywords are used for generation of correct answer along with further related options in case of multiple-choice questions. For multiple choice type of question, the user (examinee) then selects the appropriate answer, which is then verified with the correct option. In non-multiple-choice questions, the user provides a single answer generally a numerical value, whose correctness is

Fig. 1 Block diagram of the proposed system (AQPG and AVS)

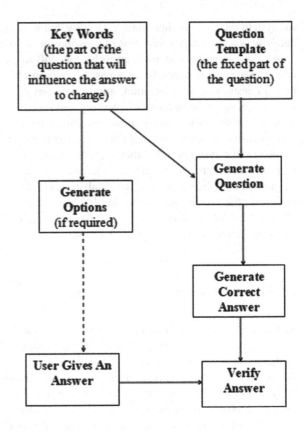

verified by the system. For every correct answer, the marks of the user add up and at the end of the test, the user's score is displayed.

4   Implementation and Performance Evaluation

Here, is a sample example of the existing type of question (blood relations) that has been deployed in the existing Android application.

Here (Fig. 2), the terms *'female'*, *'nephew'* and *'mother'* are the keywords. With simultaneous generation of questions, these keywords will change with the selection of keywords from the data set using randomization technique. As such, the answer also changes, but, it can be generated from the question itself, i.e. with each question, the correct answer is regenerated, and hence this makes our system unique. In this example, there are multiple choices that get generated in accordance to the randomly chosen keywords. Finally, the answer verification is made and the user's score gets updated automatically.

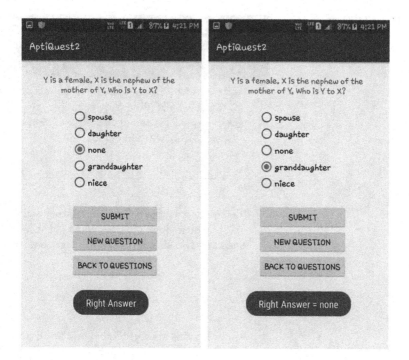

Fig. 2 Blood relationship question

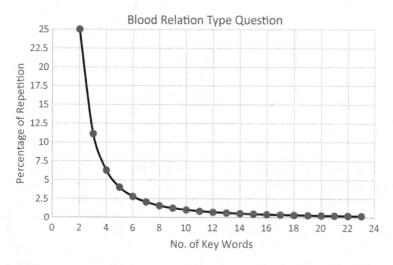

Fig. 3 Percentage of repetition of blood relationship type question

Fig. 4 Non-MCQ type
questions

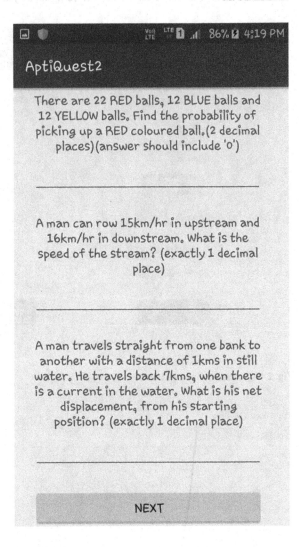

In AQPG and AVS, each category has their individual performance criteria. The performance criteria depend on the total amount of questions that is generated per category. In the above blood relation problem, there are a total of three indexes where keywords can be inserted. In the first index, two keywords (male or female) can be inserted. In the second and third indexes, the keywords (relationships like brother, father, sister, etc.) are inserted and may repeat. With the total of (say) 11 keywords, the total probable questions are 121 and as the first index offers 2 probabilities, the numbers of probable questions are 242. The probability of repeated questions will be approximately 0.82%.

Moreover, to decrease the probability of occurrence of repetitive questions and increase the performance, one has to just increase the number of keywords, and the

Fig. 5 Pattern type question

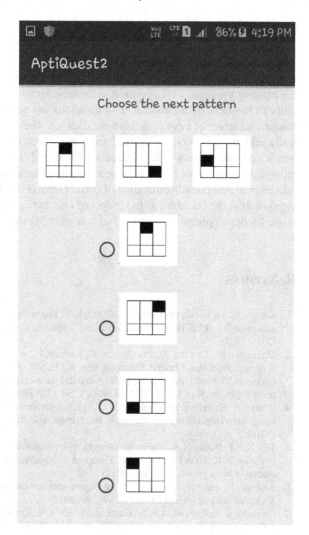

probability decreases. For example, if in the above case, the keywords would have been 12, the probability would drop to 0.69%. Likewise, if it were 10 keywords, the probability would have been 1%. Hence, we can say that according to each category of question, the performance is different. What is common is that with the increase of keywords, the number of questions that are able to be generated is higher. Hence, the probability of occurrence of repeated questions decreases and ultimately the performance and efficiency increases. The following graph shows the above performance criteria (Fig. 3).

Other question types include patterns, probability, stream speed, stream distance, sequence, etc. A few of which have already been incorporated in our existing app (tentative name AptiQuest2, which is the second prototype) (Figs. 4 and 5).

5 Conclusion

The proposed system (AQPG and AVS) has a unique feature; it requires less storage space as it does not have to maintain a huge collection of question banks like in the existing systems. This system is reliable, easy to use and eliminates the possibility of biasness by the examiner as questions are generated in real time with the random selection of keywords from the data set. The questions are created dynamically and the answers are derived from the generated questions. Moreover, the user's/examinee's answer is evaluated/verified with the correct answer. Thus, the proposed system aims to create a new dynamic question paper generator with the provision of automated verification of correct answer. Last, as there is no such existing system in the literature as the proposed one, hence there is a lot of potential and scope for development of such kind of automated system.

References

1. Kamya, S., Sachdeva, M., Dhaliwal, N., Singh, S.: Fuzzy logic based intelligent question paper generator. In: IEEE International Advance Computing Conference (IACC), pp. 1179–1183 (2014)
2. Mohandas, M., Chavan, A., Manjarekar, R., Karekar, D.: Automated question paper generator system. J. Adv. Res. Comput. Commun. Eng. 4(12), 676–678 (2015)
3. Gadge, P., Vishwakarma, R., Mestry, S.: Advanced question paper generator implemented using fuzzy logic. Int. Res. J. Eng. Technol. 4(3), 1750–1755 (2017)
4. Kamya, S., Sachdeva, M., Dhaliwal, N., Singh, S.: Automated question paper generator system using apriori algorithm and fuzzy logic. Int. J. Innovative Res. Sci. Technol. (IJIRST) 707–710 (2016)
5. Teo, N.H.I., Bakar, N.A., Karim, S.: Designing GA-based auto-generator of examination questions. In: 2012 Sixth UKSim/AMSS European Symposium on Computer Modeling and Simulation (EMS), pp. 60–64. IEEE (2012)
6. Ambole, P., Sharma, U., Deole, P.: Intelligent question paper generation system. Int. J. Sci. Tech. Advancements (IJSTA) 2(1), 257–259 (2016)
7. Choudhary, S., Waheed, A.R.A., Gawandi, S., Joshi, K.: Question paper generator system. Int. J. Comput. Sci. Trends Technol. (IJCST) 3(5), 1–3 (2015)
8. Sivanandam, S.N., Deepa, S.N.: Principles of Soft Computing, 2nd edn. Wiley Publication, Wiley India (2011)
9. Leekha, A., Barot, T., Salunke, P.: Automatic question paper generator system. Int. J. Sci. Res. Eng. Technol. (IJSRET) 6(4), 331–332 (2017)
10. Naik, K., Sule, S., Jadhav, S., Pandey, S.: Automatic question paper generation using randomization algorithm. Int. J. Eng. Tech. Res. 2(12), 192–194 (2014)
11. Shahida, N., Jamail, M., Sultan, A.B.M.: Shuffling algorithms for automatic generator question paper system. Comput. Inf. Sci. 3(2), 244–248 (2010)
12. Suresh, K.P.: An overview of randomization techniques: an unbiased assessment of outcome in clinical research. J. Hum. Reprod. Sci. 4(1) (2011)
13. Han, J., Kamber, M.: Conception and Technology of Data Mining. China Machine Press, Beijing (2007)
14. Wong, J.N.: Tutorials of Data Mining (Translated). Tsinghua University Press, Beijing (2003)
15. Wang, C., Li, R., Fan, M.: Mining Positively Correlated Frequent Itemsets. Comput. Appl. 27, 108–109 (2007)

Affinity Maturation of Homophones in Word-Level Speech Recognition

P. Ghosh, T. S. Chingtham and M. K. Ghose

Abstract Homophones are recognized in sentence-level speech recognition due to the dominating meaning and lexical analysis technique, but still it is a difficult ambiguity in word-level recognition where the uttered expression is either short or in a single string. To reduce the homophone ambiguity from the word-level recognition, an artificial immune system algorithm is proposed in the earlier part of this research where the recognition rate of a correct homophone word is minimal. In this paper, an affinity maturation technique is proposed to increase the correct recognition rate at the preliminary phase of the artificial immune algorithm.

Keywords Human–Computer Speech Interaction (HCSI) · Word-Level Speech Recognition (WLSR) · Artificial Immune System (AIS) · Affinity Maturation Artificial Immune Network (AINet)

1 Introduction

The speech recognition system is the common medium to communicate with most of the smart devices, where few online available speech recognition techniques like Google Speech [1], Apple Speech [2], Microsoft Speech [3, 4], etc. are used. The intelligent personal assistants like Google Now [5], Apple Siri [6] and Microsoft Cortana [7] are using those speech recognition techniques to assist users. The limitation

P. Ghosh (✉)
Department of Computer Science & Engineering, SIEM, MAKAUT, Siliguri,
West Bengal, India
e-mail: papri.mss@gmail.com

T. S. Chingtham
Department of Computer Science & Engineering, SMIT, SMU, Majhitar, Sikkim, India

M. K. Ghose
Department of Computer Application, Sikkim University, Tadong,
Gangtok, Sikkim, India

© Springer Nature Singapore Pte Ltd. 2019
J. Kalita et al. (eds.), *Recent Developments in Machine Learning and Data Analytics*,
Advances in Intelligent Systems and Computing 740,
https://doi.org/10.1007/978-981-13-1280-9_13

137

of recognizing the proper homophones is difficult in word-level speech recognition technique where the lexical analysis does not work.

In the earlier part of this research, an artificial immune system algorithm is proposed, where the reinforcement learning technique is approached [8–10]. Through the reward and penalty method, one system is implemented[1] where the system uses any speech recognition system to fetch the text, and if the converted text is homophone, then it extracts the appropriate homophone from the homophone dataset (H_{ds}) based on its affinity ($h_{Affinity}$) value. After a healthy amount of usage, the system is accomplished with a maturity to recognize a proper homophone word. But the accurate recognition rate was not up to the mark. The current research first applies an affinity maturation algorithm to mature the homophone dataset at the introductory stage, and second, it proposes another algorithm to compare the performance evaluation of the first algorithm.

2 Sample Homophone Dataset

During the research, a total of 2940 homophones (h) are collected from different sources [11–18] where the total number of homophone sets is 1371. Each and every homophone set (h_s) contains two or more than two homophones. Table 1 displays the sample homophone dataset where a single homophone set contains similar homophone words. Each and every homophone word has its own homophone affinity ($h_{Affinity}$) and based on the affinity value, the homophone priorities ($h_{Priority}$) are measured. In the system dataset, one denotes lower priorities and the higher value denotes higher priorities. If two homophone words have the same affinity, then their priorities are same. Before applying the affinity maturation technique, all the homophone affinities ($h_{Affinity}$) are set to zero and all the homophone priorities are set to one.

3 Affinity Maturation of Homophone Dataset

The affinity maturation follows the vaccination technique of natural immune system (*NIS*). In this method, the selected homophone dataset (H_{ds}) is being matured with a set of e-book inputs. The affinity maturation algorithm accepts an e-book as an input and examines the whole book by counting the existence of the homophone words stored in the homophone dataset (H_{ds}). The total number of occurrences of a homophone word is added to the homophone affinity ($h_{Affinity}$), and thus, it calculates the population of a homophone word within a given e-book input. Algo. 1 describes the algorithm methodically.

In the proposed affinity maturation algorithm, after each e-book input, the new and old homophone priorities are compared. If the priorities are not equal, then the

---

[1]The system is available at http://dayonmyplate.in/speech_ais/.

Table 1 Uttered and converted text in Google Speech

Set No.	Homophones	Homophone affinity [$h_{Affinity}$]	Homophone priority [$h_{Priority}$]
1	Eye	0	1
	I	0	1
2	Been	0	1
	Bin	0	1
	Bean	0	1
3	Carrot	0	1
	Carat	0	1
	Caret	0	1
4	Bands	0	1
	Banns	0	1
	Bans	0	1

homophone change count (h_{cc}) is incremented by one. Thus, the total change count (Total$_{CC}$) and the change percentage (PCT$_{Total_{CC}}$) of all homophone priorities are measured.

Algo.1. Affinity Maturation Algorithm of Homophone Dataset

Step.1. **START**

Step.2. **FOR each E-BOOK** $\bar{e}$ // e-book input

    Step.2.1. **FOR each word** $\hat{w}_i$ **from** $\bar{e}$ // Update $h_{Affinity}$

        2.1.1. **FOR each homophone** h_j **from** H_{ds}

            2.1.1.1. **IF** $\hat{w}_i = h_j$

            2.1.1.1.1. **THEN** $h_{Affinity_j}$ ++

    Step.2.2. **FOR each** h_s

        2.2.1. **Update_Priority_Sequence(h_s)**

        //Update_Priority_Sequence Algorithm rearrange the $h_{Priority}$ based on the $h_{Affinity}$

    Step.2.3. **FOR each homophone** h

        2.3.1. **IF Priority_Sequence(h_{new}) = Priority_Sequence(h_{old})**

            2.3.1.1. $h_{CC} = 0$

        2.3.2. **ELSE**

            2.3.2.1. $h_{CC} = 1$

    Step.2.4. Measurement of **$Total_{CC}$** and **$PCT_{Total_{CC}}$**

    // Total Change Count and Percentage of Total Change Count

Step.3. **STOP**

For the affinity maturation technique, a total of 100 e-books are collected from different sources and domains like engineering, medical, literature, science and fictions, etc. [19–35]. Approximately, 39 lakhs words are processed to mature the homophone dataset (H_{ds}). It is experimented after each iteration that the homophone dataset becomes matured and the percentage of total change count becomes decremented.

Table 2 Uttered and converted text in Google Speech

Set No.	Homophones	Old affinity	Old priority	New affinity	New priority	Change count
1	Arc	2	2	3	2	0
	Ark	0	1	0	1	0
2	Eye	81	2	218	2	0
	I	5789	3	13,693	3	0
	Aye	12	1	12	1	0
3	Been	1789	2	3835	3	1
	Bin	0	1	3	2	1
	Bean	0	1	1	1	0
4	Pocks	0	1	0	1	0
	Pox	0	1	2	2	1

To measure the evaluation of the total change count, Algo. 2 is proposed and implemented productively.

Algo.2.	**Algorithm for Analysis of Affinity Maturation Technique**
Step.1.	**START**
Step.2.	**FOR** each $\bar{I}$ interval
	Step.2.1. $Total_{CC} = 0$
	Step.2.2. **FOR** each h
	Step.2.2.1. **IF** $h_{Priority_{Old}} \neq h_{Priority_{New}}$
	Step.2.2.1.1. $Total_{CC}{+}{+}$
	Step.2.3. $PCT_{Total_{CC}} = {Total_{CC}}/{n_h} \times 100$ // n_h is total number of homophones in the homophone dataset
Step.3.	**STOP**

Algo. 2 calculates the percentage of total change count ($PCT_{Total_{CC}}$) at $\bar{I}$-interval. It verifies all the homophone priority ($h_{Priority}$) value and measures the percentage of the total change count based on the following equation:

$$PCT_{Total_{CC}} = Total_{CC}/n_h \times 100 \tag{1}$$

The process of calculating the homophone change count (h_{cc}) is displayed in Table 2. After each and every iteration, the homophone affinity ($h_{Affinity}$) value is updated and based on the new affinity value; the homophone priority value ($h_{Priority}$) is rearranged. After the rearrangement of homophone priority value ($h_{Priority}$), if old and new priority values are identified to be unequal, then one value is assigned to the homophone change count (h_{cc}) value. Thus, in the total homophone set (H_{ds}), all the homophone change count (h_{cc}) is assigned, and therefore, the total change count ($Total_{CC}$) and the percentage of the total change count ($PCT_{Total_{CC}}$) are calculated.

Table 3 Percentage revolution in homophone dataset after applying affinity maturity model with $\bar{I} = 20$

No. inputs	After 20 e-book inputs	After 40 e-book inputs	After 60 e-book inputs	After 80 e-book inputs	After 100 e-book inputs
31.80%					
	11.46%				
		8.57%			
			5.03%		
				2.86%	

Table 3 demonstrates the percentage evolution of the homophone dataset with $\bar{I} = 20$.

The decreased percentage of total change count indicates that with more number of e-book inputs, the homophone dataset (H_{ds}) becomes matured. Moreover, after maturing the homophone dataset (H_{ds}) with certain e-book inputs, a negligible change is observed with further input(s).

4 Conclusion and Future Scope

The affinity maturation algorithm for the homophone dataset (H_{ds}) ensures the increased recognition rate for selecting the correct homophones in word-level speech recognition. This affinity maturation technique may be used for domain-based homophone selection, i.e. if the system works for a particular domain, then during the maturation process, the input e-book may be selected from the same domain for a better result. This approach is applicable to any other research where the artificial immune system algorithm is considered as a solution strategy.

References

1. Cloud Speech API-Speech to Text Conversion Powered by Machine Learning. Available at: https://cloud.google.com/speech/. Access Date: 08 Jan 2018
2. Speech. Available at: https://developer.apple.com/documentation/speech. Access Date: 08 Jan 2018
3. Wang, Z., Stock, J., Ravi, P., Parnisari, M.I.: Microsoft Speech API Overview, Sept 2017. Available at: https://docs.microsoft.com/en-us/azure/cognitive-services/speech/home. Access Date: 08 Jan 2018
4. Bing Speech API-Speech Recognition. Available at: https://azure.microsoft.com/en-in/servic es/cognitive-services/speech/. Access Date: 08 Jan 2018
5. Google Now. Available at: https://www.androidcentral.com/google-now. Access Date: 27 Jan 2018
6. Apple Siri. Available at: https://www.apple.com/in/ios/siri/. Access Date: 27 Jan 2018

7. Cortana Is Your Truly Personal Digital Assistant. Available at: https://www.microsoft.com/en-in/windows/cortana. Access Date: 27 Jan 2018
8. Kaelbling, L.P., Littman, M.L., Moore, A.W.: Reinforcement learning: a survey. J. Artif. Intell. Res. **4**, 237–285 (1996). Available at: https://www.jair.org/media/301/live-301-1562-jair.pdf
9. Mnih, V., Badia, A.P., Mirza, M., Graves, A., Harley, T., Lillicrap, T.P., Silver, D., Kavukcuoglu, K.: Asynchronous methods for deep reinforcement learning. In: Proceedings of the 33rd International Conference on Machine Learning, New York, NY, USA, 2016, JMLR: W&CP, vol. 48. Available at: http://proceedings.mlr.press/v48/mniha16.pdf
10. Mnih, V., Kavukcuoglu, K., Silver, D., Graves, A., Antonoglou, I., Wierstra, D., Riedmille, M.: Playing Atari with Deep Reinforcement Learning. Available at: https://www.cs.toronto.edu/~vmnih/docs/dqn.pdf. Access Date: 10 Oct 2017
11. List of Homophones. Available: http://www.allaboutlearningpress.com/list-of-homophones/. Access Date: 29 Nov 2016
12. Homophone. Available: http://en.wikipedia.org/wiki/Homophone. Access Date: 29 Nov 2016
13. Homophones. Available at: http://www.singularis.ltd.uk/bifroest/misc/homophones-list.html. Access Date: 16 Dec 2016
14. Homophones. Available at: http://www.allaboutlearningpress.com/homophones/. Access Date: 16 Dec 2016
15. Common Homophones List. Available at: https://www.englishclub.com/pronunciation/homophones-list.htm. Access Date: 16 Dec 2016
16. Homophones Word List. Available at: http://www.lccc.edu/sites/www.lccc.edu/files/images/LCCC%20homophones2.pdf. Access Date: 16 Dec 2016
17. Homophones Examples with Definitions. Available at: http://www.grammarinenglish.com/homophones/. Access Date: 16 Dec 2016
18. Homophones by Letter. Available at: http://www.homophone.com/browse. Access Date: 16 Dec 2016
19. The Textfile Directory. Available at: http://textfiles.com/directory.html. Access Date: 18 July 2017
20. Free eBooks for Life!. Available at: https://www.free-ebooks.net/. Access Date: 18 July 2017
21. BookRix. Available at: https://www.bookrix.com/. Access Date: 18 July 2017
22. Free Kids Books. Available at: https://freekidsbooks.org/. Access Date: 18 July 2017
23. Free Book Centre. Available at: http://www.freebookcentre.net/. Access Date: 18 July 2017
24. Internet Archive. Available at: https://archive.org/. Access Date: 18 July 2017
25. National Digital Library of India. Available at: https://ndl.iitkgp.ac.in/. Access Date: 18 July 2017
26. Library Genesis2M. Available at: http://libgen.io/. Access Date: 18 July 2017
27. BookSC. Available at: http://booksc.org/. Access Date: 18 July 2017
28. Many Books. Available at: http://manybooks.net/. Access Date: 18 July 2017
29. Open Library. Available at: https://openlibrary.org/. Access Date: 20 July 2017
30. Popular Books. Available at: https://centslessbooks.com/. Access Date: 20 July 2017
31. Smashwords. Available at https://www.smashwords.com/. Access Date: 20 July 2017
32. bookboon.com. Available at: http://bookboon.com/. Access Date: 20 July 2017
33. goodreads—Meet Your Next Favorite Book. Available at: https://www.goodreads.com/. Access Date: 20 July 2017
34. feedbooks—Read Anywhere. Available at: http://www.feedbooks.com/. Access Date: 20 July 2017
35. Free ebooks—Project Gutenberg. Available at: https://www.gutenberg.org/. Access Date: 20 July 2017

Feature Map Reduction in CNN for Handwritten Digit Recognition

Sinjan Chakraborty, Sayantan Paul, Ram Sarkar and Mita Nasipuri

Abstract Handwritten digit recognition is a well-researched area in the field of pattern recognition that is used for distinguishing the pre-segmented handwritten digits. Deep learning is a recent research trend in this field and architectures like convolutional neural network (CNN) are being used. CNN is a computationally expensive architecture that is normally used to provide high accuracy in complex classification problems which require tuning of millions of parameters. On the contrary, less complex classification problems can be solved using considerably less number of parameters—hence using CNN on them leads to unnecessary wastage of resources. In the present work, we have proposed to reduce the feature maps that are used to train the CNN to reduce storage space and computation time. Experimental results show that the time required to train the CNN decreases with reducing the number of feature maps without notably affecting the accuracy.

Keywords Feature reduction · CNN · Digit recognition · MNIST · Feature map

1 Introduction

Handwritten digit recognition is concerned with learning models to distinguish pre-segmented handwritten digits. They are used in a variety of areas like pin code recognition on mails, handwritten numeric form entries, bank cheque processing, etc. Since handwritten digits can be of various orientations and styles, researchers

S. Chakraborty · S. Paul · R. Sarkar (✉) · M. Nasipuri
Department of Computer Science and Engineering, Jadavpur University, Kolkata, India
e-mail: raamsarkar@gmail.com

S. Chakraborty
e-mail: sinjanc@gmail.com

S. Paul
e-mail: sayantanpaul98@gmail.com

M. Nasipuri
e-mail: mitanasipuri@gmail.com

© Springer Nature Singapore Pte Ltd. 2019
J. Kalita et al. (eds.), *Recent Developments in Machine Learning and Data Analytics*,
Advances in Intelligent Systems and Computing 740,
https://doi.org/10.1007/978-981-13-1280-9_14

143

face many challenges for automated recognition of handwritten digits. The current trend of research in pattern recognition including handwritten digit recognition is deep neural networks [1–9].

One of the most difficult and expensive parts of machine learning is feature engineering, which is the process of using domain knowledge of the data to create features that make machine learning algorithms work. On the other hand, deep neural networks involve feature learning, i.e., the network extracts the features through learning by itself. Hence, the need for feature engineering is completely eliminated. Deep learning involves the use of highly over-parameterized neural networks that are time-consuming to train. Here, we attempt to reduce the overall classification time in using deep neural networks by reducing the feature space used to train the model in order to get an optimal model. The feature map reduction has been done using selecting the filter maps of a convolutional layer of the convolutional neural network (CNN) randomly.

2 Related Research

Research in deep learning has been ongoing over the past decade. But recently, deep learning architectures like CNNs are being extensively used for recognition. Handwritten digit dataset of MNIST is a benchmark dataset that has been used to evaluate many cutting-edge machine learning ideas by various researchers since its inception. A few papers on CNN-based pattern classification models that have used the MNIST dataset for conducting experiments have been discussed here.

Xu et al. have worked on alleviating the overfitting in CNNs in [1]. In [2], Singhal et al. have proposed a new framework for deep learning of hyperspectral image classification. Polania et al. have proposed a new scheme for compressed sensing using restricted Boltzmann machines and deep belief nets in [3]. Wang et al. in [4] have discussed a new optimization approach for building correlations between filters in CNNs. Ercoli et al. [5] have calculated hash codes using a multi k-means procedure and used them for retrieval of visual descriptors. Katayama and Yamane [6] have proposed a model that adapts to CNN trained by non-rotated images and even for rotated images by evaluating the feature map obtained from the convolution part of CNN. Dong et al. [7] have conducted a research on optimizing schemes for the RRAM-based implementation of CNN. Teow [8] has presented an easily understandable minimal CNN model for handwritten digit recognition. Tan et al. [9] have proposed a stochastic diagonal approximate greatest descent (SDAGD) to train the weight parameters in CNN.

3 An Overview of Convolutional Neural Networks [10]

CNNs are neural networks with every neuron possessing a learnable weight and a bias. The neuron receives some inputs, represented by an input vector and performs a dot product of the input vector with its corresponding weight vector. A nonlinearity is optionally introduced by adding the bias vector to the dot product.

The fully connected layers have neurons arranged in two dimensions—length and breadth, whereas the convolutional layers of a CNN have neurons arranged in three dimensions: length, breadth, and depth. Moreover, unlike a fully connected network where the neurons in a layer are fully connected to all the neurons in the previous layer, and the convolutional layers are connected partially to the layer preceding it.

A simple CNN is a sequence of layers of neurons. Every layer of a CNN uses a differentiable function to transform the three-dimensional activation matrix generated by the previous layer in the network to another. A CNN architecture normally consists of convolutional, pooling, and fully connected layers.

A set of learnable filters are used in the convolutional layer. These filters have limited length and breadth but there is only one filter along the entire depth of the input volume as shown in Fig. 1. The filters are slid across the length and breadth of the input volume. Then the dot product of the filter entries and the input at various positions are computed to generate the output of this layer.

Pooling layers are used periodically between successive convolutional layers. This layer is mainly used to steadily reduce the spatial size of the output of the convolutional layers. This helps to reduce the number of parameters and computations involved in the network. Also, higher level features can be learnt due to the reduction in spatial size. Moreover, it also controls overfitting by reducing the number of features involved by maintaining the range of the feature space in a flexible manner. For example, in max pooling, as only the maximum values are accepted, the feature space becomes limited to the maximum values in each pooling kernel.

ReLU has been used as the activation function for the convolutional and fully connected layers. The definition of a ReLU can be given by $h = \max(0, p)$, where $p =$

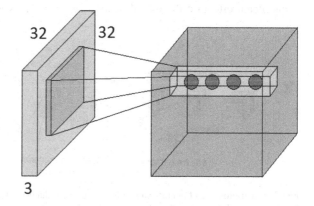

Fig. 1 The input volume has been represented in yellow. The volume of neurons in the convolutional layer has been represented in green

$Wx + q$. ReLU does not suffer from the problem of vanishing gradient unlike the sigmoid activation function which arises when $p > 0$. Moreover, ReLU produces sparse representations when $p \leq 0$, which are better than the dense representations produced by sigmoid functions. Hence, ReLU has been used as the activation function.

4 Feature Map Reduction

In pattern classification problems, often too many features are considered on the basis of which the final classification is done. Sometimes, most of these are correlated, and hence become redundant.

In the CNN defined in this paper, both the first and second convolutional layers use a kernel size of 5×5 with zero padding $= 2$ and stride $= (1,1)$. Hence, 64 feature maps each consisting of $7 \times 7 = 49$ pixels are generated as inputs for the first fully connected layer after the second convolution and max pooling layer. Out of the 64 feature maps, a fixed number (less than 64) of feature maps are selected randomly. Rest of the feature maps are then provided as input for the first fully connected layer which are used for classification. This procedure helps to reduce the storage space and computation time by removing redundant features, if any. The architecture is shown in Fig. 2.

5 Experimental Results

To evaluate the proposed feature map reduction concept, popularly used MNIST handwritten digit dataset is selected here. Experiments are performed on the MNIST dataset by varying the feature size. The size of the training dataset is chosen as the first 10,000 images of the 60,000 images provided by MNIST due to resource constraints, while the testing dataset retains all 10,000 images. From Table 1, it can be observed that the general trend of the reduction of feature maps to some extent has almost insignificant effect on the overall testing accuracy. This justifies the usefulness of

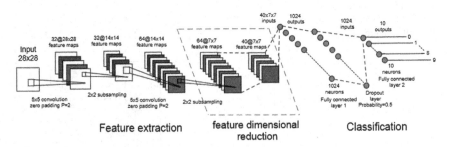

Fig. 2 Illustration of architecture showing feature map reduction used in the CNN. [Layer in which reduction is performed is enclosed in dotted lines.]

Table 1 Variation of training accuracy, testing accuracy, and execution time with number of feature maps

S. No.	Feature dimension	Number of feature maps	Training accuracy	Testing accuracy	Training time (in s)
1	$64 \times 49 = 3136$	64	1	0.9919	109.034
2	$60 \times 49 = 2940$	60	1	0.9903	108.454
3	$55 \times 49 = 2695$	55	0.98	0.9913	113.473
4	$50 \times 49 = 2450$	50	1	0.9909	110.152
5	$45 \times 49 = 2205$	45	1	0.9902	107.257
6	$40 \times 49 = 1960$	40	0.96	0.9867	101.039
7	$35 \times 49 = 1715$	35	0.94	0.9858	102.529
8	$30 \times 49 = 1470$	30	0.98	0.9809	102.278

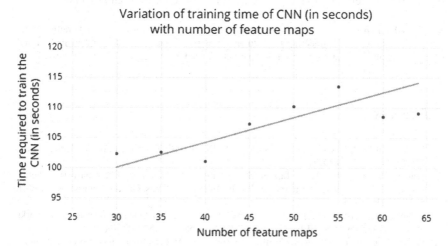

Fig. 3 Tendency of variation of training time of CNN (in s) with number of feature maps

applying the feature map reduction on CNN, as the data in the table indicates. But it is to be noted that in some cases, testing accuracy or training time does not follow the presumed trend. The reason for this is the randomness of the initial weight assignment of the kernel values. Due to this, some exceptions in the decreasing test accuracy and training time may occur. So, it may be inferred that the time required for training the CNN decreases with the decrease in the number of feature maps, which is depicted in Fig. 3.

6   Conclusion

In this paper, feature map reduction for CNN has been demonstrated. Since architectures like CNN are computationally expensive and lead to wastage of resources when used with less complex research problems, we have reduced the feature space used to train the model and compared the training accuracy, test accuracy, and execution time of the problems trained using the above procedure with that of CNNs using the entire feature space. The experiments provide conclusive evidence for the usefulness of CNNs with reduced feature space for less complex problems. The limitation of the proposed method is that the reduction in feature space has been done randomly—hence a lot of important features might get eliminated. As a future scope for this work, we can use some feature ranking procedures such as mutual information or symmetrical uncertainty to reduce the irrelevant or redundant feature maps.

Acknowledgements This work is partially supported by the CMATER research laboratory of the Computer Science and Engineering Department, Jadavpur University, India, and PURSE-II and UPE-II Jadavpur University projects. RS is partially funded by DST grant (EMR/2016/007213).

References

1. Xu Q., Pan G.: SparseConnect: regularising CNNs on fully connected layers. Electron. Lett. **53**(18), 1246–1248 (2017)
2. Singhal, V., Aggarwal, H.K., Tariyal, S., Majumdar, A.: Discriminative robust deep dictionary learning for hyperspectral image classification. IEEE Trans. Geosci. Remote Sens. **55**(9), 5274–5283 (2017)
3. Polania, L.F., Barner, K.E.: Exploiting restricted Boltzmann machines and deep belief networks in compressed sensing. IEEE Trans. Signal Process. **65**(17), 4538–4550 (2017)
4. Wang, H., Chen, P., Kwong, S.: Building correlations between filters in convolutional neural networks. IEEE Trans. Cybern. **47**(10), 3218–3229 (2017)
5. Ercoli, S., Bertini, M., Bimbo, A.D.: Compact hash codes for efficient visual descriptors retrieval in large scale databases. IEEE Trans. Multimed. **19**(11), 2521–2532 (2017)
6. Katayama, N., Yamane, S.: Recognition of rotated images by angle estimation using feature map with CNN. In: IEEE 6th Global Conference on Consumer Electronics (GCCE), pp. 1–2, 21 Dec 2017
7. Dong, Z., Zhou, Z., Li, Z.F., Liu, C., Jiang, Y.N., Huang, P., Liu, L.F., Liu, X.Y., Kang, J.F.: RRAM based convolutional neural networks for high accuracy pattern recognition and online learning tasks. In: Silicon Nanoelectronics Workshop (SNW), pp. 145–146, 4–5 June 2017
8. Teow, M.Y.W.: Understanding convolutional neural networks using a minimal model for handwritten digit recognition. In: IEEE 2nd International Conference on Automatic Control and Intelligent Systems (I2CACIS), pp. 167–172, 21 Oct 2017
9. Tan, H.H., Lim, K.H., Harno, H.G.: Stochastic diagonal approximate greatest descent in convolutional neural networks. In: IEEE International Conference on Signal and Image Processing Applications (ICSIPA), pp. 451–454, 12–14 Sept 2017
10. www.compsci6971.github.io/notes/convolutionalnetworks. Accessed 21 Jan 2018

Multi-lingual Text Localization from Camera Captured Images Based on Foreground Homogenity Analysis

Indra Narayan Dutta, Neelotpal Chakraborty, Ayatullah Faruk Mollah, Subhadip Basu and Ram Sarkar

Abstract Detecting and localizing multi-lingual text regions in natural scene images is a challenging task due to variation in texture properties of the image and geometric properties of multi-lingual text. In this work, we explore the possibility of identifying and localizing text regions based on their degree of homogeneity compared to the non-text regions of the image by binning red, green, blue channels and gray levels into bins represented individually by binary images whose connected components undergo several elimination processes and the possible text regions are distinguished and localized from non-text regions. We evaluated our proposed method on our camera captured image collection having multi-lingual texts in languages namely, English, Bangla, Hindi and Oriya and observed 0.69 as the F-measure value for best case where the image has good number of possible text regions.

Keywords Multi-lingual · Scene text · Text localization
Connected component analysis · Binning

I. N. Dutta · N. Chakraborty · S. Basu · R. Sarkar (✉)
Department of Computer Science and Engineering, Jadavpur University,
Kolkata 700032, India
e-mail: raamsarkar@gmail.com

I. N. Dutta
e-mail: indranarayandutta@gmail.com

N. Chakraborty
e-mail: neelotpal_chakraborty@yahoo.com

S. Basu
e-mail: subhadip@cse.jdvu.ac.in

A. F. Mollah
Department of Computer Science and Engineering, Aliah University,
Kolkata 700160, India
e-mail: afmollah@aliah.ac.in

© Springer Nature Singapore Pte Ltd. 2019
J. Kalita et al. (eds.), *Recent Developments in Machine Learning and Data Analytics*,
Advances in Intelligent Systems and Computing 740,
https://doi.org/10.1007/978-981-13-1280-9_15

1 Introduction

Text localization [1] in complex scene images is a domain receiving significant attention due to its importance in a number of visual-based applications like content-based image search, reading street texts, aid for visually impaired or language translation for commuters or tourists, navigation, etc. Scene texts provide relevant information in the process of image understanding and retrieval [2]. However, issues like the diversity of acquisition conditions, low resolution, font variability, complex backgrounds, different lighting conditions, blur, etc.

In this work, we proceed by binning the red, green, blue channels, and gray levels of the image. The bin components undergo elimination process where the connected components analysis (CCA) [3] is done and based on alignment, stroke width and edge properties, the possible text regions are distinguished and localized from the non-text regions.

2 Related Works

Recent research works include development of some robust methods like Stroke Width Transform (SWT), which have become extremely popular text features.

The idea of script-independence is hugely stressed in the works [2–4] where SWT features are applied and yields comparatively good results. The work in [5, 6] lists features of SWT and eHOG (Edge Histogram of Oriented Gradients) to determine the characterness of individual components in an image. The work in [7] employs baseline methods to localize the text regions. In the work [8], geometric features, intensity of the detected region and outer boundary, stroke width and the border gradients together filter out the non-text components from the multi-lingual texts. In work [9], adaptive clustering is used based on color, stroke width, and location (Compactness) differences, and in work [10], Fourier-Laplacian and Hidden Markov Model (HMM) is combined for detecting text in both scene images and video frames. Deep or convolutional neural networks have been utilized in works [11–13] to collaborate with stroke width and MSER to enhance accuracy of localizing text.

3 Proposed Method

While processing any input image, we face two major challenges: (a) The complex texture [14] of a random input image; (b) High variation in geometry and line orientation [15] due to multi-lingual nature of text (Fig. 1).

Fig. 1 Flowchart of the proposed method of text localization

The above-mentioned challenges are addressed by working on and combining the data from all the three image planes namely Red, Green, and Blue along with Gray level information.

3.1 Color Component Separation

The difference in intensity range of the foreground and background in Fig. 2 leads us to divide the entire intensity range (0–255) into bins so that the foreground and background fall under separate bins as shown in Fig. 3. High variation in geometry and line orientation [15] due to multi-lingual nature of text.

However, to determine the desired homogenous [14] regions, we address the following problems:

Multi-Binned Components At times, components share two bin, as in Fig. 2a. The binning depicted in Fig. 3a, uses two levels of Bins to cover components spread over two bins. For Bin Sizes 16 and 32, the Bin Levels will be as follows:

For Bin Size = 16 we get

Level 1 : {0 − 15, 16 − 31, 32 − 47 ... 112 − 127, 128 − 143 ... 240 − 255}
Level 2 : {8 − 23, 24 − 39, 39 − 55 ... 104 − 119, 120 − 135 ... 233 − 248}

For Bin Size = 32 we get

Level 1 : {0 − 31, 32 − 63, 64 − 95, 96 − 127, 128 − 159 ... 224 − 255}
Level 2 : {16 − 47, 48 − 79, 78 − 111, 112 − 143, 144 − 175 ... 208 − 239}

Fig. 2 Difference in intensity range of background and foreground

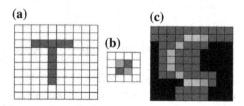

(a) (b) (c)

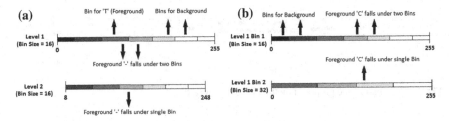

Fig. 3 Separate background and foreground bins. **a** Bin level of size 16 **b** Foreground 'C' falls under a single bin for bin size = 32 and not for bin size = 16

Fig. 4 Bin samples generated from original image

As seen in Fig. 3a, the component '-'falls under two bins at level-1, but at level-2 of, it falls under just one bin. In Fig. 2c, the intensity values of the component shows a range greater than 16. Hence, it will fall in one bin of higher size.

Though it may seem advantageous to choose bins of large size, the problem with too large bins may lead to the background and foreground falling under same bin. Hence, we choose a range of bin sizes with Bin Size(0) = 0, by applying the following equation:

$$\text{Bin Size}\,(i) = \text{Bin Size}\,(i-1) + \sqrt{(\text{Max.Intensity}) - \text{Bin Size}\,(i-1)}. \quad (1)$$

Binary Images are generated by binning from all four planes: Red, Green, Blue and the Gray level where each image represents one bin as shown in Fig. 4. Images are inverted to generate twice the number of binary images for processing.

3.2 Detecting Stable Regions

Connected Components (CCs) [16] from each bin image are extracted and small gaps within the image are filled to make components uniform, and extremely small and large components are removed. Choosing multiple bin sizes has two advantages: (1) probability of background-foreground separation increases (2) actual objects are differentiated from object parts as explained in the below section.

True Objects In some cases, parts of objects instead of whole, occur in multiple bins. On increasing the bin size to the next higher value, an even bigger object segment occurs because increase in size means an increase in intensity value range. True objects are whole objects and do not change significantly in size when bin size is increased as per Eq. (1). We perform the following algorithm on each CC of each bin to remove object segments:

Algorithm 1 1. For a particular bin b_i of size s_i, find the bins $\{x_i\}$ where $range(x_i) \cap range(b_i) \neq \phi$ and bin size(x_i) = s_{i+1} (the next higher bin size)
2. For every CC, C in bin b_i, iterate through each bin $\{x_i\}$ and find components c_i where c_i ε x_i and $region(c_i) \cap region(C) \neq \phi$
3. The following criteria to be fulfilled by c_i are:

   i. Number of pixels of c_i is not too large compared to the number of pixels of C (we said that number of pixels c_i is near C if (number of pixels of c_i) < 1.2*(number of pixels of C).
   ii. Height and width of the bounding box of c_i are calculated as:
      a. Height of Bounding Box of c_i < minimum(1.2*(height of C), (height of C) + 5)
      b. Width of Bounding Box of c_i < minimum(1.2*(width of C), (width of C) + 5)

4. If component c_i satisfies all the criteria in Step 3, we call it a true object, otherwise it is removed.

The algorithm detects the stable regions as shown in Fig. 5.

3.3 Analysis of Component Alignment

After removing object segments from the bin images, we evaluate the characterness [5] of the remaining objects by distinguishing them as:

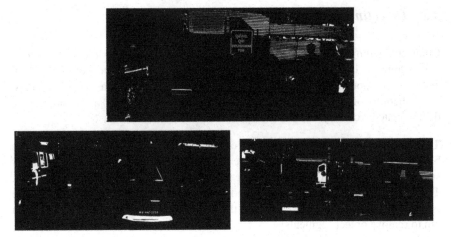

Fig. 5 Sample binary images of detected stable regions after processing the bins

Aligned components They share similar heights and baseline positions and are at a small distance from each other.

Non-aligned components Isolated components which do not satisfy the aligned components criteria (usually text is written in connected form).

The aligned component set is a single unit and determination of its text/non-text nature is done for the whole set by considering the average of characterness cue values.

3.4 Analysis of Characterness

We calculate the basic characterness cues like Solidity, Eccentricity, Extent and Euler number. For non-aligned components, solidity threshold is 0.65 and 0.8 average value for aligned components. Final characterness of the components is determined using script-independent features, namely: Stroke Width variation [5], edge Histogram of Gradients (eHOG) [5].

Stroke Width Variation This feature helps in determining if a set of aligned components or an isolated component are text regions or not and is represented as

$$SW(r) = \frac{\text{var}(1)}{E(1)^{\wedge}2}. \tag{2}$$

Here l is the distance between two edge pixels in perpendicular direction. We observe a threshold of 0.25 for $SW(r)$ to give uniform results, after conducting several experiments on various images.

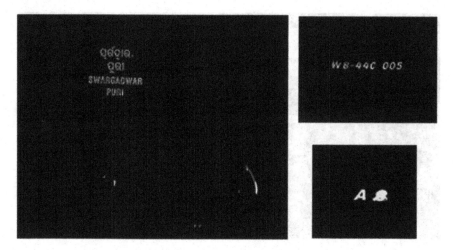

Fig. 6 Samples of components after characterness analysis

Histogram of Gradients at Edges This feature also helps in determination of Characterness of components. We find the skeleton of the connected component and determine the eHOG of the skeleton as

$$\mathrm{eHOG}(r) = \frac{\sqrt{(w(r) - y(r))^2 + (x(r) - z(r))^2}}{w(r) + x(r) + y(r) + z(r)}. \tag{3}$$

Here, $w(r)$, $x(r)$, $y(r)$, $z(r)$ are the number of gradients of particular type, where $w(r)$ is the pixel count with gradient (ϕ) as $0 < \phi \le \pi/4$ or $7\pi/4 < \phi \le 2\pi$, $x(r)$ has gradients $\pi/4 < \phi \le 3\pi/4$, $y(r)$ has gradients $3\pi/4 < \phi \le 5\pi/4$ and $z(r)$ has gradients $5\pi/4 < \phi \le 7\pi/4$; eHOG threshold was set to 0.3. Components left in bins after characterness analysis is in Fig. 6.

3.5 Combination of Information

Text regions tend to stand apart in at least two color planes. For example, a blue text in a red background stands apart more in the red and blue planes. We isolate components having occurrence in at least two planes (or having a component almost similar in a plane). After combining all the bins we get the result as depicted in Fig. 7.

Algorithm 2 1. For bin b_i of size s_i, let $\{c_i\}$ be the set of CCs found in the bin.
2. Let $\{x_i\}$, $\{y_i\}$ and $\{z_i\}$ be the set of bins in color planes other than the current plane of size s_i.

Fig. 7 The possible text regions are bounded in red after bin combination. Some false positives and false negatives are also bounded

3. If a component C is present in any of the bins $\{x_i\}$, $\{y_i\}$ or $\{z_i\}$ such that $region(c_i) \cap region(C) \neq \phi$ and the C and c_i are structurally similar (criteria same as that in Algorithm 1), then the largest of the two components above are retained for the final localization, else c_i is eliminated.

4 Experiments and Evaluation

We conducted experiments on our system with an in-house dataset consisting of more than 300 smartphone camera captured images containing multi-lingual and multi-oriented texts.

Based on number of detected text regions (sample outputs are depicted in Fig. 8), we have calculated the precision, recall and F-measure for best, worst and average cases as recorded in Table 1.

It is observed that the F-measure is 0.69 which is attributed to detection of a good True Positive (TP) count and low False Positives (FP) and False Negatives (FN).

Table 1 Proposed work performance in best, worst and average cases

Cases	Precision	Recall	F-measure
Best case	0.69	1.00	0.69
Worst case	0.43	0.39	0.46
Average case	0.48	0.84	0.58

Fig. 8 Sample outputs of text regions bounded in red, detected by the proposed method

However, the value 0.46 is due to rare text presence thus giving us less TPs. The TPs consist of Indic scripts along with its English counterpart, thus giving us the impression that we have achieved a great degree of script-independence.

5 Conclusion

The proposed work introduces a unique way of exploiting the texture properties of images by binning the foregrounds and then performing CCA to give multi-lingual texts thereby achieving a great deal of language independence. We achieve high recall for text heavy images but with decrease in textual contents, the recall diminishes. In future, improvements will be made in its robustness by employing efficient classifier methods for binning and decide the combination of the processed bins.

Acknowledgements This work is partially supported by the CMATER research laboratory of the Computer Science and Engineering Department, Jadavpur University, India, PURSE-II and UPE-II, project. SB is partially funded by DBT grant (BT/PR16356/BID/7/596/2016) and UGC Research Award (F. 30-31/2016(SA-II)). RS, SB and AFM are partially funded by DST grant (EMR/2016/007213). The heading should be treated as a 3rd level heading and should not be assigned a number.

References

1. Neumann, L., Matas, J.: Real-time scene text localization and recognition. In: 2012 IEEE Conference on Computer Vision and Pattern Recognition (CVPR). IEEE (2012)
2. Lluís, G., Karatzas, D.: Textproposals: a text-specific selective search algorithm for word spotting in the wild. Pattern Recogn. **70**, 60–74 (2017)
3. Bosamiya, J.H. et al.: Script independent scene text segmentation using fast stroke width transform and GrabCut. In: 2015 3rd IAPR Asian Conference on Pattern Recognition (ACPR). IEEE (2015)

4. Sounak, D., et al.: Script independent approach for multi-oriented text detection in scene image. Neurocomputing **242**, 96–112 (2017)
5. Yao, Li, et al.: Characterness: an indicator of text in the wild. IEEE Trans. Image Process. **23**(4), 1666–1677 (2014)
6. Gonzalez, A. et al.: Text location in complex images. In: 2012 21st International Conference on Pattern Recognition (ICPR). IEEE (2012)
7. Kumar, D., Prasad, M.N., Ramakrishnan, A.G.: Multi-script robust reading competition in ICDAR 2013. In: Proceedings of the 4th International Workshop on Multilingual OCR. ACM (2013)
8. Gomez, L., Karatzas, D.: Multi-script text extraction from natural scenes. In: 2013 12th International Conference on Document Analysis and Recognition (ICDAR). IEEE (2013)
9. Yin, X.-C., et al.: Multi-orientation scene text detection with adaptive clustering. IEEE Trans. Pattern Anal. Mach. Intell. **37**(9), 1930–1937 (2015)
10. Aneeshan, S., et al.: Multi-oriented text detection and verification in video frames and scene images. Neurocomputing **275**, 1531–1549 (2018)
11. Xu, H., Xue, L., Su, F.: Scene text detection based on robust stroke width transform and deep belief network. In: Asian Conference on Computer Vision. Springer, Cham (2014)
12. Huang, W., Qiao, Y., Tang, X.: Robust scene text detection with convolution neural network induced mser trees. In: European Conference on Computer Vision. Springer, Cham (2014)
13. He, T., et al.: Text-attentional convolutional neural network for scene text detection. IEEE Trans. Image Process. **25**(6), 2529–2541 (2016)
14. Honggang, Z., et al.: Text extraction from natural scene image: a survey. Neurocomputing **122**, 310–323 (2013)
15. Zhu, Y., Yao, C., Bai, X.: Scene text detection and recognition: recent advances and future trends. Front. Comput. Sci. **10**(1), 19–36 (2016)
16. Ye, Q., Doermann, D.: Text detection and recognition in imagery: a survey. IEEE Trans. Pattern Anal. Mach. Intell. **37**(7), 1480–1500 (2015)

Script Identification from Camera-Captured Multi-script Scene Text Components

Madhuram Jajoo, Neelotpal Chakraborty, Ayatullah Faruk Mollah, Subhadip Basu and Ram Sarkar

Abstract Identification of script from multi-script text components of camera-captured images is an emerging research field. Here, challenges are mainly twofold: (1) typical challenges of camera-captured images like blur, uneven illumination, complex background, etc., and (2) challenges related to shape, size, and orientation of the texts written in different scripts. In this work, an effective set consisting of both shape-based and texture-based features is designed for script classification. An in-house scene text data set comprising 300 text boxes written in three scripts, namely Bangla, Devanagri, and Roman is prepared. Performance of this feature set is associated with five popular classifiers and highest accuracy of 90% is achieved with Multi-layer Perceptron (MLP) classifier, which is reasonably satisfactory considering the domain complexity.

Keywords Script identification · Multi-script · Scene text · GLCM · HOG · MLP

1 Introduction

Identifying scripts of scene text images is a domain of great interest to the research community since it has a crucial usage in visual applications like content-based image searching, aiding visually hindered by reading text in the wild, or language

M. Jajoo · N. Chakraborty · S. Basu · R. Sarkar (✉)
Department of Computer Science and Engineering, Jadavpur University, Kolkata, India
e-mail: raamsarkar@gmail.com

M. Jajoo
e-mail: madhuramjajoo10@gmail.com

N. Chakraborty
e-mail: neelotpal_chakraborty@yahoo.com

S. Basu
e-mail: subhadip@cse.jdvu.ac.in

A. F. Mollah
Department of Computer Science and Engineering, Aliah University, Kolkata 700160, India
e-mail: afmollah@aliah.ac.in

© Springer Nature Singapore Pte Ltd. 2019
J. Kalita et al. (eds.), *Recent Developments in Machine Learning and Data Analytics*,
Advances in Intelligent Systems and Computing 740,
https://doi.org/10.1007/978-981-13-1280-9_16

translation for commuters/tourists [1]. Script recognition is a challenging task due to several factors like diversity of acquisition conditions, low resolution, font variability, complex backgrounds, different lighting conditions, blur, and others [2]. It becomes more complex when scene text is a multi-script one since a particular Optical Character Recognition (OCR) [14] system, in general, can handle a single script. Hence, a prerequisite is a script recognition module which helps in identifying the multiple scripts. Script recognition problem is hugely explored at OCR level but not at scene text level and hence we work on in-house multi-script scene text data comprising three popularly used scripts in Indian subcontinent, namely Bangla, Devanagri, and Roman to design a script identification model.

2   Related Works

Recent research works include the development of some shape- and texture-based features like GLCM and HOG, which have become extremely popular in determining character features which may vary scriptwise.

Some mid-level features are pooled out of local features in the work [3] to determine script. The works in [4] use HOG (Histogram of Gradients) to determine the script of individual components in an image. Multi-orientations in scripts are dealt with in the works [5] where adaptive clustering methods are used based on differences in color, stroke width, and location (Compactness), and [6] where a combination of Fourier-Laplacian and Hidden Markov Model is used in recognizing the scripts. The process proposed in [7] employs a deep belief network which is introduced for scene text script recognition. A sparse belief lexicon has been developed along with similarity to recognize scene texts in the work [8]. In the work [9], character wise analysis is done to recognize the script type. An end-to-end neural network is trained for identifying scripts in [10].

3   Proposed Method

In the proposed work, we consider text components extracted from multi-script scene text images. Word level components generated from a sample scene text image is shown in Fig. 1. Then we estimate some shape-based [11] (from binarized image) and texture-based features [11] (from grayscale image) from these text boxes. Next, feature values are fed to classifier(s) for script identification.

Fig. 1 An illustration of reducing a scene text image to word level localization

3.1 Shape-Based Feature

Here, $B_n(X_n, Y_n)$ are the boundary points of the shape in the image, $C(C_x, C_y)$ is the centroid of the shape, and n is the contour pixel count unless otherwise stated. To define the following features, we use Eq. (1) where P, Q, and R are 1×2 vector.

$$
\text{Area}(A, B, C) = \tfrac{1}{2} \begin{array}{ccc} A(1,1) & A(1,2) & 1 \\ B(1,1) & B(1,2) & 1 \\ C(1,1) & C(1,2) & 1 \end{array} \tag{1}
$$

Complex Co-ordinates. Normalizing real ($F1$) and imaginary ($F2$) parts of $S_{Z(n)}$ gives two feature values as:

$$
Z(n) = [X_n - C_x] + i\big[y_n - C_y\big]. \tag{2}
$$

$$
S_{Z(n)} = \sum_{i=1}^{n} Z(n) \tag{3}
$$

Centroid Distance Function. This feature gives the distance between nth boundary point and the centroid, and is given as:

$$
\text{CD}_n = \big[(X_n - C_X)^2 + (Y_n - C_Y)^2\big]^{1/2} \tag{4}
$$

$F3$ is the summation of CD_n values. Subtraction of C's from the B's makes $F1$, $F2$, and $F3$ translation invariant.

Angle of Tangency. For all digitally curved contours, a nominal-sized window w is introduced to determine the contour's tangential direction, Θ_n with greater accuracy. Feature $F4$ sums up positive Θ_n values.

$$
\theta_n = \tan^{-1} \frac{y(n) - y(n - w)}{x(n) - x(n - w)} \tag{5}
$$

Angle of Slope. This angle between positive x-axis and line joining points at boundaries say, B_1 and B_2, is given by:

Fig. 2 Illustration of the
area feature

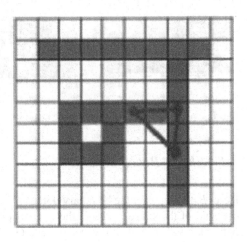

$$\text{Slope Angle} = \tan^{-1} \frac{y2 - y1}{x2 - x1} \tag{6}$$

$F5$ represents the summing of the angles of slope at every point of contour with respect to B_i's.

Centroid Area. The Centroid Area is calculated using two consecutive B_i's and the C (Fig. 2).

$$F6 = \sum \text{Area}(B_i, B_{i+1}, C) \tag{7}$$

where $i \in (1, n - 1)$ using Eq. (1).

Three Consecutive Points Area. Area between the three consecutive points is calculated as:

$$F7 = \sum \text{Area}(B_i, B_{i+1}, B_{i+2}) \tag{8}$$

where $i \in (1, n - 2)$ using Eq. (1).

3.2 Texture-Based Features

The texture of an image constitutes the distribution of color or intensity values in its special domain. Texture analysis can be done by studying the frequently occurring patterns in an image. In our work, we used two popular texture-based features, Histogram of Oriented Gradients (HOGs) and Gray-Level Co-occurrence Matrix (GLCM) which are described briefly in the following sections.

Histogram of Oriented Gradients. HOG features are used for detecting text objects by counting the number of gradient orientations in localized image parts.

$$\text{Magnitude}_{p(x,y)} = \sqrt{Gx(x, y)^2 + Gy(x, y)^2} \tag{9}$$

$$\text{Direction}_{p(x,y)} = \tan^{-1} \frac{Gy(x, y)}{Gx(x, y)} \tag{10}$$

where

$$Gx(x, y) = p(x, y + 1) - p(x, y) \tag{11}$$

$$Gy(x, y) = p(x + 1, y) - p(x, y) \tag{12}$$

Size of Image: [48, 48]; Size of Cell: [10, 10]; Size of Block: [4, 4]; Number of Bins: 8; Size of overlapping block: [2, 2]. With these values, we get HOG feature length $(N) = 128$ thereby contributing 128 more features (F8–F135).

GLCM Parameters. The texture of an image is analyzed using GLCM where the occurrence of specific-valued pixel pairs sharing a certain spatial relationship in an image, is determined. GLCMs are $i \times j \times n$ matrices, where n is the parameter count. Element (i, j) in matrix is the number of pairs (i, j) occurring in the image. Following are the parameters of GLCM where i and j are the intensity values or indices of GLCM and $G(i, j) = \text{GLCM}(i, j)$:

Contrast. Intensity contrast between a pixel and its neighbor is measured over the whole image using the following equation:

$$\text{Contrast} = \sum_{i,j} |i - j|^2 G(i, j) \tag{13}$$

Correlation. The correlation between a pixel and its neighbor is measured as:

$$\text{Correlation} = \sum_{i,j} \frac{(i - \mu i) \cdot (i - \mu j) \cdot G(i,j)}{o_i o_j} \tag{14}$$

where o_i and o_j are the intensity variances of all reference pixels in the relationships that contributed to the GLCM.

Energy. It is the sum of square of GLCM elements and is determined as:

$$\text{Energy} = \sum_{i,j} G(i, j)^2 \tag{15}$$

Homogeneity. It measures the uniformity in distribution of elements of GLCM to those in diagonal and is calculated as:

$$\text{Homogeneity} = \sum_{i,j} \frac{G(i, j)}{1 + |i - j|} \tag{16}$$

All the above features are of order [1, 2], and hence eight more features (F136–F143) are extracted from each text box.

4 Experimental Results and Discussion

A total of 300 scene text components are considered for evaluating the identification of scripts at component level. Here, we have taken same number of text images for each of the scripts namely, Bangla, Devanagri, and Roman. The input images are binarized using Otsu's thresholding to estimate $F1$–$F7$ features, and grayscale images are considered for extraction of HOG and GLCM features ($F8$–$F143$). From a total of 300, 240 words (80 per script) constitute the train set and the remaining 60 words (20 per script) constitute the test set. Our chosen feature set is fed to five well-known classifiers: Naïve Bayes, MLP, Support Vector Machine (SVM), Multi-Class Classifier, and Simple Logistic, and the results with different combination of feature vectors are tabulated using the line graphs in Fig. 3.

With the combined feature set, we obtain classification for five popular classifiers as shown in Table 1, where MLP gives the highest accuracy of 90% but with a build time of 168.67 s. Some samples of success, as well as failure cases, are shown in Fig. 4.

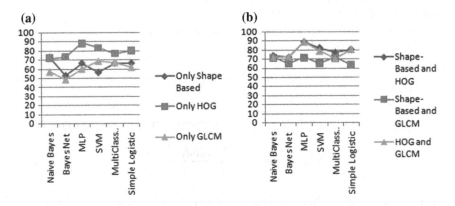

Fig. 3 Line graph showing results of **a** individual features **b** two feature combination

Table 1 Accuracies of popular classifiers for script identification using shape based, HOG, and GLCM features

Classifiers →	Naïve Bayes	MLP	SVM	Multi-class classifier	Simple logistic
Accuracy →	75.00	90.00	81.67	76.67	80.00
Build time(s) →	0.06	168.67	0.61	0.91	4.7

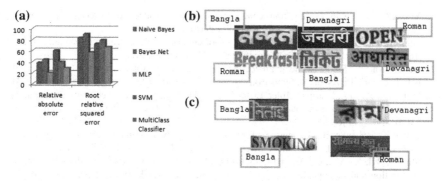

Fig. 4 a Comparative performance of various classifiers designed for script classification **b** Successfully classified scene text components **c** Scene text components with incorrect script classification

5 Conclusion

In this paper, we proposed a script identification model for scene text components by combining both shape- and texture-based features. Here, the camera-captured images contain texts in multiple scripts, namely Bangla, Devanagri, and Roman. Five popular classifiers are used and MLP is found to yield the highest accuracy of 90%, which is reasonably satisfactory considering the domain complexity. However, it may not perform well in the case of heterogeneous image backgrounds. Also, the robustness of shape-based features may be affected due to information loss in binarization. Hence, we aim to develop methods to deal with these issues in future. Increasing the size of our database along with the inclusion of more scripts are left for future works.

Acknowledgements This work is partially supported by the CMATER research laboratory of the Computer Science and Engineering Department, Jadavpur University, India, PURSE-II and UPE-II, project. SB is partially funded by DBT grant (BT/PR16356/BID/7/596/2016) and UGC Research Award (F. 30-31/2016(SA-II)). RS, SB and AFM are partially funded by DST grant (EMR/2016/007213).

References

1. Neumann, L. Matas, J.: Real-time scene text localization and recognition. In: Computer Vision and Pattern Recognition (CVPR), 2012 IEEE Conference on IEEE (2012)
2. Gómez, L., Karatzas, D.: Textproposals: a text-specific selective search algorithm for word spotting in the wild. Pattern Recogn. **70**, 60–74 (2017)
3. Singh, A.K., et al.: A simple and effective solution for script identification in the wild. In: Document Analysis Systems (DAS), 2016 12th IAPR Workshop on. IEEE (2016)
4. Li, Y., et al.: Characterness: an indicator of text in the wild. IEEE Trans. Image Process. **23**(4), 1666–1677 (2014)

5. Yin, X.C., et al.: Multi-orientation scene text detection with adaptive clustering. IEEE Trans. Pattern Anal. Mach. Intell. **37**(9), 1930–1937 (2015)
6. Sain, A., Bhunia, A.K., Roy, P.P., Pal, U.: Multi-oriented text detection and verification in video frames and scene images. Neurocomputing **275**, 1531–1549 (2018)
7. Xu, H., Xue, L., Su, F.: Scene text detection based on robust stroke width transform and deep belief network. In: Asian Conference on Computer Vision, pp. 195–209. Springer, Cham (2014)
8. Weinman, Jerod J., Learned-Miller, Erik, Hanson, Allen R.: Scene text recognition using similarity and a lexicon with sparse belief propagation. IEEE Trans. Pattern Anal. Mach. Intell. **31**(10), 1733–1746 (2009)
9. Wang, K., Babenko, B., Belongie, S.: End-to-end scene text recognition. In: Computer Vision (ICCV), 2011 IEEE International Conference on, (pp. 1457–1464). IEEE (2011)
10. Shi, B., Bai, X., Yao, C.: An end-to-end trainable neural network for image-based sequence recognition and its application to scene text recognition. IEEE Trans. Pattern Anal. Mach. Intell. **39**(11), 2298–2304 (2017)
11. Ye, Qixiang, Doermann, David: Text detection and recognition in imagery: a survey. IEEE Trans. Pattern Anal. Mach. Intell. **37**(7), 1480–1500 (2015)

Implementation of BFS-NB Hybrid Model in Intrusion Detection System

Sushruta Mishra, Chandrakanta Mahanty, Shreela Dash
and Brojo Kishore Mishra

Abstract Recently due to rise in technology and connectivity among various components of a system, various new cybersecurity issues are emerging. To handle this rapid growth in computer attacks, intrusion detection system models are developed and used, which are characterized with optimum accuracy and minimum complexity. These systems are the models used to detect and identify any suspicious event occurring on a system. Based on the information sent by it, all sorts of activity can be analyzed and distinguished among normal or malicious activities. Our proposed work constitutes developing a data mining-based intrusion detection model which uses a classification algorithm as well as a feature selection technique. This proposed method is named as BFS-NB and is used to identify various network-based attacks and thereby analyzing the performance of our network. The dimensionality reduction is performed using best first search method, and then naïve Bayes classifier is used to predict the type of attacks. The performance is evaluated using some parameters like accuracy, sensitivity, and time delay. The results indicate that our developed technique gives an optimal performance and may be used as an effective intrusion detection system by professionals.

Keywords Best first search · Data mining · Intrusion detection system
Accuracy · Sensitivity · Naïve Bayes

S. Mishra (✉) · S. Dash
KIIT University, Bhubaneswar, India
e-mail: mishra.sushruta@gmail.com

S. Dash
e-mail: shreelamamadash@gmail.com

C. Mahanty · B. K. Mishra
CVRCE, Bhubaneswar, India
e-mail: chandra.mahanty@gmail.com

B. K. Mishra
e-mail: brojokishoremishra@gmail.com

© Springer Nature Singapore Pte Ltd. 2019
J. Kalita et al. (eds.), *Recent Developments in Machine Learning and Data Analytics*,
Advances in Intelligent Systems and Computing 740,
https://doi.org/10.1007/978-981-13-1280-9_17

1 Introduction

Issue of data security is predominant with the rise of networking complexity in computers. Several network-oriented applications have surfaced due to rapid development of networked system-based resources. This rise in networking has given rise to several unauthorized events which include both internal and external threats [1]. With the rapid rise in computer-based events, there is a constant growth of networking-dependent attacks, which impacts integrity and availability of crucial data. Hence, it is very important for a networked system model to make use of secure tools like antivirus and firewall. A firewall is not completely secure from all external attacks. Similarly, tools like honey pot and antivirus too are not entirely resistant to all sorts of attacks. Hence to deal with such networking attacks, a real-time intrusion detection system model is the need of the hour. It alarms the computer system when any malicious or suspicious event occurs in a system by monitoring any hazardous activities [2]. A secure intrusion detection model can be successfully implemented as both software and hardware [3, 4]. Hardware-based system is used to deal with enormous amount of network traffic but quite costly while software-based model is simple to update and easily configurable. Hence, it is important to verify the availability of an efficient intrusion detection system [5].

2 Intrusion Detection System Categories

Masud et al. [6] used two different techniques of intrusion detection system which can be visualized from Table 1. Taxonomy of IDS is constructed with the use of three basic concepts which include signature detection, anomaly detection, and hybrid detection. Various IDS are categorized using system features, detection interval, and response to identifying intrusions.

Table 1 General IDs taxonomy

Anomaly	Self-learning	Non-time series
		Time series
	Programmed	Descriptive stats
		Default deny
Signature	Programmed	State modeling
		Expert system
	Self-learning	String matching
		Simple rule-based
Signature inspired	Self-learning	Automatic feature selection

3   Intrusion Detection System

Detection of different kinds of malicious attacks and unwanted network traffic can be monitored through an intrusion detection system. Based on any suspicious event occurring on system, an alert signal is conveyed to the network administrator. General intrusion detection model is depicted in Fig. 1. Various malicious and sensitive activities involve information-driven attacks on system applications, permissions and privilege services, confidential access to organizational data, and malware. Some intrusion-based systems simply monitor the network activity and alert for any suspicious event. Few others are associated with performing actions in response to some threat detection.

Different kinds of intrusion detection system are as follows:

NIDS: Network intrusion detection system is arranged in some strategic zone inside the network so that it can easily monitor network traffic on all devices on the system. It acts as a bottleneck to impact the overall rate of the network.

HIDS: Host intrusion detection system is executed on individual devices on the network. Here, the inbound and outbound packets are monitored on the devices and accordingly an alert signal is sent to the user if any malicious event is observed.

Signature-oriented IDS: Here, traffic monitoring occurs on the network and is compared with a dataset of attributes from known threats. There is a matching of events between the observed threat and the recorded threat in database and accordingly dealt with.

Anomaly based IDS: Here, the network traffic is monitored and compared with a predefined baseline. The baseline detects the threat type and determines the bandwidth used in the network and decides whether it is normal activity or not [7].

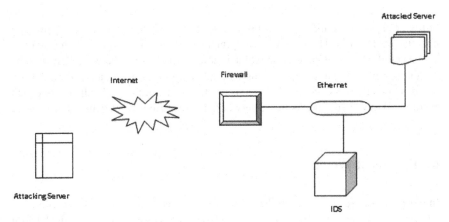

Fig. 1 An intrusion detection system

4 Data Mining in Intrusion Detection System

Data mining is a very constructive and intelligent technique that enables various organizations to give emphasis on important aspects in their databases. Algorithms of data mining help in prediction of future behavior based on predefined data samples permitting business to be more active and decision-oriented. It can also be utilized for misuse and inconsistent intrusion identification. In an intrusion detection model, the information deals from multiple sources such as network traffic or logs, system logs, application logs, alarm messages, etc. Due to varied data source and format, the complexity increased in auditing and analysis of data [7]. Data Mining has a huge advantage in data extraction from large volumes of data that are noisy and dynamic; thus, it is of great importance in intrusion detection system. Every data tuple of a data record is tagged as "normal" or "intrusion" in misuse detection while a learning classifier is used for training over the labeled data. Anomaly generally refers to an event that might potentially turn out to be an intrusion. A great level of accuracy rate for identifying known attacks and their variants is one vital benefit of misuse identification methods [8, 9]. An intrusion detection system model is useful in guarding a network against anonymous attacks from external as well as internal environment. It also helps in determining the category of attack. It helps in detecting intrusion in several platforms like college network and industries, etc.; in past, some data mining techniques have been used for this purpose and their performance is evaluated. In [10], a hybrid model of intrusion detection model is developed and implemented for DDOS attacks successfully.

5 Dataset

Software used to identify system intrusions prevents a computer model from any unauthorized access. The task of developing an intrusion detection system is to construct a model based on prediction which is able to differentiate between connections with attack potential and connections without any attacks. Here, a standard dataset for local area network of US Air Force is used, which comprises 38 attributes which correspond to various attacks in the network. These attributes are denoted in Table 2.

6 Proposed Work

In our proposed research model BFS-NB, we have used naïve Bayes as a classifier while best first search is the attribute optimization method applied. The KDD data records are gathered from US Air force. The dataset constitutes 38 distinct types of attacks which are categorized into four different types. After analyzing various features in the dataset only the requisite attributes are retained while others are

Table 2 Features of KDD dataset

Duration	Protocol	Service
Scr bytes	Des bytes	Land
Urgent	Hot	Num failed logins
Num compromised	Root shell	Su attempted
Num file creations	Num shell	Num access files
Is host login	In guest login	Count
Serror rate	Svr serror rate	Rerror rate
Same srv rate	Diff srv rate	Srv diff host rate
Dst host srv count	Dst host same srv rate	Dst host diff srv rate
Dst host srv diff host rate	Dst host serror rate	Dst host srv serror rate
Dst host srv diff host rate	Normal or attack	

eliminated. Then, data preprocessing takes place where missing values are handled with consistent figures. Best first search is used as a feature selection tool to optimize the dataset. Among the data samples, 20% of them are selected with labeled data for training using naïve Bayes while rest with unlabelled data are used for testing phase. A hierarchical structure is built which is similar to decision tree that determines the category of attacks in the system. A tuple is assumed to be normal if it lacks any attack. But if a potential attack is found, then its category is identified. This is applicable in an active system scenario where packets are permitted to pass through the system if it is normal while an alert is sent to the user if an attack is detected.

6.1 Best First Search

Classifier used in our study is best first search to determine the classification accuracy. Features with higher value of merit are termed as potential features which are further used to classify the tuples. Here, the entire feature subset is searched by adding a backtracking procedure. It may initiate with an empty feature set and traverse in forward direction or it may start with complete feature set searching backward or even traverse in both direction of search space.

7 Results and Analysis

Our intrusion detection system was evaluated using WEKA data mining tool for analysis. NSL-KDD data samples are used as the dataset which is an upgraded version of KDDcup99 data. Cross-validation method is used to access the results of the evaluation. This technique is applied in scenarios where the main objective is to predict an outcome and to determine its effectiveness in reality.

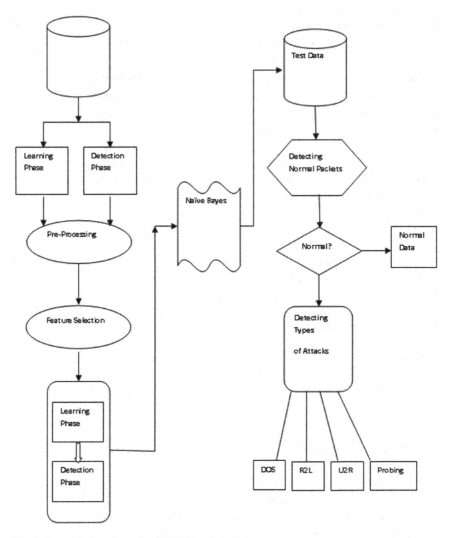

Fig. 2 Intrusion detection using BFS-NB technique

In this section, our proposed algorithm (BFS-NB) was evaluated with some performance metrics like classification accuracy, delay, and sensitivity. Anomaly detection is performed with the help of this data mining model developed. Maximum instances used for the evaluation was 1183. Evaluation of performance starts using feature optimization of the raw data samples using best first search. Experimental evaluation depicts that the efficiency of the optimal attribute set predicts the classification. It was demonstrated that BFS-NB yields an optimal accuracy with 92.5%. The time taken for prediction of attack is also less compared to our algorithm with 1.23 s. The sensitivity analysis with BFS-NB is 97% while specificity value is 97.6%.

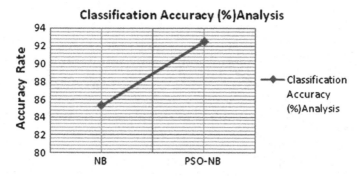

Fig. 3 Accuracy analysis in IDS using BFS-NB

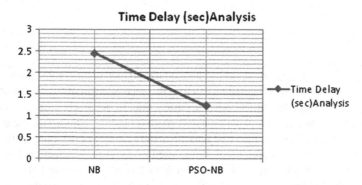

Fig. 4 Time delay analysis in IDS using BFS-NB

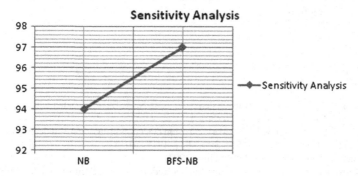

Fig. 5 Sensitivity analysis in IDS using BFS-NB

The result obtained is quite optimal compared to implementing it with naïve Bayes classifier without using BFS algorithm for feature selection. The results are depicted in Figs. 2, 3, 4, 5 and 6.

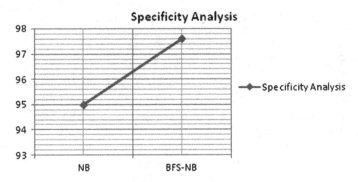

Fig. 6 Specificity analysis in IDS using BFS-NB

8 Conclusion

Recently, intrusion is having a vital impact on the privacy of an organization. Therefore, an efficient intrusion detection system is the need of the hour which can identify
several potential attacks with a higher accuracy rate. In our work, a strong intrusion
detection model was proposed with naïve Bayes as classifier to minimize the delay in
building model for classification and maximize the accuracy of detecting intrusion in
organization. Best first search method was used as an attribute selection technique.
Results show that our proposed algorithm BFS-NB is an effective hybrid model,
which is helpful in precise detection of types of attacks in an intrusion detection system. Our future work includes the application of extreme learning machine (ELM)
in intrusion detection.

References

1. Mukherjee, S., Sharma, N.: Intrusion detection using Naive Bayes classifier with feature reduction. Procedia Technol. **4**, 119–128 (2012)
2. Swamy, K.V., Lakshmi, K.V.: Network intrusion detection using improved decision tree algorithm. Int. J. Comput. Sci. Inf. Sec. **3**(5), 4971–4975 (2012)
3. Wallner, R.: Intrusion Detection Systems (2007)
4. Di Pietro, R., Mancini, L.V. (ed.) Intrusion Detection Systems. Springer Science & Business
   Media, (June 12 2008)
5. Alhomoud, A., Munir, R., Disso, J.P., Awan, I., Al-Dhelaan, A.: Performance evaluation study
   of intrusion detection systems. Procedia Comput. Sci. **1**(5), 173–180 (2011)
6. Masud, M., Khan, L., Thuraisingham, B., in Data Mining Tools for Malware Detection. CRC
   Press, (Dec 7 2011)
7. Anderson, J.P.: Computer security threat monitoring and surveillance, Washington, Pennsylvania. J. Comput. Sci. Netw. Sec. USA (1980)
8. Dokas, P., Ertoz, L., Kumar, V., Lazarevic, A., Srivastava, J., Tan, P.N.: Data mining for network intrusion detection. In: Proceedings of NSF Workshop on Next Generation Data Mining,
   pp. 21–30, Nov (2002)

9. Javitz, H.S., Valdes, A., NRaD, C.: The NIDES statistical component: description and justification. Contract. **39**(92-C), 0015 (1993)
10. Cepheli, Ö., Büyükçorak, S., Karabulut Kurt, G.: Hybrid intrusion detection system for ddos attacks. J. Electr. Comput. Eng. 1–8 (2016)
11. Patil, P.R., Sharma, Y., Kshirasagar, M.: Performance analysis of intrusion detection systems implemented using hybrid machine learning techniques. Int. J. Comput. Appl. (0975–8887) **133**(8), 35–38 (2016)

Context-Sensitive Spelling Checker for Assamese Language

Ranjan Choudhury, Nabamita Deb and Kishore Kashyap

Abstract The task of finding real-word error in sentences is a complex problem in the field of natural language processing. In this paper, we are presenting a new method for context-sensitive spell checking for Assamese language, a technique to tackle the issue of real-word error detection and correction. The emphasis is concentrated on n-gram language model for Assamese language to develop the spelling checker. The purpose is to detect spelling mistake according to the context of the sentence and suggest the most probable word out of a confusion set generated for each of the misspelled words in the sentence. The system was tested for spell checking and word suggestion on the basis of precision and recall of correct and incorrect words. An overall performance of 76% was attained. It can be inferred that the system has the potentiality for further research and other software development for Assamese language.

Keywords Context-sensitive spell checking · n-gram · Edit distance · Assamese language · Confusion set · Spelling mistake

1 Introduction

Two types of spelling mistake that occur in a sentence are real-word error and nonword error [1]. The nonword spelling mistake is an error in which the sentence consists of a word that is not present in the language. For example,

বিশ্বকাপৰ সময়ত দলটোক সমৰ্থন কনৰাৰ বাবে প্ৰাক্তন খেলুৱৈসকলক সমালোচনা কৰিছে।

R. Choudhury (✉) · N. Deb · K. Kashyap
Department of IT, Gauhati University, Guwahati, India
e-mail: choudhury.ranjan22@gmail.com

N. Deb
e-mail: nd@gauhati.ac.in

K. Kashyap
e-mail: kk@gauhati.ac.in

© Springer Nature Singapore Pte Ltd. 2019
J. Kalita et al. (eds.), *Recent Developments in Machine Learning and Data Analytics*,
Advances in Intelligent Systems and Computing 740,
https://doi.org/10.1007/978-981-13-1280-9_18

The word কনৰাৰ is not a valid word of Assamese language. Hence, it is termed as a nonword spelling mistake. Detection and correction of this type of spelling mistake are relatively easy than the real-word spelling mistake.

Real-word spelling mistake is an error in a sentence in which one or some of the words do not fit in the context of the sentence. In this case, the erroneous word belongs to a valid set of words of the language. For example,

বিশ্বকাপৰ সময়ত দলটোক সমৰ্থন মকৰাৰ বাবে প্ৰাক্তন খেলুৱৈসকলক সমালোচনা কৰিছে।

The word মকৰাৰ is a valid Assamese word but does not fit in the context of the sentence. It might be a typing mistake of the word নকৰাৰ.

Context-sensitive spell checking is a technique which identifies the real-word spelling errors and provides a correction of the erroneous word. We have tried to achieve this using a statistical machine learning approach using n-gram language model for Assamese language.

The process involves finding a confusion set for each of the testing words of the sentence and determining the most probable word out of the confusion set which gives the maximum probability as per the context of the sentence. The n-gram language model generated from a standard Assamese language corpus helps us here to determine the context of a word in a sentence.

1.1 Trigram of a Sentence

The trigram is a sequence of three tokens $\{w_i, w_{i+1}, w_{i+2}\}$ of a sentence $\{w_1, w_2 \ldots w_k\}$. Given a sentence likeবিশ্বকাপৰ সময়ত দলটোক সমৰ্থন নকৰাৰ বাবে প্ৰাক্তন খেলুৱৈসকলক সমালোচনা কৰিছে, trigram of the sentence would be {বিশ্বকাপৰ সময়ত দলটোক}, {সময়ত দলটোক সমৰ্থন}, {দলটোক সমৰ্থন নকৰাৰ}, {সমৰ্থন নকৰাৰ বাবে}, {নকৰাৰ বাবে প্ৰাক্তন}, {বাবে প্ৰাক্তন খেলুৱৈসকলক}, {প্ৰাক্তন খেলুৱৈসকলক সমালোচনা} and {খেলুৱৈসকলক সমালোচনা কৰিছে}.

A trigram language model stores the probability of each trigram sequence calculated from a language corpus.

1.2 Confusion Set

The set of words which are often mistyped for another word in a sentence is called a confusion set. It may happen due to deletion of a letter from the word, transposition of the adjacent letters of the word, and replacement or insertion of some letters into the word.

1.3 Statistical Trigram Approach for Spell Checking

The prime objective here is to find the most suitable word out of a confusion set of the testing word in a sentence. Let w_i be the testing word of the sentence $w_1 \ w_2 \ldots w_i \ldots w_n$ and w_i^c is the confusion set of w_i. The contextual accuracy of the word w_i would be determined by calculating the trigram probability of sentence $w_1\ w_2 \ldots w_i \ldots w_n$ where w_i is substituted by each word from the confusion set w^c. For a word w_i^c from w^c, if $probability(w_1\ w_2 \ldots w_i^c \ldots w_n) > probability(w_1\ w_2 \ldots w_i \ldots w_n)$, the word w_i^c is contextually more accurate than word w_i in sentence $w_1\ w_2 \ldots w_i \ldots w_n$.

If the likelihood of a confused word is more suitable in the context of the sentence, the testing word will be detected as a spelling mistake.

2 Related Study

Many scholars have proposed different approaches to tackle the issue [2]. This is probably the first approach that has been tried for Assamese language. Some of the approaches for other languages have been discussed in this section.

Rokaya et al. [3] concentrated on field association terms dictionaries to tackle the issue of real-word errors for context-sensitive spell checking. They have used machine learning and statistical methods that depend on F and external dictionaries based on the concepts of field association terms and the power links. The method utilizes the advantage of the resource-based methods and statistical and machine learning methods. The precision and recall of their system were found to be 90 and 70%, respectively.

Hirst and Budanitsky [4] proposed a method for correction of real-word spelling errors by restoring lexical cohesion. They identify the tokens which are semantically unrelated to the context and are spelling variations of the word that is more appropriate in the context of the sentence. The tokens which do not fit in the context are detected as spelling error and the corresponding spelling variation matching the context is suggested as a correction. The measure of semantic distance determines the relatedness to the context.

Wilcox O'Hearn et al. [5] considered the problem of real-word error detection and correction based on trigrams. They have utilized the method of Hirst and Budanitsky to analyze the limitation and advantages of the trigram-based noisy channel model of real-word spelling error correction presented by Mays and Mercer in 1991.

Fossati and Eugenio [1] have used mixed trigram approach to address the issue of real-word spell checking. Here, they have considered language trigram model along with parts of speech tagging in the mixed trigram. They have implemented, trained, and tested their system with data taken from the Penn Treebank. The experiment has shown good results considering hit rates of both detection and correction.

Carlson and Fette [6] presented a memory-based approach for context-sensitive spell checking at web scale. The training data is a very large database of token n-gram occurrences in the web. They have designed and implemented an algorithm that suggests a word out of a confusion set for each spelling mistake.

Brill and Moore [7] have considered the problem of spelling correction based on generic string to string edits in a noisy channel model. Source model and channel model are the two components of the noisy channel. They have worked on automatically training a system to detect and correct single-word spelling errors considering probabilities of each of these string to string edits. This model has shown significant improvements in accuracy and performance.

3 Methodology

The block diagram of the system is as follows (Fig. 1).

The process starts with extraction of sentences from the user input and tokenizing it into words. The tokenized sequence of words is analyzed in the spell checker to find the sequence of words having maximum trigram probability using the n-gram language model.

The system was developed in Java for the core processing and Java applet for the user interface part.

3.1 Generation of the Confusion Set

The following edits are considered while generating the confusion set:

- Deletion: Removal one letter from the word.
- Transposition: Swapping of adjacent letters of the word.
- Replacement: Changing of a letter into another letter.
- Insertion: Adding a letter into the word.

All those edit operations when applied on a word produce a large number of probable confusion words some of which are not valid words of the language. So those words which are not found in the vocabulary of the language are discarded and the distinct words from the set produced a confusion set for the input word. The confusion set is prepared for each token of the input sentence.

For example, for the word মকৰাৰ,

- The deletion operation produces the output as [কৰাৰ, মৰাৰ, মকাৰ, মকৰৰ, মকৰা], where one of the letters from the original word is deleted.
- The transposition operation produces the output as [কমৰাৰ, মৰকাৰ, মকাৰৰ, মকৰৰা], where the letters of the original words are swapped with adjacent letters of the word.

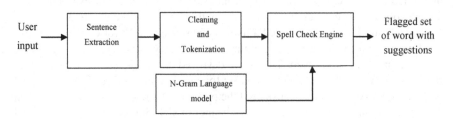

Fig. 1 Block diagram of the spell checker

- The replacement operation on মকৰাৱ produces the following output: [ককৰাৱ, থকৰাৱ, গকৰাৱ,... মকৰাৱ, মথৱাৱ, মগৱাৱ,... মকৱাক, মকৱাথ, মকৱাগ,... মকৱা৮, মকৱা৯, মকৱা০]

Here, each letter of the original word is replaced by each letter from the alphabet of the language. Therefore for each letter of the word, it produces 73 different words. For a word having five letters, the replacement operation produces 365 words.

- The insertion operation produces the following output: [কমকৱাৱ, থমকৱাৱ, গমকৱাৱ, ঘমকৱাৱ,... মককৱাৱ, মথকৱাৱ,... মকৱাৱক, মকৱাৱথ, মকৱাৱগ, মকৱাৱঘ,... মকৱাৱ৮, মকৱাৱ৯, মকৱাৱ০]

Here, each letter from the alphabet is inserted into the original word in each position of the letters of the word. Hence, the length of the word becomes $n + 1$, where n is the length of the original word. For each position of the letters of the word, it produces 73 different words. Therefore for a word having five letters, the word length becomes six, and it produces 438 different words.

It is evident that all the four edit operations produce words all of which do not belong to the language. So the produced words are checked against the vocabulary of the language. For all words generated by the edit operation, if a word is present in the vocabulary, then it is added to the candidate set for the confusion set; otherwise, the word is discarded. Now, the candidate set contains all the edited words which belong to the language. Now, the candidate set is pruned if any word has occurred more than once in it so that a distinct set of word is obtained. This set of distinct word is added to the confusion set.

Finally, the confusion set generated for the word মকৱাৱ is [কৱাৱ, মৱাৱ, মকৱৱ, মকৱা, মকাৱৱ, নকৱাৱ, মকৱাৱ, মাৱাৱ, মকৱাই].

3.2 Preprocessing of the Input Sentences

Cleaning is an important task before a language corpus could be used for modeling. It includes

- Removal of non-textual items such as smileys, hyperlinks, special symbols, etc.
- Removal of words or characters which does not belong to Assamese language.

The removal was done by comparing the tokens to a list of predefined regular expressions that need to be cleaned.

After the input is cleaned, it is tokenized. The tokenization is done at two levels.

- Sentence level: Sentences are extracted from the user input sequence by tokenization using the end of sentence character, "|" for Assamese sentences. After the sentences are tokenized, each sentence is passed on for the word level tokenization.
- Word level: Individual words are extracted from the user input sequence based on the white space character. Each token at this level contains a word of the user input word sequence.

The cleaned and tokenized sequences of words are passed on for spell checking.

3.3 Spell Checking

For each sentence of the input sequence, words of the sentence are tested for spelling mistake. If the testing word is not found in the vocabulary, it is marked as a nonword error and a suggestion set is generated using the context of the sentence. If the word is present in the vocabulary, the segment of the sentence starting from the beginning word to the testing word is passed on for checking the context. For this segment of the sentence, the best fit against the testing word is suggested by calculating the sentence trigram value by replacing the testing word with each of the confused words in the sentence.

The conditional probability using trigram language model of a word sequence $W = \{w_1, w_2, w_3, \ldots, w_n\}$ is calculated by

$$p(W) = p(w_1, w_2, w_3, \ldots, w_n) = \prod_{i=1}^{n} p(w_i | w_{i-2}, w_{i-1})$$

If a particular n-gram is not listed in the language model, then its probability can be calculated by smoothing the count for the n-gram. Since the n-gram is not listed, its count would be 0. So by Add-One smoothing techniques, we get smoothed count as

$$sc(w_1, w_2 \ldots w_N) = \frac{[c(w_1, w_2 \ldots w_N) + 1] * c(w_1, w_2 \ldots w_{N-1})}{c(w_1, w_2 \ldots w_{N-1}) + V}$$

Again, the smoothed probability for the n-gram is

$$sp(w_N | w_1, \ldots w_{N-2}, w_{N-1}) = \frac{sc(w_1, w_2 \ldots w_N)}{c(w_1, w_2 \ldots w_{N-1})}$$

Here,

$c(w_1, w_2 \ldots w_N)$ is the count for the n-gram.
$sc(w_1, w_2 \ldots w_N)$ is the smoothed count for the n-gram.
$sp(w_N | w_1, \ldots w_{N-2}, w_{N-1})$ is the probability for the word w_N after the sequence $w_1, \ldots w_{N-2}, w_{N-1}$.
V is the size of the vocabulary.

For example, taking W= (বিশ্বকাপৰ সময়ত দলটোক সমৰ্থন নকৰাৰ বাবে প্ৰাক্তন খেলুৱৈসকলক সমালোচনা কৰিছে) the trigram probability for the sentence is

$p(W) = p($বিশ্বকাপৰ$| < s >, < s >) * p($সময়ত$| < s >,$বিশ্বকাপৰ$) * p($দলটোক $|$বিশ্বকাপৰ, সময়ত$)$
$\qquad * p($সমৰ্থন$|$সময়ত, দলটোক$) * p($নকৰাৰ$|$দলটোক, সমৰ্থন$) * p($বাবে$|$সমৰ্থন, নকৰাৰ$)$
$\qquad * p($প্ৰাক্তন$|$নকৰাৰ, বাবে$) * p($খেলুৱৈসকলক$|$বাবে, প্ৰাক্তন$)$
$\qquad * p($সমালোচনা$|$প্ৰাক্তন, খেলুৱৈসকলক$) * p($কৰিছে$|$খেলুৱৈসকলক, সমালোচনা$)$
$\qquad * p(</s > |$সমালোচনা, কৰিছে$)$

where <s> is the start of sequence symbol and </s> is the end of sequence mark.

If a trigram probability, say (বাবে|সমৰ্থন, নকৰাব) is not found in the language model, then the smoothed probability of that trigram is created by performing Add-One smoothing over the unknown trigram.

$$sp(বাবে \mid সমৰ্থন, নকৰাব) = \frac{sc(সমৰ্থন, নকৰাব, বাবে)}{c(সমৰ্থন, নকৰাব)}$$

Again, smoothed count, $sc()$, is calculated as follows:

$$sc(সমৰ্থন, নকৰাব, বাবে) = \frac{[c(সমৰ্থন, নকৰাব, বাবে) + 1] * c(সমৰ্থন, নকৰাব)}{c(সমৰ্থন, নকৰাব) + V}$$

Since the trigram সমৰ্থন, নকৰাব, বাবে is not present in the language model, its count would be zero. So, the formula for its smoothed count becomes

$$sc(সমৰ্থন, নকৰাব, বাবে) = \frac{c(সমৰ্থন, নকৰাব)}{c(সমৰ্থন, নকৰাব) + V}$$

Again if the count of bigram সমৰ্থন, নকৰাব is also not present in the language model, we smooth that count using the unigramনকৰাব as follows.

$$sc(সমৰ্থন, নকৰাব) = \frac{c(নকৰাব)}{c(নকৰাব) + V}$$

In the example shown above, **মকৰাব** is the testing word which is being tested for spelling mistake and the words in the box is the confusion set generated for the word **মকৰাব**. The conditional probability for the sentence is calculated by replacing the testing word by each word of the confusion set (Fig. 2). For each substituted string, if the trigrams are present in the language model, its sequence value is calculated and is added to the suggestion set. Otherwise, $n - 1$ grams are tested for the same. The final suggestion set is then sorted and if the probability of the original sequence is found less than the probability of any sequence replacing the testing word by a confused word, then it will be marked as a spelling mistake. The word for which the probability is found greater than the original sequence will be added to the suggestion list. Suggestion list will be presented to the user sorted by the conditional probability of the sequence.

The output of the spelling checker shows the misspelled word along with the probable suggestion for that word checking the context of that word in the sentence.

In Fig. 3, the system has been tested for a sentence in three cases

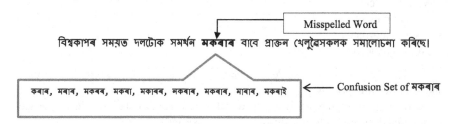

Fig. 2 Use of confusion set in spell checking

```
                    Enter the sentence to check spelling mistake or enter 0 to exit:
বিশ্বকাপৰ সময়ত দলটোক সমৰ্থন কৰবাৰ বাবে প্ৰাক্তন খেলুৱৈসকলক সমালোচনা কৰিছে।
The word কনৱাৰ at location 5 is not in the dictionary.
 Did you mean any from the below sorted suggestion
নকৰাৰ, কৱাৰ, কনৱ, কনৱাড, কনৱাদ, নৱাৰ,
                    Enter the sentence to check spelling mistake or enter 0 to exit:
বিশ্বকাপৰ সময়ত দলটোক সমৰ্থন নকৱাৰ বাবে প্ৰাক্তন খেলুৱৈসকলক সমালোচনা কৰিছে।
You typed নকৱাৰ at location 5. Did you mean নকৱাৰ
                    Enter the sentence to check spelling mistake or enter 0 to exit:
বিশ্বকাপৰ সময়ত দলটোক সমৰ্থন নকৱাৰ বাবে প্ৰাক্তন খেলুৱৈসকলক সমালোচনা কৰিছে।
                    Enter the sentence to check spelling mistake or enter 0 to exit:
```

Fig. 3 The final output of the spellchecker for testing a sentence

- Nonword spelling mistake: The following sentence containing a nonword was fed to the system:

বিশ্বকাপৰ সময়ত দলটোক সমৰ্থন **কনৱাৰ** বাবে প্ৰাক্তন খেলুৱৈসকলক সমালোচনা কৰিছে।

Here, the word কনৱাৰ is not a valid Assamese word. So, a set of words [নকৱাৰ, কৱাৰ, কনৱা, কনৱাড, কনৱাদ, নৱাৰ] is suggested against the misspelled word, which is sorted according to the context of the sentence.

- Real-word spelling mistake: The following sentence was fed to the system:

বিশ্বকাপৰ সময়ত দলটোক সমৰ্থন **মকৰাৰ** বাবে প্ৰাক্তন খেলুৱৈসকলক সমালোচনা কৰিছে।

In the above sentence, the word মকৰাৰ is a valid Assamese word which means the name of a creature. So the word seldom fits in the context of the sentence. Hence, another word নকৰাৰ, which is more probable in the context, is suggested against that word.

In the third case the statement, বিশ্বকাপৰ সময়ত দলটোক সমৰ্থন নকৰাৰ বাবে প্ৰাক্তন খেলুৱৈসকলক সমালোচনা কৰিছে। was fed to the system that has no erroneous words in the sentence. Hence the statement remains unchanged.

4 Results and Discussion

The performance of the system was evaluated based on the following measures:

- Precision correct,
- Recall correct,
- Precision incorrect,
- Recall incorrect, and
- Predictive accuracy.

The recall and precision are defined in the terms of a set of the **recognized words** and a set of **valid words**.

By distinguishing the precision and recall on erroneous and correct forms, we get more refined measures for the evaluation of the system.

Following test cases were considered for the evaluation:

- **True Positives (TP)**: Valid words recognized by the spell checker.
- **True Negatives (TN)**: Invalid words recognized by the spell checker.
- **False Negatives (FN)**: Valid words not recognized by the spell checker.
- **False Positives (FP)**: Invalid words not recognized by the spell checker.

Some of the data validated are given below

	Word Index	Testing Word	Suggested Word Set	Result
S E N T E N C E 1	1	তেও	[তেওঁ[	TN
	2	আহি	[আমি[	TN
	3	ডাঙৰ	[ডাঙৰ[	TP
	4	কান	[কাম[	TN
	5	কৰোঁৰ	[কৰোঁৰ[	TP
	6	ওপৰত	[ওপৰত[	TP
	7	গুৰুত্ব	[গুৰুত্ব[	TP
	8	নিদিল	[দিদিল[	TN
	9	জন্তু	[জিন্তু, সন্তু, জন্তুও, জন্তুক, তন্তু, জন্তু, জন্তুৰ[	FP
	10	একেসময়তে	[এনেসময়তে, একেসময়তে[	TP
	11	বাটি	[ঘাটি[	TN
	12	ত্ৰাস	[ত্ৰাস[	TP
	13	মৰিবলৈ	[কৰিবলৈ[	TN
	14	কোনো	[কোনো[	TP
	15	শক্তিশালী	[শক্তিশালী[	TP
	16	তথা	[তথা[	TP
	17	কার্যকৰী	[কার্যকৰী[	TP
	18	নাতি	[নীতি[	TN
	19	আগবঢ়োৱাত	[আগবঢ়োৱাত[	TP
	20	ব্যৰ্থ	[ব্যৰ্থ[	TP
	21	হৈছিল	[হৈছিল[	TP
S E T E N C E 2	1	নতুন	[নতুন[	TP
	2	নিযুক্তি	[নিযুক্তি, নিযুক্তিৰ, নিযুক্ত[	TP
	3	সমষ্টি	[সৃষ্টি[	TN
	4	ভও	[থণ্ড[	TN
	5	সম্পৰ্কে	[সম্পৰ্কে[	TP
	6	তেওঁৰ	[তেওঁৰ[	TP
	7	কোনো	[কোনো[	TP
	8	ধাৰণা	[ধাৰণা[	TP
	9	মাই	[নাই[	TP

On testing a document of 500 words, where 360 cases were found as true positive, 24 were found as false positive, 93 were found as false negative, and 23 were found true negative, the results were obtained as follows:

– Precision correct for the test set was

$$p_c = \frac{\text{TP}}{\text{TP} + \text{FP}} = \frac{360}{360 + 24} = 0.9375$$

– Precision Incorrect for the test set was

$$p_i = \frac{TN}{TN + FN} = \frac{23}{23 + 93} = 0.1982$$

– Recall correct for the test set was

$$R_c = \frac{TP}{TP + FN} = \frac{360}{360 + 93} = 0.7947$$

– Recall Incorrect for the test set was

$$R_i = \frac{TN}{TN + FP} = \frac{23}{23 + 24} = 0.4893$$

The overall performance measure, predictive accuracy, was calculated as follows:

– Predictive accuracy

$$PA = \frac{TP + TN}{TP + TN + FP + FN} = \frac{360 + 23}{360 + 23 + 24 + 93} = 0.7660$$

The approach used here is based on the statistics generated from the language corpus. For every input sentence, it generates the corresponding trigram, bigram, or unigram and calculates the sentence trigram value taking the statistics generated from the corpus.

Some conditions arise when the trigram for some sentences is not found in the pre-calculated statistics of the corpus. In that case, Laplace smoothing technique is done to eliminate the problem of zero probability in the sentences. The advantage of taking the smoothed trigram is that it eliminates the chance of occurring zero probability in the n-gram and assigns a value to the n-gram based on its $n - 1$ gram. Let us consider a sequence "$a\ b\ c$". If the trigram sequence for "$a\ b\ c$" is not found in the corpus, we would take the smoothed trigram for "$a\ b\ c$" which is based on the bigram of "$b\ c$".

The spell checker performs better if the trigrams of the user input word sequences match some of the trigram or bigram of the sequence in the language model as spell checking is performed based on the n-gram language model developed from the language corpus. Again, the quality of the language model is dependent on the linguistic corpus that has been chosen for the purpose. Though the spell checker has been found to be recognizing many words, in some cases, some words are suggested which have resulted in false negatives or false positives. One of the reasons for this is the unavailability of trigram or bigram sequences of the input sentence in the language model generated from the training corpus.

Let us consider a sentence, $= \{w_1, w_2, w_3, w_4\}$. Checking the context for a word say w_3 in the sentence, each word of the confusion set of the testing word, w_3, is tested for the best fit in the sentence S. Say w_{31}, w_{32}, w_{33} are the confused word for w_3. For finding the best fit for the word w_3, the spell checker finds the trigram w_1, w_2, w_c (w_c is the set of confused words including w_3) to calculate the sequence trigram of S. If the trigram for this sequence is not found in the language model, the spell checker will take the count of $w_2 w_c$ to calculate the smoothed trigram for the sequence w_1, w_2, w_c using Add-One smoothing technique. At this level, it may happen that the count for some combination w_2, w_i ($i \in c$) is more than w_2, w_3 which will result in greater probability for the sequence having w_2, w_i in it. So the spell checker will suggest w_i as the best fit in place of w_3. If w_3 was a valid word according to the context of the sentence, this will result in a false negative case.

A probable solution for overcoming this issue is to choose a large corpus specific to the particular domain of the problem. Choosing a very large corpus on the other hand will also increase the complexity of the system, which remains a matter of further research.

5 Conclusion

The problem of detecting real-word error and generating suggestions for the same is an interesting and complex issue. Though a lot of work has been done in this regard in other languages, a lot more needs to be done specially for the Assamese language. In the project, n-gram language modeling technique has been studied and implemented for context-sensitive spell checking for Assamese language. The n-gram language model up to trigram was developed out of an Assamese text corpus containing 220,743 words. A system was developed for using the language model to check the validity of the words in the user input sentences. The spell checker is tested on the basis of recall and precision on various test sets within the corpus as well as out of the corpus of Assamese Language. The overall performance of the spell checker has found to be around 76%. Though an attempt is made to handle the unknown sentences for which trigrams or bigrams are not available in the language model by performing Laplace smoothing, further research is required in this regard. The results have shown that a lot of suggested words can be filtered out if parts of speech tagging is used along with n-gram language modeling. Hence, the accuracy of the spell checker can be further improved by incorporating parts of speech tagging along with the n-gram language model. It can be inferred that the work has the potential for further research and software development for spell checking in Assamese language.

References

1. Fossati, D., Di Eugenio, B.: A mixed trigrams approach for context sensitive spell checking. Awards N00014-00-1-0640 from the Office of Naval Research, and awards IIS-0133123 and ALT-0536968 from the National Science Foundation
2. Choudhury, R. et al.: A survey on the different approaches of context sensitive spell checking. Int. J. Eng. Sci. Comput. 6(6) (June 2016). https://doi.org/10.4010/2016.1642
3. Rokaya, M., Nahla, A., Aljahdali, S.: Context-sensitive spell checking based on field association terms dictionaries. IJCSNS Int. J. Comput. Sci. Netw. Sec. 12(5) (May 2012)

4. Hirst, G., Budanitsky, A.: Correcting real-word spelling errors by restoring lexical cohesion. Nat. Lang. Eng. **11**(1), 87–111. (c) 2005 Cambridge University Press. https://doi.org/10.1017/s 135132490400356

5. Wilcox O'Hearn, A. et al.: Real-word spelling correction with trigrams: a reconsideration of the Mays, Damerau, and Mercer model, Department of Computer Science, University of Toronto, Toronto, Ontario, Canada M5S 3G4 (2008)

6. Carlson, A., Fette, I.: Memory-based context-sensitive spelling correction at web scale, NSF Cyber Trust initiative (Grant#0524189), under ARO research grant DAAD19-02-1-0389

7. Brill, E., Moore, R.C.: An improved error model for noisy channel spelling correction, Microsoft Research One Microsoft Way Redmond, Wa. 98052

Distance Transform-Based Stroke Feature Descriptor for Text Non-text Classification

Tauseef Khan and Ayatullah Faruk Mollah

Abstract Natural scene or document images captured from camera devices containing text are the most informative region for communication. Extraction of text regions from such images is the primary and fundamental task of obtaining textual content present in images. Classifying foreground objects as text/non-text elements is one of the significant modules in scene text localization. Stroke width is an important discriminating feature of text blocks. In this paper, a distance transform-based stroke feature descriptor is reported for component level classification of foreground components obtained from input images. Potential stroke pixels are identified from distance map of a component using strict staircase method, and distribution of distance values of such pixels is used for designing the feature descriptors. Finally, we classify the components using a neural network-based classifier. Experimental result shows that component classification accuracy is more than 88%, which is much impressive in practical scenario.

Keywords Stroke potential pixel (SPP) · Stroke potential value (SPV)
Equidistance pixel (EP) · Distance transform (DT) · Multilayer perceptron (MLP)

1 Introduction

Text region detection and localization from scene images captured by camera held mobile device are still considered as an unsolved research area in computer vision and image processing. Though researchers have reported a number of algorithms for text region retrievals from such images, automated text detection and recognition from the natural image are still a challenging task due to unconstrained environment. An effective approach toward text localization is to extract the foreground objects from an image and then to classify the extracted components as text or non-text. Following

T. Khan (✉) · A. F. Mollah
Department of Computer Science and Engineering, Aliah University,
Kolkata 700160, India
e-mail: tauseef.hit2013@gmail.com

© Springer Nature Singapore Pte Ltd. 2019
J. Kalita et al. (eds.), *Recent Developments in Machine Learning and Data Analytics*,
Advances in Intelligent Systems and Computing 740,
https://doi.org/10.1007/978-981-13-1280-9_19

this approach, several methods have been developed to suppress the background and identify text components in the recent past. Stroke is one of the discriminating features for component level classification. Usually, stroke profile of text component reflects certain degree of uniformity whereas for non-text, it is arbitrary in nature. This property can be utilized to extract regions that are likely to contain text information. It has been observed that most of the previously implemented algorithms for scene text retrieval are focused on horizontally or vertically aligned text-likely object analysis [1–6]. But, for arbitrary oriented text, a very few methods have been reported till now.

Dalal et al. [7] implemented a histogram of gradient (HOG) of different orientations for target object recognition. Outdoor scene images containing text line detection is implemented using advanced HOG-based texture descriptor (T-HOG) [8]. Tian et al. [9] proposed a convolutional HOG descriptor with co-occurrence matrix (ConvCo-HOG) for script invariant text recognition from scene images. Texture pattern is often uniform in nature. A local binary pattern (LBP) for texture classification of grayscale image is proposed in [10]. Texture pattern for facial image is analyzed using multi-scale LBP (MS-LBP) [11]. A number of different texture-based feature descriptor techniques are also reported in [12, 13].

The stroke-based approach is effective to detect text regardless of its scale, size, font, and language variation. Epstein et al. [14] proposed a stroke width transform (SWT) based approach is used for scene text retrieval by transforming the pixel intensity to its stroke width values. An end-to-end text localization method is presented for natural images by extracting candidate regions using MSER and analyzing them with stroke area ratio to identify text from non-text ones [15].

Subramanian et al. [16] proposed a method using character strokes to localize text regions from natural images. Though several methods have been designed to exploit the strength of stroke width profile, it has remained an under-explored task due to the fact that accurate computation of stroke is still an unsolved problem. Considering the invariant nature of stroke profile information, efforts should be made to accurately compute such profile for using in text/non-text classification problem.

In this paper, distance transform (DT) based approach is presented to generate stroke distributions of candidate components. Then, a set of feature descriptors are prepared from extracted foreground objects and are used for component level classification. The detailed explanation of this work is given in Sect. 2. In Sect. 3, a brief experimental analysis is given, and finally in Sect. 4, a conclusive discussion is drawn.

2 Present Work

First, the color image is converted into grayscale intensity space by weighted average of RGB channels as explained in [17]. Then, foreground objects are extracted from grayscale image using region growing algorithm with single seed point motivated by our previous work [18]. Now, a global binarization technique is applied on foreground components to convert the binary image component into distance-transformed image. After that, we identify pixels having regional maximum distance values positioned in the middle of strokes. These are considered to be the stroke potential pixels (SPPs). Finally, a discretization grouping technique is applied on the occurrence of SPPs, from which values of each group are taken as features that are fed into a neural network-based classifier for component level classification.

2.1 Distance Transformation of Foreground Components

Foreground candidate components obtained from camera images using a background expansion approach [18] are binarized with Otsu's global binarization method. Some of these binary components are shown in Fig. 1. Then, we applied Euclidean distance transformation function on each binary component as shown in Fig. 2. It may be observed that for text regions distance value is high across the center of text strokes and it gradually decreases while moving across edges.

2.2 Identification of Stroke Potential Pixels

It has been observed that distance value is high across middle of stroke width of each foreground component whereas distance values gradually decrease while moving away from center in both directions. In this work, to identify stroke potential pixels (SPPs) a horizontal kernel of 1×5 is applied on the image and in order to identify center pixel as SPP, kernel is rotated anticlockwise. Here, to identify SPP, a strict staircase method is applied on DT map, where three cases have been considered while rotating the kernel. By rotating the kernel in anticlockwise direction, all the four possible directions are considered. Now in every direction, we consider three cases and if any of the case among three cases is triggered, then target pixel is considered as

(a)   **(b)**   **(c)**   **(d)**

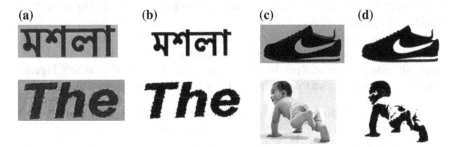

Fig. 1 Extracted foreground components: **a** text components, **b** binarized text components, **c** non-text components, and **d** binarized non-text components

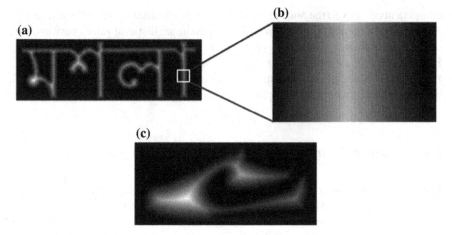

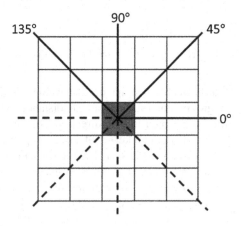

Fig. 2 DT on foreground components. **a** DT view of a text component, **b** magnified view of a portion along text stroke, and **c** DT view of a non-text component

Fig. 3 Stroke potential pixel (SPP) selection by anticlockwise rotation of horizontal kernel

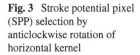

SPP. Figure 3 shows how the kernel rotates anticlockwise and considers all possible directions to identify SPPs.

It has been observed that for text components stroke can be straight or curvy in nature. So, for each stroke single pixel is selected as SPP, whereas in some scenario stroke shaving two potential pixels with same distance values marked as SPP. Figure 4 shows the different scenarios for single SPP and equidistance double SPPs.

Strict Staircase Method To consider a stroke pixel as SPP, we consider a strict staircase method. There can be the following three cases:

Case 1 If distance value of the center pixel (CP) is greater than all the left half pixels (LHP) and also all the right half pixels (RHP), then the CP is marked as SPP. Here,

(a)

2.0	3.0	4.0	3.0	2.0
2.0	3.0	4.0	3.0	2.0
2.0	3.0	4.0	3.0	2.0
2.0	3.0	4.0	3.0	2.0
2.0	3.0	4.0	3.0	2.0
2.0	3.0	4.0	3.0	2.0
2.0	3.0	4.0	3.0	2.0

(b)

2.0	3.0	3.0	2.0	1.0
2.0	3.0	3.0	2.0	1.0
2.0	3.0	3.0	2.0	1.0
2.0	3.0	3.0	2.0	1.0
2.0	3.0	3.0	2.0	1.0
2.0	3.0	3.0	2.0	1.0
2.0	3.0	3.0	2.0	1.0

Fig. 4 Stroke width calculation for each stroke of text component: **a** for single SPP and **b** for double SPP in equidistance scenario (SPP marked in red) (color figure online)

Fig. 5 Scenario of case 1 for considering a CP as SPP

we followed a strict staircase rule. Equation 1 shows the staircase rule for identifying SPP. Figure 5 shows the scenario of case 1.

$$\text{if}((CP > LHP_1 > LHP_2)\&\&(CP > RHP_1 > RHP_2))$$
$$CP \text{ marked as SPP} \tag{1}$$

Case 2 If the distance value of a pixel is same as that of a CP, it is called Equidistance Pixel (EP). Here, EP is positioned left of CP. Equation 2 shows the rule for identifying SPP. Figure 6 shows the scenario of case 2.

$$\text{if}((CP == EP)\&\&(CP > LHP)\&\&(CP > RHP_1 > RHP_2))$$
$$CP \text{ marked as SPP} \tag{2}$$

Case 3 In this case, EP is positioned right of CP, and the remaining rule is same for identifying SPP. Equation 3 shows the rule for selecting SPPs. Figure 7 shows the scenario of case 3.

Fig. 6 Scenario of case 2 for identifying stroke potential pixel, where equidistance pixel is marked in red and CP in black (color figure online)

Fig. 7 Scenario of case 3 for identifying SPP, where EP marked in red and CP identified as SPP is marked in black (color figure online)

$$if((CP == EP)\&\&(CP > LHP_1 > LHP_2)\&\&(CP > RPH))$$
$$CP \text{ marked as SPP} \qquad\qquad (3)$$

In order to generate stroke value distribution, profiles of foreground object components' SPPs are identified first. As SPPs are normally positioned at the middle of strokes, the width of stroke is just twice the value of SPPs. It is assumed that stroke width is near uniform across the text components, whereas for non-text components stroke width varies arbitrarily. Selection of SPP considers all the four directions of kernel by rotating anticlockwise. The number of SPPs may vary with the number of directions. Different combination of directions is chosen out of all four possible directions. Figure 8 shows identified potential stroke pixels of some text components.

In this paper, an experiment is carried out by choosing a different combination of directions to select SPPs. In Fig. 9, SPPs obtained with a different combination of directions are shown.

In the above figure, it has been observed that SPPs are very less when considering all four directions. While considering any three directions and any two directions, numbers of identified SPPs have increased significantly. But, it has been seen that

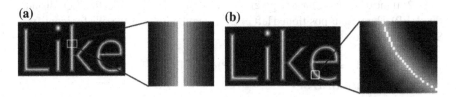

Fig. 8 Identification of SPPs after considering any three directions: **a** text component and corresponding equidistance SPPs across straight stroke, and **b** SPPs across curvy stroke of text component

Fig. 9 Identified SPPs after choosing different combination of directions: **a** any two directions, **b** any three directions, and **c** all four directions (the number of SPPs decreases with number of directions)

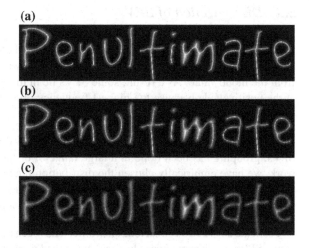

some unwanted pixels are also identified as SPPs in the corner side of strokes when considering any two directions whereas these unwanted pixels do not appear in any three directions. However, very few numbers of pixels that should be identified as SPPs are not identified as SPPs while considering any three directions shown in Fig. 10.

Now, after identifying potential stroke pixels from the DT map, we extract only distinct stroke potential values (SPVs). Then, we compute their occurrence corresponding to each distinct SPV.

Fig. 10 SPPs identification at corner of strokes: **a** considering any two directions and **b** considering any three directions

2.3 Discretization of SPV

In this step, distribution of SPV is generated and from distinct SPVs of foreground components, we compute mode of all SPVs. Then, we subtract each distinct SPV from the mode value. As a result, some SPVs lie in the negative side and some fall at the positive side because the highest frequency of SPV falls in the center of the distribution, i.e., zeroth position.

Discretization In order to discretize the frequency values of each corresponding SPVs, grouping has been done in a large range of distance values of SPPs. In this work, we have empirically chosen the distribution range of SPVs with a fixed step size. After that, we compute the cumulative frequencies for each group. Figure 11 shows the stroke distribution of SPVs of foreground components after binning.

From the above stroke distribution profile, it has been observed that for text components, SPPs are dispersed very less and converged very closely whereas for non-text components SPPs are dispersed in a wide range shown in Fig. 12.

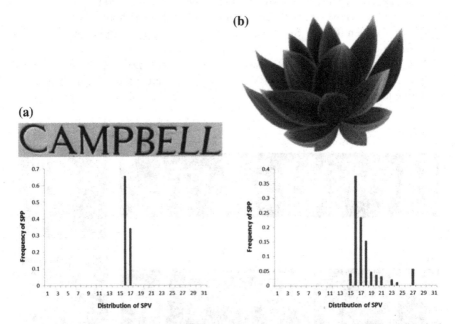

Fig. 11 Effect of binning on frequency of SPVs: **a** text component and corresponding stroke pixel distribution profile, and **b** non-text component and corresponding stroke pixel distribution profile

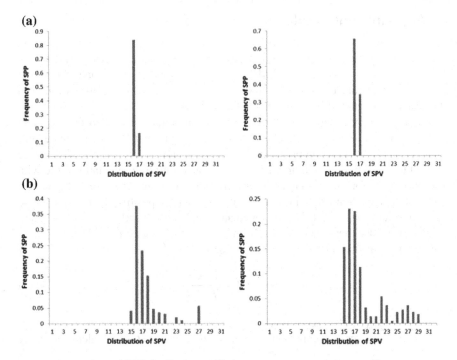

Fig. 12 Comparison of SPV distribution profile for **a** text component images and **b** non-text component images

2.4 Feature Descriptor Generation and Component Classification

Classification of foreground objects as text or non-text, a feature file is generated which is input to a binary classifier for component classification. Here, the cumulative frequency value of each group is treated as a feature. In this work, the number of features, i.e., 31, is empirically chosen. After generating feature map of extracted candidate blocks, the feature space is fed into a neural network-based learning classifier. Here, multilayer perceptron (MLP) is used for classification. This neural classifier consists of three layers, viz., the input layer, hidden layer, and the output layer. The number of extracted features from components is treated as a number of neurons in the input layer. Here, output layer neurons are 2 (i.e., text and non-text). Hidden layer neuron is chosen empirically.

3 Experimental Result

Experimental process is applied to a database of 200 camera images captured from handheld mobile device having 13 MP resolutions. A total of 91 text and 76 non-text elements are extracted for classification. Some mixed mode components found in this test set are ignored for component classification.

Figure 13 shows some sample foreground components extracted from camera captured images for component classification. To train the MLP classifier Text Component and Non-text Component are labeled as 0 and 1, respectively. Now, these labeled text and non-text components are separated into train file and test file with the ratio of 2:1 for binary classification. After that from these labeled components, all the statistical parameters, i.e., true positive (TP), false positive (FP), true negative (TN), and false negative (FN), are calculated, and subsequently, precision (P) and recall (R) are also measured. Finally, balanced F-score/F-measure (FM) is computed with the help of P and R as mentioned in Eq. 4.

$$FM = 2 \times \frac{P \times R}{P + R} \tag{4}$$

where $R = \frac{TP}{TP+FN}$, $P = \frac{TP}{TP+FP}$.

Finally, the overall component classification accuracy (C_A) is calculated using Eq. 5.

$$C_A = \frac{TP + TN}{TP + TN + FP + FN} \tag{5}$$

In Table 1, a comparative analysis is shown by considering different direction for identification of potential stroke pixels. It has been observed that classification accuracy is highest for any three directions, whereas any two directions produce less classification accuracy and all four directions lead to poor classification accuracy.

Fig. 13 Sample images for component classification: **a** text components and **b** non-text components

Table 1 A comparative analysis using statistical parameters

No. of direction	Precision (P)	Recall (R)	F-measure (FM)	Classification accuracy (C_A) (%)
Any two directions	89.24	83.83	86.45	86.28
Any three directions	89.58	86.81	88.16	88.12
Any four directions	76.85	83.83	80.18	80.00

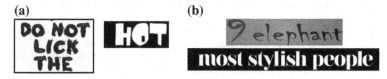

Fig. 14 Some text components with arbitrary stroke size: **a** properly classified as text components, and **b** fails to classify as text components

Some foreground text components are misclassified as non-text in any two or all four directions, whereas consideration of any three directions for identification of SPPs generates much better classification accuracy. In Fig. 14, it is shown that some text components having arbitrary strokes are still correctly classified as text components in any three directions, though for some complex text components, it fails to classify as text components.

4 Conclusion

In this paper, we present distance transform-based features for text/non-text classification of foreground objects extracted from camera captured images. Distribution of unique distance values of the distance-transformed component image is used for computing the features. As discussed in Sect. 2, strokes of binary text components have a certain degree of uniformity. Hence, the designed features can effectively distinguish uniform/near-uniform stroked components from arbitrary stroked components. Thus, this method is found to be robust enough to classify texts with near-uniform stroke width. However, it is sensitive to certain fonts having nonuniform strokes. For high-resolution handwritten component classification, it may be very effective. As this method is applicable for binary components, an appropriate binarization technique is required. Application of such a binarization technique will certainly enhance the effectiveness of the developed feature set.

Acknowledgements This work is carried out in the research lab of Computer Science & Engineering Department of Aliah University. The first author is grateful to Maulana Azad National Fellowship (MANF) for the financial support.

References

1. Zhang, Z., Zhang, C., Shen, W., Yao, C., Liu, W., Bai, X.: Multi-oriented text detection with fully convolutional networks. In: IEEE Conference on Computer Vision and Pattern Recognition, pp. 4159—4167, IEEE (2016)
2. Chen, X., Yuille, A.L.: Detecting and reading text in natural scenes. In: IEEE Conference on Computer Vision and Pattern Recognition, pp. II–II (2004)
3. Yao, C., Bai, X., Liu, W., Ma, Y., Tu, Z.: Detecting texts of arbitrary orientations in natural images. In: IEEE Conference on Computer Vision and Pattern Recognition, pp. 1083–1090, IEEE (2012)
4. Yi, C., Tian, Y.: Text string detection from natural scenes by structure-based partition and grouping. In: IEEE Transactions on Image Processing, pp. 2594–2605, IEEE (2011)
5. Neumann, L., Matas, J.: Real-time scene text localization and recognition. In. IEEE Conference on Computer Vision and Pattern Recognition, pp. 3538–3545, IEEE (2012)
6. Huang, W., Lin, Z., Yang, J., Wang, J.: Text localization in natural images using stroke feature transform and text covariance descriptors. In: IEEE International Conference on Computer Vision, pp. 1241–1248, IEEE (2013)
7. Dalal, N., Triggs, B.: Histograms of oriented gradients for human detection. In. IEEE Computer Society Conference on Computer Vision and Pattern Recognition, pp. 886–893, IEEE (2005)
8. Minetto, R., Thome, N., Cord, M., Leite, N.J., Stolfi, J.: T-HOG: an effective gradient-based descriptor for single line text regions. Pattern Recognit., 1078–1090 (2013). Elsevier
9. Tian, S., Bhattacharya, U., Lu, S., Su, B., Wang, Q., Wei, X., Lu, Y., Tan, C.L.: Multilingual scene character recognition with co-occurrence of histogram of oriented gradients. Pattern Recognit. **51**, 125–134 (2016). Elsevier
10. Ojala, T., Pietikäinen, M., Harwood, D.: A comparative study of texture measures with classification based on featured distributions. Pattern Recognit., 51–59 (1996). Elsevier
11. Mäenpää, T., Pietikäinen, M.: Multi-scale binary patterns for texture analysis. Image Anal., 267–275 (2003). Springer
12. Goto, H., Tanaka, M.: Text-tracking wearable camera system for the blind. In: 10th International Conference on Document Analysis and Recognition, pp. 141–145, IEEE (2009)
13. Ye, Q., Huang, Q., Gao, W., Zhao, D.: Fast and robust text detection in images and video frames. Imag. Vision Comput. **23**(6), 565–576 (2005). Elsevier
14. Epshtein, B., Ofek, E., Wexler, Y: Detecting text in natural scenes with stroke width transform. In: IEEE Conference on Computer Vision and Pattern Recognition (CVPR), pp. 2963–2970, IEEE (2010)
15. Neumann, L., Matas, J.: Efficient scene text localization and recognition with local character refinement. In: 13th International Conference on Document Analysis and Recognition, pp. 746–750, IEEE (2015)
16. Subramanian, K., Natarajan, P., Decerbo, M., Castanon, D.: Character-stroke detection for text-localization and extraction. In: 9th International Conference on Document Analysis and Recognition, ICDAR, pp. 33–37, IEEE (2007)
17. Mollah, A.F., Basu, S., Nasipuri, M.: Text detection from camera captured images using a novel fuzzy-based technique. In: 3rd International Conference on Emerging Applications of Information Technology (EAIT), pp. 291–294, IEEE (2012)
18. Khan, T., Mollah, A.F.: A novel text localization scheme for camera captured document images. In: 2nd International Conference on Computer Vision & Image Processing (CVIP), pp. 253–264, Springer Nature (2017)

A Hybrid Approach to Analyze the Morphology of an Assamese Word

Mirzanur Rahman and Shikhar Kumar Sarma

Abstract Analyzing morphology of a word is a crucial task and may be varied based on Language Grammar for different languages. Assamese is a language spoken by the people of Assam, the northeastern part of country India, located in south of the eastern Himalayas. Assamese is the major language spoken in Assam and it is served as a bridge language among different speech communities in the whole area of the state. We are using Assamese language as a domain of research where we are trying to find out the morphology of Assamese word using machine learning technique. In our work, we proposed a hybrid technique for analyzing an Assamese word by merging the advantages of statistical method and finite state transducer method. As a result, we get approximately 90.1% correct result for our test case.

Keywords Machine learning technique · Morphological analyzer · Finite state transducer · Assamese language · SFST · Hidden Markov model

1 Introduction

Morphology is the branch of linguistics that studies the structure of words and word formation of that language. Morphological analysis is a process by which we can find the basic unit of word. The basic unit of the word is known as morpheme. Automatic processing for language needs digitized information/morpheme information (morphology of the word) about the language. So, morphological analysis is a required process for machine learning.

Assamese is the major dialect or official language used in the state of Assam for day-to-day conversation and the state is situated at the northeastern part of the country, India. Assamese language always placed as the first choice while commu-

M. Rahman (✉) · S. K. Sarma
Department of Information Technology, Gauhati University, Guwahati, Assam, India
e-mail: mirzanurrahman@gmail.com

S. K. Sarma
e-mail: sks001@gmail.com

© Springer Nature Singapore Pte Ltd. 2019
J. Kalita et al. (eds.), *Recent Developments in Machine Learning and Data Analytics*,
Advances in Intelligent Systems and Computing 740,
https://doi.org/10.1007/978-981-13-1280-9_20

nicating among various discourse groups of the state. The language Assamese is an Indo-Aryan language originated from Vedic dialects. The language as it stands today passes through tremendous modifications in all the component, viz., phonology, morphology, conjunction, etc. There are two variations of Assamese language according to dialectical regions, i.e., Eastern Assamese and Western Assamese languages [1]. Both are different in terms of phonology and morphology. But still the written text is same for all the regions.

Morphological information of a word is an important component of any language. In language processing techniques such as machine translation [2], parsing, Parts of Speech (POS) tagger, text summarization, etc. require morphological analyzers to find out the lexical component of a word. And lexical components are the very important parts of a grammar of a language.

In our work, only standard written Assamese literary data is considered for processing the language. We have used hidden Markov model as statistical method and SFST tool (SFST tool is developed by Institute of Natural Language Processing, University of Stuttgart) as finite state transducer for implementing the hybrid analyzer.

In this paper, we will discuss how we merge the advantage of statistical and FST method to increase the correctness of the result.

2 Proposed Hybrid Morphological Analyzer

In our previous works [2, 3], we present a finite state transducer model for analyzing morphology of Assamese word using SFST (Stuttgart finite state transducer) [3] and Apertium Lttoolbox [2].

We found some of the drawbacks with FST-based model that it gives good result for known words if all the word is properly added in rule transducer. An FST-based model is unable to find the proper result for unknown words.

To increase the number of words analyzed by an analyzer, we add a statistical model for analyzing unknown words with our proposed model. In our work, we are using hidden Markov model (HMM) and Viterbi algorithm and suffix stripping method for analyzing unknown words.

There are numbers of research works [1, 4] for finding root word using suffix stripping and dictionary-based suffix stripping method. In suffix tripping method, a known suffix is trimmed down to get root word. In dictionary-based method, a root word dictionary is used to find the root word form inflected word by suffix tripping method.

Our main aim of the work is to find out the morphological information of Assamese words with our proposed hybrid model. We present Assamese word morphological structure as

$$\varnothing/\text{prefix}|root|\varnothing/(\text{suffix})^*$$

A word may contain null prefix, prefix, null suffix, single suffix, or multiple suffixes. A suffix tripping method can give good result for single suffix word, but get confused for the word with multiple suffixes and the word containing characters' form suffix list, but actually not a suffix.

Researchers in [5, 6] describe nicely about different word categories of Assamese words based on suffix added on a word. In our work, we are using these word categories [5, 6] for categorizing different words in Assamese sentences. According to them, Assamese word in a sentence may be a

a. Root word (mark as W_r) [Ex. মানুহ =মানুহ +∅],
b. Root word with character from suffix list, i.e., pseudo-single suffix (mark as W_{ps}) [Ex. গাখীৰ= গাখীৰ+ ∅],
c. Single suffix (mark as W_s) [Ex মানুহৰ = মানুহ + ৰ] and
d. multiple suffix (mark as W_m) [Ex. মানুহৰপৰা=মানুহ + ৰ + পৰা].

W_r category contains root words, which does not contain any suffix and the character(s) from the suffix list. W_{ps} category contains root words which does not contain any suffix but contains character(s) from the suffix list. W_s category contains inflected words which contain single suffix from the suffix list. W_m category contains inflected words which contain multiple suffixes from the suffix list.

In our proposed approach, known words are processed by SFST tool and unknown words are handled by HMM module. We know that HMM is good for sequence labeling, but it cannot directly find out root word. So, we use HMM module for finding the word category W_r, W_{ps}, W_s, and W_m and then send it to different Java modules (For suffix stripping) based on word category to find out the root word along with morphological information.

We proposed our morphological analyzer as two-phase morphological analyzer: First phase is basically an SFST module and second phase is basically an HMM module mix with few Java modules to analyze the unprocessed words (unable to process by SFST module). Figure 1 shows complete system architecture of our proposed module.

Input module is for inputting single word or sentence. Tokenizer and cleaner module divide a sentence to individual token and clean unwanted characters. We select token/word based on delimiter like space, (comma), " (double cote), . (dot), - (hyphen), and | (Assamese stop).

SFST module: This module is taking input from previous module and analyzes the inputted word as per the rules added to the FST processor. An FST-based morphological analyzer can be divided into three phases: (1) Generation of lexicon file, (2) Creation of rule list for inflection and derivational morphology, and (3) Morphological processor (Fig. 2).

The lexicon file was generated using a raw corpus collected and manually classified into various classes according to their inflection and derivation types. FST rule file also generated manually for inflected words and added to the morphological processor.

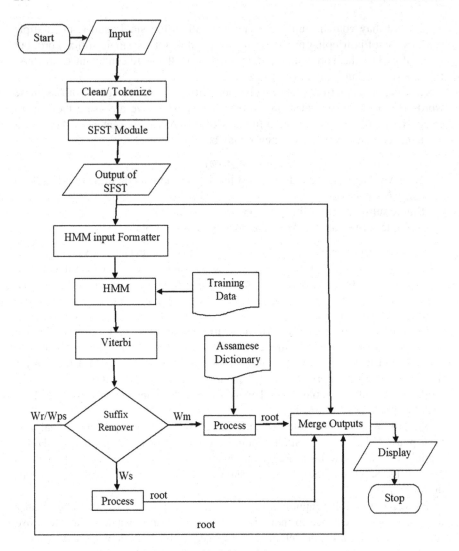

Fig. 1 System architecture of the proposed hybrid model

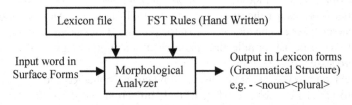

Fig. 2 Block diagram for SFST-based morphological analyzer

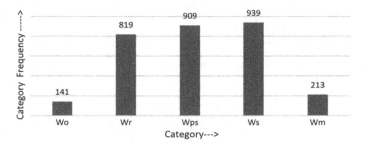

Fig. 3 Number of words in different categories

Morphological processor is the main processing unit of an FST-based analyzer for finding out the lexical form of an inputted surface form of a word. Morphological processor analyzes an inputted word with the help of lexical file and FST rule.

HMM module: Since a finite state transducer based analyzer is unable to analyze unknown word properly, we add a statistical module (Hidden Markov model) with the proposed architecture.

We have used the following steps for training our system:

1. We have collected 141 different sentences which contain 2880 words. We have tagged the corpus four different categories (W_r, W_{ps}, W_s, and W_m) and calculated the HMM parameter manually (Fig. 3).
2. We are using the following formula for calculating the probability of a sequence [7] required for training our system:

$$\text{PROB}(C_1, \ldots, C_T | W_1, \ldots, W_T) = \prod_{i=1,T} \text{PROB}(C_i | C_{i-1}) * \text{PROB}(w_i | C_i)$$

$w_1, w_2, \ldots, w_T$ be a sequence of words and $C_1, \ldots, C_T$ is a sequence of categories
Say a sentence X Y Z and word category $W_m\ W_r\ W_s$

$$\text{Probability} = \text{Prob}(W_m | W_o) \times \text{Prob}(W_r | W_m) \times \text{Prob}(W_s | W_r)$$
$$\times \text{Prob}(X | W_m) \times \text{Prob}(Y | W_r) \times \text{Prob}(Z | W_s)$$

3. Calculating Bigram probabilities [7] is as follows:
   Bigram model can provide good estimation for finding word categories. We are using the following formula for estimating bigrams.
   Say we want to calculate bigram probabilities of W_r (Root word) follows W_m (Multiple Suffix):

$$\text{PROB}(C_i | C_{i-1}) \equiv \frac{\text{Count}(W_m \text{ at position } i - 1 \text{ and } W_r \text{ at } i)]}{\text{Count}(W_m \text{ at position } i - 1)}$$

where $C_i = W_r$ and $C_{i-1} = W_m$.

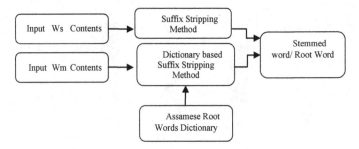

Fig. 4 Suffix remover module

Fig. 5 Merging the final output by suffix remover module

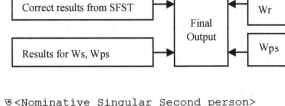

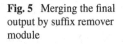

এখন= এখন <root>
দেশত= দেশ <root> + ত <Nominative Singular Second person>
লটকন= লটকন<root>
নামেৰে= নাম<N>+ (ওৰে <Declensor>
এটা= এটা<root>
বৰ= বৰ <JJ><root>
দুখীয়া= দুখীয়া <root>
বামুণ= বামুণ<root>
আছিল= আছিল<root>

Fig. 6 Sample output of hybrid model

4. We are using Viterbi algorithm for finding the best sequence provided by hidden Markov model. HMM cannot identify the best sequence from different existing sequences. For example, in a sentence of three words with four different word categories, there will be $3^4 = 81$ possible path for that sentence. So whenever the number of words in a sentence is increased, the number of possible sequence will also increase.

Suffix remover module: This module uses suffix stripping method for finding out the root word from an inflected word. In HMM module, the final output is stored in four different text files based on the word categories identified by the HMM module. From these four output files, only two file W_s and W_m will be processed by the suffix remover module. Other two files for W_r and W_{ps} will be directly used for output as root word (Fig. 4).

After analyzing W_m and W_s contents by the suffix remover module, it starts processing the final output. The final output is a combination of SFST module output, content of W_{ps}, W_r file, and the result analyzed by suffix remover module (Fig. 5).

When merging the result for final display, the program will first check for analyzed and not analyzed output of SFST module. If a word is analyzed by the SFST module, then it will send for display in a required format. If a word is not analyzed by the SFST module, it will start searching for the result in W_r, W_{ps}, and the output file of suffix remover module and display the result according to the proper format (Fig. 6).

3 Result and Analysis

To show the efficiency of the test result, we are using standard notations for computations, i.e., precision and recall and F-measure. Precision is calculated as number of correct analysis divided by number of analyzed word. Recall counts as number of words correctly analyzed divided by total number of words and F-measure

Table 1 Test result for SFST tool and hybrid model

		Prefix/suffix	No of words	Percentage
	Total words	–	2213	
SFST module	Correctly analyze	Prefix word	34	1.54%
		Suffix word	1839	83.1%
	Not analyze	Prefix word	11	0.49%
		Suffix word	186	8.40%
	Wrongly analyze	Prefix word	13	0.59%
		Suffix word	130	5.87%
HMM and suffix stripping method	Correctly analyze	Prefix word	0	0%
		Suffix word	122	61.93%
	Not analyze	Prefix word	0	0%
		Suffix word	15	7.61%
	Wrongly analyze	Prefix word	11	5.58%
		Suffix word	49	24.87%
Hybrid model (merging the result of SFST & HMM)	Correctly analyze	Prefix word	34	1.54%
		Suffix word	1961	88.61%
	Not analyze	Prefix word	0	0%
		Suffix word	15	0.68%
	Wrongly analyze	Prefix word	24	1.08%
		Suffix word	179	8.09%

Table 2 Precision, recall, and F-measure for SFST tool

Precision	Recall	F-measure
0.93	0.85	0.89

Table 3 Precision, recall, and F-measure for the hybrid model

Precision	Recall	F-measure
0.907	0.901	0.903

is calculating as $2PR/(P + R)$, where P and R are represented as precision and recall of the program.

For testing our system, we have collected around 2213 (After cleaning) words from wikisource.org (source: https://wikisource.org/wiki/Category:বুঢ়ী_আইৰ_সাধু).

After analyzing the test result, we have observed that, for known word, FST-based system provides best results but for unknown word it is unable to determine the word structure. Table 1 shows the result found in our test case.

The test result shows that hybrid model increases the number of correctly analyzed word and then SFST-based module. An SFST module solely analyzed 1873 words correctly from 2213 words. But after adding HMM module as the second phase of hybrid model, the system correctly analyzes 1995 words from 2213 words and the recall also increases by 0.05 for hybrid model (Table 2).

The results of the morphological analysis of Assamese word processed with the help of finite state techniques using SFST tool prove very efficient and speedy. SFST tool is capable of handling and analyzing very long input strings, which is normal in agglutinative languages. An FST-based system gives the best performance depending on the knowledge provided to the system (Table 3).

A statistical model requires more and more information about that language to train the system for providing good performance. Estimates can be particularly unreliable, if only a small number of samples are involved. So system accuracy depends on the trained knowledge. Our statistical model HMM (Hidden Markov model) tune up the efficiency of the hybrid architecture by increasing the analysis result set.

Though or source lexicon is not of an impressive size, our sample of tokens for evaluation of the model indeed is good enough to test the model for coverage. Also as per our small evaluation conducted in the and looking at the reasons for failure on some accounts, we can say that merging of finite state techniques with statistical techniques is ideally suited to cover a wide range of possible word forms in a morphologically rich and complex agglutinative language like Assamese.

References

1. Saharia, N., Sharma, U., Kalita, J.: Analysis and evaluation of stemming algorithms: a case study with Assamese. In Proceedings of the International Conference on Advances in Computing, Communications and Informatics, Chennai, pp. 842–846 (2012)
2. Rahman, M., Sarma, S.K.: An implementation of Apertium based Assamese morphological analyzer. Int. J. Nat. Lang. Comput. **4**(1) (2015). ISSN: 2319-4111 (Print); 2278-1307 (online); https://doi.org/10.5121/ijnlc.2015.4102

3. Rahman, M., Sarma, S.K.: Analysing morphology of Assamese words using finite state transducer. Int. J. Innov. Res. Comput. Commun. Eng. **4** (12) (2016). ISSN (Print): 2320-9798; ISSN (Online): 2320-9801; https://doi.org/10.15680/ijircce.2016.041208
4. Parakh, M., Rajesha, N.: Developing morphological analyzer for four Indian languages using a rule based affix stripping approach. In: Linguistic Data Consortium for Indian Languages. CIIL, Mysore (2011)
5. Saharia, N., Konwar, K.M., Sharma, U., Kalita, J.: An improved stemming approach using hmm for a highly inflectional language. In: Proceedings of 14th International Conference on Intelligent Text Processing and Computational Linguistics (CICLing), pp. 164-173. Samos, Greece (2013)
6. Saharia, N.: Computational morphology and syntax for a resource-poor inflectional language. Ph.D. Thesis (chapter 3). http://shodhganga.inflibnet.ac.in/bitstream/10603/48741/13/13_chapter%203.pdf
7. Allen, J.: Natural Language Understanding (2007). ISBN 978-81-317-0895-8

Netizen's Perspective on a Recent Scam in India—An Emotion Mining Approach

Aviroop Mukherjee, Agnivo Ghosh and Tanmay Bhattacharya

Abstract Every few days, there are reports of someone scamming the government or the bank through one way or another. So, the question arises: Why should ordinary people bear the burden of freeloaders who keep indulging their desires by dipping into public savings through fraudulent activities? Due to the easy availability of Internet, there are a large number of netizens who share their views and opinions on social networking and video sharing websites, making it a valuable platform for tracking and analyzing emotions. Thus, to analyze how the common people feel about these scams, we examine how netizens voice their opinions about a recent scam—the PNB scam, on social media websites. Since Twitter emotion mining is common nowadays, we will use two more popular social networking websites—Facebook and YouTube to analyze this data. Such tracking and analysis can provide an insight into the emotions and opinions of the people regarding the PNB scam.

1 Introduction

We have analyzed the PNB scam from a neutral point of view. The PNB scam involved billionaire Nirav Modi and his relatives scamming the second largest PSU bank of India using fake Letter of Credit or Letter of Undertaking (LoU). The bank was scammed of ₹11,600 crores ($1.77 billion). If we compare the amount of money involved in the scam with the financial performance of the bank, we can find that the amount involved in the scam was 94.1% of the bank's FY2016's operating income (earnings before interests and taxes or EBIT). This gives us a clear picture of the magnitude of the scam.

Twitter analysis has become increasingly common nowadays [1]. So, we have used Facebook and YouTube to analyze data. YouTube has over 1 billion monthly active users and Facebook has over 2.2 billion monthly active users. Because of this huge number of user base, we can achieve an accurate reflection of the opinions of a

A. Mukherjee (✉) · A. Ghosh · T. Bhattacharya
Techno India, Salt Lake, West Bengal, India
e-mail: aviroopmukherjee96@gmail.com

© Springer Nature Singapore Pte Ltd. 2019
J. Kalita et al. (eds.), *Recent Developments in Machine Learning and Data Analytics*,
Advances in Intelligent Systems and Computing 740,
https://doi.org/10.1007/978-981-13-1280-9_21

211

large proportion of the population by analyzing the comments left on YouTube videos and posts by news pages. These provide us a huge set of opinionated data which we can mine to analyze the general opinion, emotion, or attitude of a person. Thereafter, we have represented our analysis in some visual formats like word clouds and various forms of graphs, a quick glance at which helped us lead to many significant inferences.

2 Literature Review

Mohammad and Kiritchenko [2] in their paper targeted answering simple questions such as whether something good or bad is being said about an object or person and what the sentiment of the individual is toward that person or object. The NRC taking advantage of their advanced expertise in text analytics has developed lexicons that enable the creation of software for automatic emotion analysis and hence provide a deeper understanding of the sentiments and emotions.

Neri [3] performed sentiment analysis on about 1000 Facebook posts. He proposed two techniques of sentiment analysis:

- Supervised machine learning—Involves first creating a training series of user-defined polarity values.
- Unsupervised machine learning—Uses a lexicon with words scored for polarity values such as neutral, positive, or negative.

Bakliwal and Foster [4] achieved an accuracy of 61.6% using supervised learning on a corpus of the Irish elections to determine if the emotions toward a political group or individual are positive, negative, or neutral. Shashank and Bhattacharyya [5] used similar techniques for the emotion analysis of online chats. They categorized the datasets into positive and negative classifications using elaborate annotation and feature engineering. Singh [6] in his paper examined the sentiment analysis of a prominent political figure by analyzing a dataset gathered from Twitter. The dataset was obtained from Twitter during the Prime Minister Narendra Modi's visit to the United States of America. He assigned a sentiment score to the words (S_w) to those words that are not present in AFINN111 through this simple formula:

$$S_w = \frac{\sum_1^n S_t(w \in t)}{n}$$

Here,
$n =$ number of tweets that contain the word and
S_t is the sentiment score of the tweet containing the word w.

The sentiment of a tweet is then determined by the higher value among s_+ or s_-. If the values are equal, the tweets are considered neutral [7].

$$s_t = \begin{cases} \text{positive} & \text{if } s_+ > s_- \\ \text{negative} & \text{if } s_+ > s_- \end{cases}$$

3   Tools and Resources

R—*R* is an open-source programming language widely used in statistics and data analytics. *R* was chosen because it was open source and had a large number of packages. There are also packages that helped connect with the Facebook and YouTube APIs.

YouTube Data API v3—It provides access to YouTube data, such as videos, playlists, and channels.

Facebook Graph Explorer—It is a low-level HTTP-based API that can query the Facebook API for various information. We used cURL which is a free client-side URL transfer library to transfer comments of Facebook posts to *R*.

4   Methodology

Datasets—The datasets used from Facebook were two public posts by a prominent English newspaper group on their public Facebook page. The datasets consist of about 89 comments and 700 comments. The dataset used from YouTube was about 3363 comments left on a video by a Hindi news channel reporting the scam.

Corpus collection—For collection of data, we first opened a Facebook developer account. A developer account allows us to access the Facebook Graph API Explorer with Rfacebook package with the help of cURL. Then, we collected data using inbuilt functions from the above datasets.

For YouTube, we used the YouTube Data API v3 to access the comments of a video by a news channel video of the scam. Around 3363 comments of that video were extracted.

Processing data—The data was converted into a corpus and the package "**tm**" was used to process the corpus. The processing involved converting the encoding to a more portable one, removing white spaces, converting entire corpus to lower case to maintain uniformity, removing numbers, emoticons, hyperlinks, special characters, and punctuation, and stemming the document.

Analyzing and Visualizing—We have then visualized the data using wordcloud, syuzhet, and ggplot2 packages in *R*.

5   Experimental Results

Tables 1 and 2 show us that the most used words are **bank, Modi, BJP, corrupt,** and **scam**. Therefore, we can see people are more vocal about the involvement of political parties in the PNB scam.

Table 1 Most used words among 89 comments

Word	Frequency
Bank	44
Modi	30
Scam	23
Congress	18
People	15
BJP	14
Fraud	14
Money	11

Table 2 Most used words among 700 comments

Word	Frequency
Modi	402
BJP	353
Corrupt	341
Scam	333
Congress	301
Loan	293
Money	286
Man	167

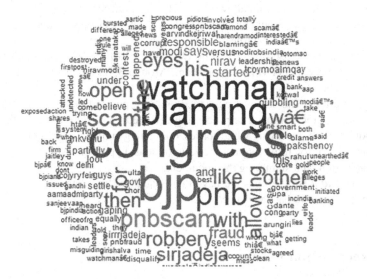

Fig. 1 Word cloud of the 3363 comments on the YouTube video

Figure 1 shows the most used words among YouTube comments of a news video on the PNB scam to be "**Congress**" and "**BJP**" followed by "**blaming**", "**pnb**", and "**watchman**".

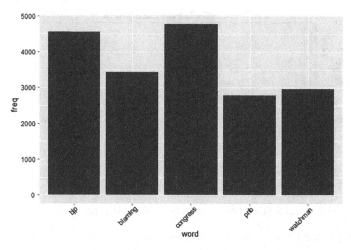

Fig. 2 Bar graph representing words plotted against their frequency for the YouTube video (parameter—frequency > 2500)

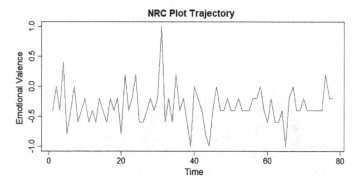

Fig. 3 The NRC plot trajectory of the emotional valence

Figure 2 shows us a graph of the most used words which we found in Fig. 3. From here, we can summarize that the most used words are "**BJP**", "**blaming**", and "**Congress**".

Next, we have plotted a graph of the emotional valence of the YouTube video with 3363 comments as it is the largest dataset. We plotted the emotional valence of the comments. Emotional valence refers to the intrinsic attractiveness or averseness of a target [8].

Figure 3 shows the NRC plot trajectory of the emotional valence. The emotional valence is rescaled from −1 to +1. We can see although for most of the text, the emotional valence is below 0, more positive words are present in the dataset. This is due to the different categorizations of the various emotions.

Figure 4 shows us the NRC plot of the emotional valence after Fourier transform. The transformation is basically a percentage-based sentiment comparison using

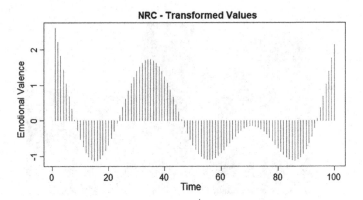

Fig. 4 The NRC plot of the transformed values of emotional valence

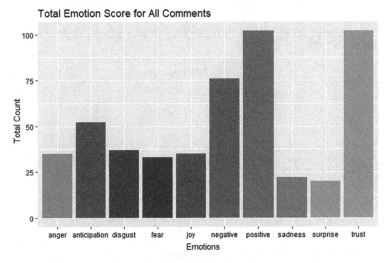

Fig. 5 Bar graph showing the total emotion scores for all comments

Fourier transformation and low-pass filter. The scaled plot is obtained by deducting the means and dividing by their standard deviation. The Fourier transform uses an approximation on the graph with sinusoidal waves. In the above graph, we have used decimation with 100 as the number of values to return and 3 as the number of components to retain.

We now divide the emotions into **sentiments and emotions**. The sentiments are negative and positive, while the emotions are anger, anticipation, disgust, fear, joy, sadness, surprise, and trust. Then, we calculate the emotion score for all the comments of the YouTube video.

Figure 5 shows us that the most expressed is emotion about trust. Also, the positive sentiments expressed in the comments are more than the negative sentiments.

6 Conclusion

In this paper, we have shown how opinion mining is done by utilizing the comments on social networking sites—Facebook and YouTube. This helped us to understand the people's emotion and opinion during the aftermath of the scam. We also generated various statistical facts and figures.

The most expressed sentiment was positive overall. The most expressed emotion was that regarding trust. This indicates people talking about trust in the government or in the banking system, which further indicates that the question of trusting the aforementioned entities is being raised. The most used words were BJP, Congress, and blaming. These show that most people in the comments are blaming either of the parties.

By studying the experimental results and statistics, it is quite evident that people are more concerned about blaming the respective political parties who are at fault for the scam according to their reason and logic. The results also show us that people are more interested in talking about politics after the scam. We do not find any mention of any banking terms among the most frequent words; or any other word that shows the moral incorrectness of what happened and the miseries and hardships that the nation will have to face. Instead, the most frequent words are those which are related to the parties. Although the anger of people is quite obvious, it will be taxpayers' money from whom the RBI will pay the bank if the bank does have enough assets to meet such a liability, but the continuous squabbling among political parties will lead to people losing faith in the banking system, which will lead to a bank run and hence, a financial crisis. People blame political parties for many reasons, but they do not think that they are the one who have chosen such politicians. We, the people of the nation, are responsible for the backwardness of the society in many ways such as voting the inappropriate political parties. We cannot blame the political parties for the corruption as we have all the powers to vote for a right candidate in an election.

Therefore, we should educate people and instill moral ethics and values in them. Educating the people about corruption is much more important and how to take up steps to counter various forms of corruption in every possible way. Humanitarian values and moral codes must be imbibed in each and every person of the nation. In order to tackle and fight against bigger national scams and corruption, we have to start annihilating it from the very grass root level. Only then can we have a nation with minimum corruption and people with the right set of values and ethics to think and act as responsible citizens of the nation.

References

1. Nielsen, F.A.: A new ANEW: evaluation of a word list for sentiment analysis in microblogs
2. Mohammad S.M., Kiritchenko, S., Zhu, X.: NRC-Canada: building the state-of-the-art in sentiment analysis of tweets
3. Neri, F., Aliprandi, C., Capeci, F., Cuadros, M., By, T.: Sentiment analysis on social media

4. Bakliwal, A., Foster, J., van der Puil, J. et al.: Sentiment analysis of political tweets: towards an accurate classifier
5. Shashank, Bhattacharyya, P.: Emotion analysis of online chats. IIT Bombay, India (2010)
6. Singh, A.P.: Sentiment analysis on political tweets
7. Stefano, B., Esuli, A., Sebastiani, F.: SENTIWORDNET 3.0: an enhanced lexical resource for sentiment analysis and opinion mining
8. Mohammad, S.M.: Sentiment analysis: detecting valence, emotions, and other affectual states from text

Finger Spelling Recognition for Nepali Sign Language

Vivek Thapa, Jhuma Sunuwar and Ratika Pradhan

Abstract The use of Human–Computer Interaction (HCI) has improved day by day. Hand gesture recognition system [1–3] can be used to interface computer with humans using hand gesture. It is applicable in the areas like virtual environment, smart surveillance, sign language translation, medical system, robot control, etc [4, 5]. The paper elaborated the mechanism to identify the Nepali Sign Language with the help of hand gesture using shape information. It uses radial approach for dividing the segmented image and to obtain the sampled points making the process scaling invariant. The feature set is obtained by using the Freeman chain code and Vertex Chain Code (VCC) techniques. Skin color model has been used to identify the hand from simple background. Further processing is done in order to remove the unwanted noise and areas. Blob analysis is then carried out in order to extract the hand gesture from the image considering the largest blob in the image. The centroid of the image is identified, and the image is divided equally by plotting a line at certain angle from the centroid of the image. For sampling of the image, the point of intersection of line and the boundary of image is taken. For the classification purpose, k-NN classification technique is used. Confusion matrix approach is used to authenticate its accuracy. The accuracy using sampling for radial approach and use of Freeman chain code for feature extraction was found to be efficient than sampling using radial approach and use of transcribing vertex chain code technique and sampling using grid and use of Freeman chain code for feature extraction.

Keywords Nepali sign language · Freeman chain code · Vertex chain code
Sign language recognition

V. Thapa · J. Sunuwar (✉) · R. Pradhan
Sikkim Manipal University, Sikkim, India
e-mail: jhuma.s@smit.smu.edu.in

V. Thapa
e-mail: vivekthapa27@live.in

R. Pradhan
e-mail: ratika.p@smit.smu.edu.in

1 Introduction

Human–Computer Interaction (HCI) [6] can be implemented using sensor-based technology, computer vision, digital image processing methods or using devices like data gloves, color cue, and markers. As HCI is mainly used to replace the input devices like mouse and keyboard, it allows the user to interact with computer system as they communicate with the other person owning several challenges. Use of sensor-based technology to perform HCI requires using additional hardware like data gloves or sensors which gives extra burden to the person that tries to use this technology. It may be concluded that computer vision and digital image process is better suited for implementing human–computer interact applications rather than using sensor-based technology. Computer vision has advantages over other method as it reduces the cost of implementing HCI and also provides simple and best approach for interaction between humans and computer in a more natural form [4, 5]. Manual alphabet recognition or static gesture recognition is one of the application areas of HCI. Gesture can be either static, which does not pose any movement or it can be a dynamic gesturer, which involves some movement. Real-time systems deal with the static as well as the dynamic gesture. The hand gesture recognition systems comprise various components to interact with human being and are used in applications such as interactive entertainments and augmented reality which requires more natural and intuitive interface. Hand gesture recognition system captures hand gesture using camera as the input devices, recognizes the gesture, and then unfolds its meaning. Sign language recognition is one of the areas of application for hand gesture recognition. Sign language is the representation of the letters of a writing system and sometimes the numeral systems, using gestures. These manual alphabets have often been used in deaf education and have subsequently been adopted as a distinct part of the number of sign languages nearly about 40 manual alphabets around the world. Historically, manual alphabets have had a number of additional applications—being used as a cipher, mnemonics, and silent religious setting. Finger spelling for consonant set of Nepali Sign language [7] is represented using simple hand gesture. Nepali Sign language is used to test the static gesture recognition that uses a single hand to perform the manual alphabets in this paper. The mapping algorithm consists of a cell representation algorithm that represents a thinned binary image into triangular cells, and a transcribing algorithm that transcribes the cells into Vertex Chain Code (VCC). The algorithms have been tested and validated by using three thinned binary images: L-block, hexagon, and pentagon. The result shows that this algorithm is capable of visualizing and transcribing them into VCC; it can also be improved by testing on more thinned binary images. Grid approach is used to make the system scale variant and accuracy depends on the thickness of the contour [8]. Representation of shape is based on a new boundary chain code and using the chain code to recognize the object as they preserve information and allow considerable data reduction. VCC recognizes images better than the classical methods but it lacks its accuracy in terms of scaling with the Freeman chain code technique [9].

2 Design Methodology

Image is acquired by using a USB camera and is preprocessed to reduce the size of the captured image. The image is resized by cropping some part of the image to remove any nearby object that is not required. Reduction the size of the image also makes the processing faster as the number of pixels representing the image is less. Enhancement is done for improving the quality of the image so that the analysis of the image is reliable. The brightness and contrast of the image are also adjusted to enhance the quality of the image and later may be made automatic. All the components of the acquired image may not be important during processing of the image as stated above, so image segmentation is done to extract only the region of interest for further analysis. The process of segmentation [3] is achieved by the use of skin color detection in RGB image. To identify the gesture in the image, two phases are initiated. First phase is the learning phase where the set of images provided to the system is used to find the feature of the input and is stored in database. The next phase is the recognition phase where the input given to the system is classified successfully by the system by using the feature learned during the first phases as represented in Fig. 1. The feature that has been considered here is the shape information for gesture as each of the gestures has unique shape.

For the purpose of generating feature set of five different gestures, each of six types is captured from different people. Those gestures are equivalent to Nepali Sign language consonant MA, BA, GHA, SA, and CHHA. The recognition system should be robust enough to classify the gesture. Therefore, the training set should contain all the orientations of the gestures. The feature table or record that is constructed during the learning phase is then used to perform the classification. In the classification phase, the feature is extracted from the image using the same technique as used

Fig. 1 Phases of gesture recognition

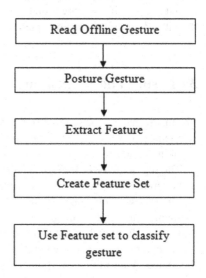

Fig. 2 Vertex chain code
from rectangular cells

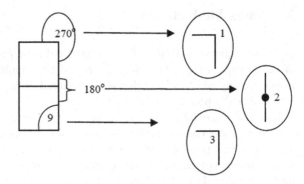

during the learning phase. After the feature is extracted successfully, then the task
of classification needs to be done. The system aims at identifying the hand gesture
uniquely using only one feature to make it simple and fast, so classification processes
should be made reliable to do so. Classification is done by finding the most nearly
matching feature in the existing feature database using k-NN classification technique.
The transcribing algorithm [8] transcribes the rectangular cells into VCC. In fact,
VCC shows the number of rectangle that in touch with a corner of the rectangle.
But for the transcribing algorithm, it proposes to see the degree of either rectangular
corner or relation between two edges of the rectangular. Rectangular cell has three
kinds of VCC: 1, 2, and 3. It represents three degrees: 270°, 180°, and 90° (Fig. 2).
The sample point was selected based on the point of intersection between the line and
image boundary which is not similar always like in cell boundary or grid approach.
So to obtain the above mentioned degree is not always possible. In order to overcome
that problem if the angle value is 0°, then it is considered as 2. If the value is on the
right side of the image and value obtained is negative, then it is 1 and if it is positive
then it is 3. But if the value is on left side of the image and value obtained is negative,
then it is 3 and if it is positive then the value is 1.

3   Results and Discussion

The image needs to be segmented first to extract the hand gesture from the image. The
segmentation was done using the skin color segmentation using the RGB component
for skin color (Fig. 3).

The segmented RGB image is then converted to binary image so that morphological operators can be used to remove any type of noise occurred during the segmentation process. Blob analysis is done in order to remove any unwanted blob from the
segmented image. The image is then resized and cropped to a standard size of 320 ×
320. Then, the center of the image is identified and the image is divided by drawing
line at certain angle and the point of intersection between line and the boundary of
the image. After finding the point of intersection, the points are connected.

After finding the point of intersection, the chain code is obtained using the Freeman chain code method. Chain code for the above gesture is 1111111107777777777775544443333211222. The first difference obtained for the above chain code is 00000007700000000000060700070007701007. Then, the shape number is found out, that is, 00000000006070007000770100700000000077 (Fig. 4).

Usually, the gesture contains equal length of shape number depending on angle which is used to divide the image (length of shape number (LEN) = 360/angle). But in some case it will produce more than that length. It is because the line can meet the boundary of the image more than one time, thereby obtaining the length of shape number more than the expected shape number length (LEN).

Table 2 represents the corresponding chain code and the shape number for inputs of Table 1. During the learning phase, the shape information of the manual alphabet are found and using the data structure a feature set is created. After the learning phase is over, the classification is performed using the feature set created during the learning phase (Table 3).

Classification is done by comparing the obtained shape number of the manual alphabets with the shape number in database. As two same gestures performed by two different people may not be exactly same, the classification compares all possible occurrence of similar sequence in the shape number. The shape number, whose distance between them is minimum, is considered to be the similar manual alphabet and is said to be classified (Table 4).

To verify the accuracy of the used approach, confusion matrix was used. To generate a training set and for the testing purpose, five sets of gesture of each type are used. The calculation of accuracy using Freeman chain code technique for feature extraction can be evaluated using confusion matrix is given in Table 5.

Accuracy using this approach is calculated by calculating the total number correctly identified gesture and total number of gesture tested. The total number of correctly identified gesture can be found out by adding diagonal values in the matrix.

$$\text{Accuracy percentage}(\%) = \frac{\text{Total Number of Identified Objects}}{\text{Total Number of Test Objects}} \times 100$$

$$= \frac{(4+2+2+4+2)}{(5+5+5+5+5)} \times 100 = \frac{14}{25} \times 100 = 56\%$$

Similarly for calculation of accuracy using grid line approach for sampling and Freeman chain code 48%, using radial approach for sampling and VCC technique 44% after checking the accuracy using confusion matrix on the given training set, it was found that the use of radial approach for sampling and Freeman chain code for extracting feature set is appropriate followed by the use of radial approach for sampling and VCC for extracting feature set of the gesture. It was found that the sampling using grid approach was less efficient than both techniques by fewer margins (Fig. 5).

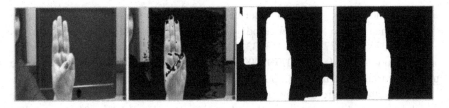

Fig. 3 Input image and its processing

(a) **(b)**

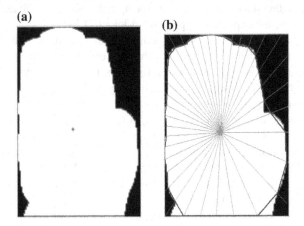

Fig. 4 **a** Biggest blog resized and cropped, **b** Line drawn from center and image boundary

Table 1 Output images of each step during the learning phases

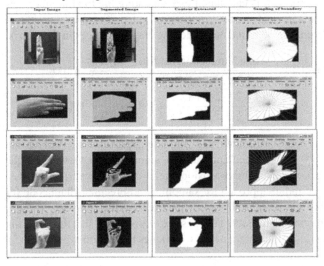

Table 2 Freeman chain code and shape number of gesture

Srl. No	Freeman chain code	First difference code	Shape number
1	11111111077777777777 5544443333211222	00000007700000000 0060700070007701007	000000000060 70007000770100700000007
2	11111111007771110 7775657554444333333	0000000707002007 70061726070007000006	0000000707002 00770061726070007000006
3	11077777775575444433 332332111111111112	077000006026700 0700071077000000000017	00000000001707 70000060267000700071077
4	111111000711311077 77677777754443333333	00000700720260 77000710000067007000006	00000060000070 07202607700071000067007

Table 3 Vertex chain code and shape number of gesture

Gesture	Vertex chain code	Shape number
1	3133331333333311223311112223331133133	1111222333113313331333313333333112233
2	3311333231331111113111313222222222233	1111111311131322222222233331133323133
3	3113133133331313111313111113131332213	1111131313322133113133133331313111313
4	1332131133333333113133113111323133333	1113231333331332131133333333113133113

Table 4 Classification of manual alphabet

S. No.	Test image	Contour extracted	Equivalent alphabet of NSL
1			MA
2			GHA

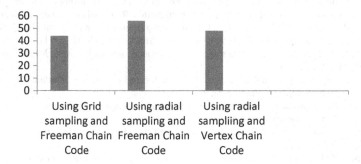

Fig. 5 Comparison between accuracy of three different techniques

Table 5 Confusion matrix for result obtained using radial approach for sampling and Freeman chain code for extracting feature set

Classification data	Training set data	Error of omission						
	Gesture	1	2	3	4	5	Total	
	1	4	0	0	1	0	5	(1/5 × 100) = 20%
	2	0	2	1	1	1	5	(3/5 × 100) = 60%
	3	0	0	2	0	3	5	(3/5 × 100) = 60%
	4	1	0	0	4	0	5	(1/5 × 100) = 20%
	5	0	2	0	1	2	5	(3/5 × 100) = 60%
	Total	5	4	3	7	6	25	
Error of commission	(1/5 × 100) = 20%	(2/4 × 100) = 50%	(1/3 × 100) = 33.3%	(3/7 × 100) = 42.86%	(4/6 × 100) = 66.67%			

4 Conclusion

The use of object description representing hand has made the classification easy and simple by transforming the gesture to shape number. The paper has highlighted the radial approach for sampling boundary points from which the shape number is obtained using Freeman chain code and VCC technique. Using this shape number, the gesture is classified successfully.

References

1. Bhuyan, M.K., Bora, P.K., Ghosh, D.: Trajectory guided recognition of hand gestures having only global motions. World Acad. Sci. Eng. Technol. **21**, 753–764 (2008)
2. Hassan Ahmed, S.M., Alexander, T.C., Anagnostopoulos, G.C.: Real-time, static and dynamic hand gesture recognition for human-computer interaction. Retrieved December 22 (2008), 2014
3. Jeon, M.-J., Lee, S.W., Bien, Z.: Hand gesture recognition using multivariate fuzzy decision tree and user adaptation. Int. J. Fuzzy Syst. Appl. (IJFSA) **1**(3), 15–31 (2011)
4. Ahad, M.A.R. et al.: Motion history image: its variants and applications. Mach. Vis. Appl. **23**(2), 255–281 (2012)
5. Aly, A.A., Deris, S.B., Zaki, N.: Research review for digital image segmentation techniques. Int. J. Comput. Sci. Inf. Technol. **3**(5), 99 (2011)
6. Pradhan, R., Sunuwar, J., Pradhan, R.: Gesture extraction using depth information. Int. J. Adv. Res. Comput. Sci. Softw. Eng. (2014)
7. Khan, R.Z., Ibraheem, N.A.: Hand gesture recognition: a literature review. Int. J. Artif. Intell. Appl. **3**(4), 161 (2012)
8. Davis, J.W.: Hierarchical motion history images for recognizing human motion. Detection and Recognition of Events in Video, 2001. Proceedings IEEE Workshop on IEEE (2001)
9. Wang, A.W.H., Tung, C.L.,: Dynamic gesture recognition based on dynamic Bayesian networks. WSEAS Trans. Bus. Econ. **4**(11), 168–173 (2008)
10. Venkatesh, B.R., Ramakrishnan K.R.: Recognition of human actions using motion history information extracted from the compressed video. Image Vis. Comput. **22**(8), 597–607 (2004)

11. Hedi, A., Gouiffes, M., Lacassagne, L.: Motion histogram quantification for human action recognition. Pattern Recognition (ICPR), 2012 21st International Conference on IEEE (2012)
12. Bobick, Aaron F., and James W. Davis. "The recognition of human movement using temporal templates." IEEE Transactions on pattern analysis and machine intelligence 23.3 (2001): 257-267.

Parsing in Nepali Language Using Linear Programming Problem

Archit Yajnik, Furkim Bhutia and Samarjeet Borah

Abstract Linear Programming Problem (LPP) is one of the powerful techniques for parsing in terms of NLP. Various works are available in the literature using optimization and soft computing techniques. In this paper, the emphasis is given on the fundamental technique of optimization, i.e., simplex method by which the parsing evolves. First, the input is tested with the Big M Method (Penalty approach). Second, the same input is fed to the two-phase simplex method. The concept is tested theoretically using integer programing problem yielding desired results. The test is conducted for ten annotated sentences in Nepali language and observed that two-phase simplex method is found to be the better one.

Keywords NLP · Nepali · Parsing · Linear programming problem · Language

1 Introduction

Natural Language Processing (NLP) is the discipline of computer science dealing with interactions between computer and human languages. It covers computer understanding and manipulation of human language. More precisely, it can be said that human languages can be analyzed and understood by computers with the help of NLP only. It also helps in deriving meaning in a smart and useful way. There is a cognitive and linguistic motivation to gain a better insight into how human communicate using natural language. Parsing is one of the important tools in the field of NLP.

A. Yajnik · F. Bhutia · S. Borah (✉)
Sikkim Manipal Institute of Technology, Sikkim Manipal University, Majhitar, East Sikkim, Sikkim 737136, India
e-mail: samarjeetborah@gmail.com

A. Yajnik
e-mail: archit.yajnik@gmail.com

F. Bhutia
e-mail: furkinbhutia@gmail.com

© Springer Nature Singapore Pte Ltd. 2019
J. Kalita et al. (eds.), *Recent Developments in Machine Learning and Data Analytics*,
Advances in Intelligent Systems and Computing 740,
https://doi.org/10.1007/978-981-13-1280-9_23

A parser can briefly be described as a tool to extract the relationships among the words of a sentence. It is a method of understanding the grammatical structure of a sentence or word, which can be solved with the help of Linear Programming Problem (LPP), bipartite matching problem, assignment problem, etc. It makes use of two components: a parser, which is a procedural component (a computer program) and grammar, which is declarative. It uses the notion of karaka relations between verbs and nouns in a sentence. A parser breaks data into smaller elements (chunks), according to a set of rules that describe its structure sequence of token (i.e., words), to determine its grammatical structure with respect to a given grammar. Various efforts are available in the literature [1–3] using optimization and soft computing techniques.

In Nepal, a small Himalayan country, Nepali is recognized as the national language. Nepali is also spoken in India, Myanmar, and Thailand partially. As per records, around the world, Nepali is spoken by approximately 45 million people. Devanagari script is used to write in Nepali language. It was developed from the Brahmi script in the eleventh century A.D. Recently, activities on Nepali NLP have been started, and few applications and tools have been developed [5–7].

In this research work, an attempt has been made to analyze parsing in Nepali language using LPP. It uses two approaches of LPP, namely, Big M and two-phase simplex methods. Several works are found in literature which uses LPP in resolving NLP issues [8–10]. It is basically a theoretical analysis using ten sentences of Nepali language.

2 Methodology

Since the aim of the work is to theoretically analyze the usefulness of LPP in the development of parser for Nepali language, the following well-known methods of LPP are used.

2.1 Big M Method

The Big M method (also known as method of penalties) contains four basic steps. First, LPP is expressed in the standard form. It includes slack and/or surplus variables (if any). Second, nonnegative variables are introduced. For constraints of ($\geq$ or $=$) type, it is always in the left side. Third, modified LPP is solved by simplex method. The computational procedure is continued until

- Optimal BFS is obtained or
- Existence of an unbounded solution.

2.2 Two-Phase Simplex Method

The two-phase method is a way to resolve a given LPP. The problem should involve some artificial variable. The phase one involves in the application of simplex method into a specially constructed LPP. It yields a final simplex table containing an elementary possible resolution to the original problem.

The second phase starts with the elementary possible resolution. It gives an optimum elementary possible resolution, if any. It also applies the simplex method.

3 Analysis

3.1 Using Big M Method

Constraint graph is used to obtain a parse in Big M method. Then, the graph is converted into LPP. To convert this, variable x is introduced for an *arc* from node i to j. It is labeled by *karaka k* in the constraint graph. There should be a variable for every *arc*. The variable takes the value as *0* and *1*. The variables, whose corresponding *arcs* are in the parse subgraph, are assigned 1 (parse), otherwise 0. The constraint rules are formulated into constraint equations described in [1, 4]. Sum of all the variables forms the cost function, which is to be minimized. This can be illustrated by forming inequalities, and word groups in the sentence are referred by (a, b, c, A):

मोहन हात ले सुन्तला न्म्न ।

Mohan   haath le   suntalakhancha.

a   b   c   A

(Mohan eats orange with his hand)

Constraint $c1$ generates the following equalities (for mandatory karakas $k1$ and $k2$):

$$M_{A,k1}:x_{A,k1,a} + x_{A,k1,c} = 1$$
$$M_{A,k2}:x_{A,k2,a} + x_{A,k2,c} = 1$$

Constraint $c2$ generates

$$O_{A,k3}:x_{A,k3,b} \leq 1$$

Constraint $c3$ generates

$$S_a:x_{A,k1,a} + x_{A,k2,a} = 1$$
$$S_b:x_{A,k3,b} = 1$$

$$S_c : x_{A,k1,c} + x_{A,k2,c} = 1$$

For ease to readability, we rename x's as follows:

$$x_{A,k1,a} = y_1$$
$$x_{A,k1,c} = y_2$$
$$x_{A,k2,a} = y_3$$
$$x_{A,k2,c} = y_4$$
$$x_{A,k3,b} = y_5$$

Now we get

$$M_{A,k1} : y_1 + y_2 = 1$$
$$M_{A,k2} : y_3 + y_4 = 1$$
$$O_{A,k3} : y_5 \leq 1$$
$$S_a : y_1 + y_3 = 1$$
$$S_b : y_5 = 1$$
$$S_c : y_2 + y_4 = 1$$

The cost function to be minimized is

$$Z = y_1 + y_2 + y_3 + y_4 + y_5 \tag{1}$$

Subject to the constraints:

$$\left.\begin{array}{l} y_1 + y_2 = 1 \\ y_3 + y_4 = 1 \\ y_5 \leq 1 \\ y_1 + y_3 = 1 \\ y_5 = 1 \\ y_2 + y_4 = 1 \end{array}\right\} \tag{2}$$

Solution Introducing slack variable y_6 and artificial variable a_i, $i = 1, 2, \ldots, 5$.

Equations (1) and (2) can be rewritten as

$$\text{Maximize } Z* = -a_1 - a_2 - a_3 - a_4 - a_5$$

Subject to the constraints:

$$y_1 + y_2 + 0.y_3 + 0.y_4 + 0.y_5 + 0.y_6 + a_1 + 0.y_8 + 0.y_9 + 0.y_{10} + 0.y_{11} = 1$$
$$0.y_1 + 0.y_2 + y_3 + y_4 + 0.y_5 + 0.y_6 + 0.y_7 + a_2 + 0.y_9 + 0.y_{10} + 0.y_{11} = 1$$
$$0.y_1 + 0.y_2 + 0.y_3 + 0.y_4 + y_5 + y_6 + 0.y_7 + 0.y_8 + 0.y_9 + 0.y_{10} + 0.y_{11} = 1$$
$$y_1 + 0.y_2 + y_3 + 0.y_4 + 0.y_5 + 0.y_6 + 0.y_7 + 0.y_8 + a_3 + 0.y_{10} + 0.y_{11} = 1$$
$$0.y_1 + 0.y_2 + 0.y_3 + 0.y_4 + y_5 + 0.y_6 + 0.y_7 + 0.y_8 + 0.y_9 + a_4 + 0.y_{11} = 1$$
$$0.y_1 + y_2 + 0.y_3 + y_4 + 0.y_5 + 0.y_6 + 0.y_7 + 0.y_8 + 0.y_9 + 0.y_{10} + a_5 = 1$$

Initial basic feasible solution is $\mathbf{B} = [\mathbf{A_7, A_8, A_6, A_9, A_{10}, A_{11}}]$ and $\mathbf{A_7 = A_8 = A_6 = A_9 = A_{10} = A_{11} = 1}$.

y_1, y_2, y_3, y_4, y_5 can enter the basis since $z_j - c_j = c_B y_j - c_j, j = 1, 2, \ldots, 4$ is the most negative evaluation so we choose arbitrary y_1.

Again, y_7 and y_9 can leave the basis since $\min\left\{\frac{x_{Bi}}{y_{i1}}, y_{i1} > 0, i = 1, \ldots, 6\right\} = \min\{1, 1\} = 1$ so, we choose arbitrary y_7.

Now applying the row operation, i.e., $R_4 \rightarrow R_4 - R_1$ and $R_7 \rightarrow R_7 + (2M - 1)R_1$. We get Table 8.

Applying the similar argument of Tables 1 in 2 here, y_3 enter the basis and y_8 leaves the basis and applying $R_4 \rightarrow R_4 - R_2, R_7 \rightarrow R_7 + (2M - 1)R_2$ we get Table 3.

Similarly, here y_5 enter the basis and y_{10} leaves the basis and applying $R_3 \rightarrow R_3 - R_5, R_7 \rightarrow R_7 + (2M - 1)R_5$ we get Table 4.

Since, $z_j - c_j \geq 0$, for all j, so it appears that the present basic solution, an optimum solution to the LPP is attained. But, due to the existence of y_{11} and y_9 in the basis, and moreover the value of XB for y_9 is -1 (<0), the current solution, i.e., $(y_1 = 1, y_3 = 1, y_5 = 1)$, is not possible.

$y_1 = 1, y_3 = 1, y_5 = 1$ means $x_{A,k1,a} = 1, x_{A,k2,a} = 1$ and $x_{A,k3,b} = 1$.

i.e., मोहन (Mohan) is Karta and Karma as well, which is not feasible. In this regard, the same experiment will be conducted with the two-phase simplex method.

3.2 Using Two-Phase Simplex Method

As discussed in Sect. 3.1, constraint graph is used to obtain a parse in Big M method. Then, the graph is converted into LPP. To convert this, variable x is introduced for an arc from node i to j. It is labeled by karaka k in the constraint graph. There should be a variable for every arc. The variable takes the value as 0 and 1. The variables, whose corresponding arcs are in the parse subgraph, are assigned 1 (parse), otherwise 0.

Example

Mohan   haath   le   suntala   khancha.
a   b   c   A
(Mohan eats orange with his hand)

Table 1 Row operation on R_1

C	B	X_B	-1	-1	-1	-1	-1	0	$-M$	$-M$	$-M$	$-M$	$-M$
			y_1	y_2	y_3	y_4	y_5	y_6	y_7	y_8	y_9	y_{10}	y_{11}
$-M$	y_7	1	*1*	1	0	0	0	0	1	0	0	0	0
$-M$	y_8	1	0	0	1	1	0	0	0	1	0	0	0
0	y_6	1	0	0	0	0	1	1	0	0	0	0	0
$-M$	y_9	1	1	0	1	0	0	0	0	0	1	0	0
$-M$	y_{10}	1	0	0	0	0	1	0	0	0	0	1	0
$-M$	y_{11}	1	0	1	0	1	0	0	0	0	0	0	1
	$Z^* = -5M$		$-2M+1$	$-2M+1$	$-2M+1$	$-2M+1$	$-2M+1$	0	0	0	0	0	0

Table 2 Row operation on R_4 and R_7 using R_1

C	B	X_B	-1	-1	-1	-1	-1	0	$-M$	$-M$	$-M$	$-M$
			y_1	y_2	y_3	y_4	y_5	y_6	y_8	y_9	y_{10}	y_{11}
-1	y_1	1	1	1	0	0	0	0	0	0	0	0
$-M$	y_8	1	0	0	1	1	0	0	1	0	0	0
0	y_6	1	0	0	0	0	1	1	0	0	0	0
$-M$	y_9	0	0	-1	1	0	0	0	0	1	0	0
$-M$	y_{10}	1	0	0	0	0	1	0	0	0	1	0
$-M$	y_{11}	1	0	1	0	1	0	0	0	0	0	1
	$Z^* = -3M - 1$		0	0	$-2M+1$	$-2M+1$	$-2M+1$	0	0	0	0	0

Table 3 Row operation on R_4 and R_7 using R_2

C			-1	-1	-1	-1	-1	0	$-M$	$-M$	$-M$
C	B	X_B	y_1	y_2	y_3	y_4	y_5	y_6	y_9	y_{10}	y_{11}
-1	y_1	1	1	1	0	0	0	0	0	0	0
-1	y_3	1	0	0	1	-1	0	0	0	0	0
0	y_6	-1	0	0	0	0	-1	1	-1	0	0
$-M$	y_9	1	0	-1	0	0	1	0	1	0	0
$-M$	y_{10}	1	0	0	0	1	0	0	0	1	0
$-M$	y_{11}	1	0	1	0	0	0	0	0	0	1
	$Z^* = -M - 2$		0	0	0	0	$-2M+1$	0	0	0	0

Table 4 Row operation on R_3 and R_7

C	B	X_B	-1	-1	-1	-1	-1	0	$-M$	$-M$
			y_1	y_2	y_3	y_4	y_5	y_6	y_9	y_{11}
$-M$	y_1	1	1	1	0	0	0	0	0	0
$-M$	y_3	1	0	0	1	1	0	0	0	0
0	y_6	0	0	0	0	0	0	1	0	0
$-M$	y_9	-1	0	-1	0	-1	0	0	1	0
-1	y_5	1	0	0	0	0	1	0	0	0
$-M$	y_{11}	1	0	1	0	1	0	0	0	1
		$Z^* = -2M - 1$	0	0	0	0	0	0	0	0

Constraint $c1$ generates the following equalities (for mandatory karakas k1 and k2):

$$M_{A,k1}:x_{A,k1,a} + x_{A,k1,c} = 1$$
$$M_{A,k2}:x_{A,k2,a} + x_{A,k2,c} = 1$$

Constraint $c2$ generates

$$O_{A,k3}:x_{A,k3,b} \leq 1$$

Constraint $c3$ generates

$$S_a:x_{A,k1,a} + x_{A,k2,a} = 1$$
$$S_b:x_{A,k3,b} = 1$$
$$S_c:x_{A,k1,c} + x_{A,k2,c} = 1$$

For ease to readability, we rename x's as follows:

$$x_{A,k1,a} = y_1$$
$$x_{A,k1,c} = y_2$$
$$x_{A,k2,a} = y_3$$
$$x_{A,k2,c} = y_4$$
$$x_{A,k3,b} = y_5$$

Now we get

$$M_{A,k1}:y_1 + y_2 = 1$$
$$M_{A,k2}:y_3 + y_4 = 1$$
$$O_{A,k3}:y_5 \leq 1$$
$$S_a:y_1 + y_3 = 1$$
$$S_b:y_5 = 1$$
$$S_c:y_2 + y_4 = 1$$

The cost function to be minimized is

$$\text{Minimize } Z = y_1 + y_2 + y_3 + y_4 + y_5 \tag{3}$$

Subject to the constraints:

$$\left.\begin{array}{l} y_1 + y_2 = 1 \\ y_3 + y_4 = 1 \\ y_5 \leq 1 \\ y_1 + y_3 = 1 \\ y_5 = 1 \\ y_2 + y_4 = 1 \end{array}\right\} \qquad (4)$$

Solution Introducing slack variable y_6 and artificial variable a_i, $i = 1, 2, ..., 5$.

Equations (3) and (4) can be rewritten as follows:
Minimize $Z* = -a_1 - a_2 - a_3 - a_4 - a_5$
Subject to the constraints:

$y_1 + y_2 + 0.y_3 + 0.y_4 + 0.y_5 + 0.y_6 + a_1 + 0.y_8 + 0.y_9 + 0.y_{10} + 0.y_{11} = 1$
$0.y_1 + 0.y_2 + y_3 + y_4 + 0.y_5 + 0.y_6 + 0.y_7 + a_2 + 0.y_9 + 0.y_{10} + 0.y_{11} = 1$
$0.y_1 + 0.y_2 + 0.y_3 + 0.y_4 + y_5 + y_6 + 0.y_7 + 0.y_8 + 0.y_9 + 0.y_{10} + 0.y_{11} = 1$
$y_1 + 0.y_2 + y_3 + 0.y_4 + 0.y_5 + 0.y_6 + 0.y_7 + 0.y_8 + a_3 + 0.y_{10} + 0.y_{11} = 1$
$0.y_1 + 0.y_2 + 0.y_3 + 0.y_4 + y_5 + 0.y_6 + 0.y_7 + 0.y_8 + 0.y_9 + a_4 + 0.y_{11} = 1$
$0.y_1 + y_2 + 0.y_3 + y_4 + 0.y_5 + 0.y_6 + 0.y_7 + 0.y_8 + 0.y_9 + 0.y_{10} + a_5 = 1$

Initial basic feasible solution is $\mathbf{B} = [\mathbf{A_7, A_8, A_6, A_9, A_{10}, A_{11}}]$ and $\mathbf{A_7 = A_8 = A_6 = A_9 = A_{10} = A_{11} = 1}$.

Phase 1

y_1, y_2, y_3, y_4 can enter the basis since $z_j - c_j = c_B y_j - c_j, j = 1, 2, ..., 4$ is the most negative evaluation so we choose arbitrary y_1 (Table 5).

Again y_7 and y_9 can leave the basis since $\min\left\{\frac{x_{Bi}}{y_{i1}}, y_{i1} > 0, i = 1, ..., 6\right\} = \min\{1, 1\} = 1$ so, we choose arbitrary y_7. Now applying the row operation, i.e., $R_4 \rightarrow R_4 - R_1$ and $R_7 \rightarrow R_7 + 2R_1$. We get Table 6.

Applying the similar argument of Table 7 in Table 8 here, y_4 *enter the basis and* y_8 *leaves the basis* and applying $R_6 \rightarrow R_6 - R_2, R_7 \rightarrow R_7 + 2R_2$ we get Table 7. Similarly, here y_5 enters the basis and y_{10} leaves the basis and applying $R_3 \rightarrow R_3 - R_5, R_7 \rightarrow R_7 + R_5$ we get Table 8.

Similarly, here y_5 enters the basis and y_{10} leaves the basis and applying $R_3 \rightarrow R_3 - R_5, R_7 \rightarrow R_7 + R_5$ we get Table 8.

Since $z_j - c_j \geq 0$, for all j, so the present Basic Feasible Solution (BFS) is an optimum solution to LPP. Since max $Z* = 0$ and artificial variable y_9, y_{11} are in the optimal basis at zero level, we proceed to second phase. Here, use the optimum BFS obtained in phase 1 as a starting BFS for the original LPP. Assign the actual cost to the original variable and a cost zero to the entire artificial variable in the objective function. Apply simplex method till an optimum BFS (if any) is obtained.

NOTE: Before initializing phase 2, remove all artificial variables from the table which was nonbasic at the end of phase 1.

Table 5 Row operation on R_1

C	B	X_B	0	0	0	0	0	0	−1	−1	−1	−1	−1
			y_1	y_2	y_3	y_4	y_5	y_6	y_7	y_8	y_9	y_{10}	y_{11}
−1	y_7	1	*1*	1	0	0	0	0	1	0	0	0	0
−1	y_8	1	0	0	1	1	0	0	0	1	0	0	0
0	y_6	1	0	0	0	0	1	1	0	0	0	0	0
−1	y_9	1	1	0	1	0	0	0	0	0	1	0	0
−1	y_{10}	1	0	0	0	0	1	0	0	0	0	1	0
−1	y_{11}	1	0	1	0	1	0	0	0	0	0	0	1
	$Z^* = -5$		−2	−2	−2	−2	−1	0	0	0	0	0	0

Table 6 Row operation on R_4 and R_7

C			0	0	0	0	0	0	−1	−1	−1	−1	−1
C	B	X_B	y_1	y_2	y_3	y_4	y_5	y_6	y_7	y_8	y_9	y_{10}	y_{11}
0	y_1	1	1	1	0	0	0	0	1	0	0	0	0
−1	y_8	1	0	0	1	1	0	0	0	1	0	0	0
0	y_6	1	0	0	0	0	1	1	0	0	0	0	0
−1	y_9	0	0	−1	1	0	0	0	−1	0	1	0	0
−1	y_{10}	1	0	0	0	0	1	0	0	0	0	1	0
−1	y_{11}	1	0	1	0	1	0	0	0	0	0	0	1
	$Z^* = -3$		0	0	−2	−2	−1	0	2	0	0	0	0

Table 7 Row operation on R_6 and R_7

C	B	X_B	0	0	0	0	0	0	-1	-1	-1	-1	-1
			y_1	y_2	y_3	y_4	y_5	y_6	y_7	y_8	y_9	y_{10}	y_{11}
0	y_1	1	1	1	0	0	0	0	1	0	0	0	0
0	y_4	1	0	0	1	1	0	0	0	1	0	0	0
0	y_6	1	0	0	0	0	1	1	0	0	0	0	0
-1	y_9	0	0	-1	1	0	0	0	-1	0	1	0	0
-1	y_{10}	1	0	0	0	0	1	0	0	0	0	1	0
-1	y_{11}	0	0	1	-1	0	0	0	0	-1	0	0	1
	$Z^* = -1$		0	0	0	0	-1	0	2	2	0	0	0

Table 8 Row operation on R_3 and R_7

C			0	0	0	0	0	0	−1	−1	−1	−1	−1
C	B	X_B	y_1	y_2	y_3	y_4	y_5	y_6	y_7	y_8	y_9	y_{10}	y_{11}
0	y_1	1	1	1	0	0	0	0	1	0	0	0	0
0	y_4	1	0	0	1	1	0	0	0	1	0	0	0
0	y_6	0	0	0	0	0	0	1	0	0	0	−1	0
−1	y_9	0	0	−1	1	0	0	0	−1	0	1	0	0
0	y_5	1	0	0	0	0	1	0	0	0	0	−1	0
−1	y_{11}	0	0	−1	−1	0	0	0	0	−1	0	0	1
	$Z^* = 0$		0	0	0	0	0	0	2	2	0	1	0

Fig. 1 Solution graph (corresponding to "Mohan eats orange with hand")

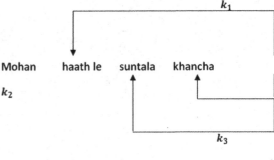

Fig. 2 Solution graph (corresponding to the meaning "orange eats Mohan with hand")

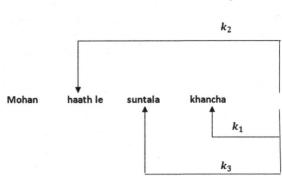

Phase 2 The starting BFS is $y_1 = 1$, $y_4 = 1$, $y_5 = 1$, $y_6 = 0$, $y_{11} = 0$, $y_9 = 0$ is depicted in Table 9.

Since $z_j - c_j \geq 0$ for all j; therefore, the present BFS is an optimum solution. And optimum solution is $y_1 = 1$; $y_4 = 1$; $y_5 = 1$ and min $Z = -\max Z^* = 3$, that is, an optimum solution is shown in Fig. 1.

In place of y_1 in Table 1 if we choose another variable like y_2 and proceeding in a similar way, we get another solution, i.e., $y_2 = 1$; $y_3 = 1$; $y_5 = 1$ as shown in Fig. 2.

In this way, all the possible parsing can be obtained using LPP approach. The correct parsed sentence can be obtained using Karak Vibhakti chart. Another analysis can be carried out using the following sentence:

रामNNP लेPLE खानाNN खाएरVBF मोहनNNP लाईPLAI पुस्तकNN दियोVBF

| a | b | B | c | d | A |
| Ram | food | eat | Mohan | book | gave |

English meaning is "After eating food Ram gave book to Mohan"
For Verb दिन्छ (A), the following expression is considered:

For Mandatory Demand (Rule 1)

For Optional Demand (Rule 2)

$M\{\{A, k1\}\} : x\{\{A, k1, a\}\} = 1$

$M\{\{A, k2\}\} : x\{\{A, k2, b\}\} + x\{\{A, k2, d\}\} = 1$ $O_{A,k4} : x_{A,k4,c} \leq 1$

For Verb खान्छ (B), the following expression is considered:

Table 9 Optimal basic feasible solution

C			-1	-1	-1	-1	-1	0	0	0
C	B	X_B	y_1	y_2	y_3	y_4	y_5	y_6	y_9	y_{11}
-1	y_1	1	1	1	0	0	0	0	0	0
-1	y_4	1	0	0	1	1	0	0	0	0
0	y_6	0	0	0	0	0	0	1	0	0
0	y_9	0	0	-1	1	0	0	0	1	0
-1	y_5	1	0	0	0	0	1	0	0	0
0	y_{11}	0	0	1	-1	0	0	0	0	1
	$Z^* = -3$		0	0	0	0	0	0	0	0

For Mandatory Demand (Rule 1)   For Optional Demand (Rule 2)

$M\{$ B,k2$\}$:x$\{$ B,k2,b$\}$ + x$\{$ B,k2,d$\}$ = 1 $S_a : x_{A,k1,a} = 1$

$S_b : x_{A,k2,b} + x_{B,k2,b} = 1$

$S_c : x_{A,k2,c} + x_{A,k4,c} = 1$

$S_d : x_{A,k2,d} = 1$

$y1 = x_{A,k1,a}, \; y2 = x_{A,k2,b}, \; y3 = x_{A,k2,d}, \; y4 = x_{A,k4,c}, \; y5 = x_{B,k2,b}, \; y6 = x_{B,k2,d}, \; y7 = x_{A,k2,c}$

$y1 = x_{A,k1,a} = 1, \; y2 = x_{A,k2,b} = 0, \; y3 = x_{A,k2,d} = 1, \; y4 = x_{A,k4,c} = 1,$

$y5 = x_{B,k2,b} = 1, \; y6 = x_{B,k2,d} = 0, \; y7 = x_{A,k2,c} = 0$

राम (Ram) is Karta, पुस्तक (Pustak) is Karma and मोहन (Mohan) is Sampradan with respect to the verb दियो (Diyo).

4 Conclusion

This paper exhibits the performance of two-phase simplex method over the Big M method for successful generation of *Paninian* parsing. The experiment has been carried out for ten sentences, and the flaws of the Big M method are observed in four sentences. Only six sentences are parsed correctly, whereas the other method could be successfully parsed in all attempts, even though the LPP-based techniques are computationally costly but these are powerful enough to capture context of the sentence. The experiment may be carried out for larger dataset for getting a clear picture.

Acknowledgements This work is a part of DST-CSRI, Ministry of Science & Technology, Government of India funded project on Study and Develop A Natural Language Parser for Nepali Language (vide sanction order SR/CSRI/28/2015 dated 28-Nov-2015).

References

1. Bharati, A., Caitanya, V., Sangal, R.: Natural Language Processing—A Paninian Perspective. Prentice-Hall of India, New Delhi (1995)
2. Amita, A.J.: An annotation scheme for English language using Paninian framework. IJISET **2**(1), 616–619 (2015)
3. Plank, B.: Natural Language Processing: Introduction to Syntactic Parsing. NLP+IR course, Spring (2012)
4. Bhattacharyya, P.: Natural language processing: a perspective from computation in presence of ambiguity, resource constraint and multilinguality. CSI J. Comput. **1**(2), 1–11 (2012)
5. Bal, B.K., Shrestha, P.: A Morphological Analyzer and a Stemmer for Nepali. Madan Puraskar Pustakalaya, Nepal, pp. 1–8 (2007)
6. Chhetri, I.,Dey, G., Das, S.K., Borah, S.: Development of a morph analyser for Nepali noun token. In: Proceedings of International Conference on Advances in Computer Engineering and Applications (ICACEA-2015), March 2015. IEEE, New York (2015). ISBN: 978-1-4673-6911-4

7. Borah, S., Choden, U., Lepcha, N.: Design of a morph analyzer for non-declinable adjectives of Nepali language. In: Proceedings of the 2017 International Conference on Machine Learning and Soft Computing, pp. 126–130, 13–16 Jan 2017. ACM, New York (2017). https://doi.org/10.1145/3036290.3036307, ISBN: 978-1-4503-4828-7

8. Martins, A.F.T., Smith, N.A., Xing, E.P.: Concise integer linear programming formulations for dependency parsing. In: Proceedings of the Joint Conference of the 47th Annual Meeting of the ACL and the 4th International Joint Conference on Natural Language Processing of the AFNLP: Volume 1, Aug 02–07, 2009, Suntec, Singapore (2009)

9. Perret, J., Afantenos, S., Asher, N., Morey, M.: Integer linear programming for discourse parsing. In: Proceedings of the 2016 Conference of the North American Chapter of the Association for Computational Linguistics: Human Language Technologies, pp. 99–109, San Diego, CA (2016)

10. Aparnna, T., Raji, P.G., Soman, K.P.: Integer linear programming approach to dependency parsing for Malayalam. In: Proceedings of International Conference on Recent Trends in Information, Telecommunication and Computing (2010)

MaNaDAC: An Effective Alert Correlation Method

Manaswita Saikia, Nazrul Hoque and Dhruba Kumar Bhattacharyya

Abstract This paper presents an effective alert correlation method referred to as MaNaDAC to support network intrusion detection. The method includes several modules such as feature ranking and selection, clustering and fusion to process low-level alerts and uses the concept of causality to discover relations among attacks. The method has been validated using DARPA 2000 intrusion dataset.

Keywords Alert correlation · Granger causality test · Intrusion detection system
Attack scenario · Feature ranking · Feature selection

1 Introduction

In the recent years, there have been extensive works done on Intrusion Detection Systems (IDS). A modern IDS applies a wide range of methods for misuse and anomaly detection and have the capability of detecting attacks in several environments. However, IDSs usually tend to generate a large amount of false alerts and often fail to detect new attacks or variations of known attacks. Moreover, they tend to focus on low-level attacks and raise alerts independently without considering any logical connections among these alerts. Consequently, the IDSs usually generate a large volume of alerts. In situations where there are intensive intrusion actions, not only will actual alerts be mixed with false alerts but these also become unmanageable due to the sheer volume. As a result, it is difficult for human users or Intrusion Response Systems (IRS) to understand the intrusions behind the alerts and take necessary actions. Alert correlation creates a high-level view or scenario of the attacks by understanding the

M. Saikia (✉) · N. Hoque · D. Bhattacharyya
Tezpur University, Tezpur, India
e-mail: manaswitasaikia@gmail.com

N. Hoque
e-mail: tonazrul@gmail.com

D. Bhattacharyya
e-mail: dkb@tezu.ernet.in

© Springer Nature Singapore Pte Ltd. 2019
J. Kalita et al. (eds.), *Recent Developments in Machine Learning and Data Analytics*,
Advances in Intelligent Systems and Computing 740,
https://doi.org/10.1007/978-981-13-1280-9_24

relationship among the alerts so as to distinguish between the alarms due to real attacks and the false alarms due to legitimate traffic and provide a coherent response to attack. There are many techniques proposed towards alert correlation and most of them are based on the observation that intrusions are not isolated but connected, as an attack has different stages comprising a sequence of intrusions with the early stages preparing for the later ones.

An IDS generates alerts of suspicion or unusual events. The major problems of alert analysis are processing of a large number of alerts, their redundancy and heterogeneous behaviour. For effective analysis of alerts, an alert correlation framework is needed that can handle seer volume of alerts and can generate high-level alerts from the low-level alerts during attack graph generation. The major contribution of this paper is development of a statistical alert correlation method. The method consists of a number of components and each component aims to achieve the specific goal of alert analysis. The components are designed in such a way that they can contribute the following during alert analysis. They can handle voluminous alert data over non-redundant space towards construction of appropriate attack scenario graph of multistep attack based on temporal and statistical properties among the alerts, and MaNaDAC was validated using benchmark and real-life dataset, and the performance has been found satisfactory.

2 Related Work

Alert correlation methods can be categorized [1] as similarity-based, knowledge-based, statistical, model-based [2] and other approaches. Similarity-based correlation techniques [3–8] use distance and probability functions to cluster alerts with the closest similarity values between alert attributes. These techniques have the ability to group alerts with high similarity into meta-alerts while ignoring the attack type. However, such techniques are incapable of detecting causality links among alerts and cannot provide any information regarding the alert's root cause.

Knowledge-based correlation [9–13] in a way is similar to misuse correlation as it matches the alerts with a prior knowledge and search for fixed patterns of alerts. The most obvious shortcoming in all knowledge-based algorithms [14] is that they fail to correlate alerts of previously unknown attacks. They must be updated frequently to detect new attacks as they are discovered. In case of pre- and post-condition [9] based approaches, the limitation is that there is the requirement for manually defining conditions for all alerts as operations such as thread reconstruction or session reconstruction does not take place automatically. Furthermore, it is impossible to monitor the evolution of a particular scenario from state to state in real time when only dependencies between alerts have been modelled. Scenario-based alert correlation techniques [10–13, 15] base itself on prior knowledge about attacker's behaviour through and in turn generate predefined attack scenarios. This approach automatically models alert correlation scenarios through predefined pre- and post-conditions.

The major limitation of this technique lies in the fact that it is not capable of detecting unknown attack scenarios.

Purely statistical alert correlation techniques [16] have the capability to detect new attack scenarios as there is no requirement for predefined knowledge. However, this approach tends to indicate high causality rate thus adding the requirement for specialized domain expertise. Statistical techniques are additionally more time consuming when compared to knowledge-based techniques and are highly sensitive to deliberate delays or planned noisy alerts [17].

3   Proposed Framework

The proposed MaNaDAC includes a number of modules to achieve specific goals towards attack identification. As shown in Fig. 1, it includes seven major components (similar to [16]) and the purpose of each component is discussed next.

(a) Alert Preprocessing: Different IDSs generate alerts in different formats and this step aims to convert the alerts into a unique format for better understanding as well as for interoperability. Moreover, an alert may have large number of features and all the features may not be relevant for alert analysis. So, a feature ranking technique is used to support selection of features with high relevance from a large feature space.

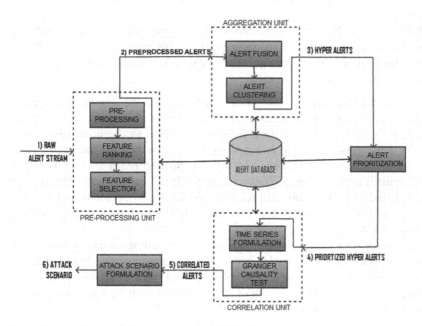

Fig. 1 Generic framework for the proposed MaNaDAC

(b) Alert Aggregation: An IDS usually generates a large number of raw alerts during network traffic analysis and often a significant portion of the raw alerts are found redundant. Analysis of redundant alerts is meaningless and time consuming. So, this component is responsible for removal of redundant alerts and combines the alerts that represent independent detection of same attack occurrence by different intrusion detection systems basing the aggregation decision on temporal difference and enclosed information.

(b.1) Fusion Process: Alert fusion has two steps, i.e. aggregation of alerts of the same IDS and aggregation of alerts generated by different IDSs. First, alerts that have the same attributes except time stamps are combined. In other words, this step intends to aggregate alerts that are output by the same IDS and are corresponding to the same attack but have a small delay. Second, the alerts from previous step that vary on time stamp are fused together, if they are close enough to fall in a predefined time window. This step aggregates alerts that have the same attributes but are reported from different heterogeneous sensors. Thus, the complete process of alert fusion drastically reduces the volume of alerts as it reduces redundancy of alerts.

(b.2) Clustering Process: Alert clustering is used to further group alerts after alert fusion. Currently, based on the results of alert fusion, alerts that have same attribute values except time stamps are further grouped into one cluster. After this step, redundancy of alerts is further reduced. Each cluster of alerts thus has a set of attribute values common among them but they occur at different time instances. Thus, a cluster of alerts can be represented as a hyper-alert [18] which in turn can be defined as follows:

Definition 1: Hyper-alert A represents alerts belonging to the same cluster as a time-ordered sequence. A series of aggregated alert instances, $a_1, a_2, \ldots, a_n$, in one cluster that have the same attributes along the time axis would be represented as a hyper-alert A such that $A = a_1, a_2, \ldots, a_n$.

(c) Alert Prioritization: The clustering process reduces the amount of raw alerts generated by grouping them into alert clusters (hyper-alerts). However, there might be a large number of such clusters. Finding correlation among each pair of such clusters is infeasible as well as unnecessary. It might be the case that some attacks are irrelevant in a context of a particular network. So as to assign priorities to hyper-alerts, prior information on the protected network is a requirement. The priority model should take into account the relevance of the alert to the configuration of the protected network and hosts as well as severity of the assessed attack while computing the priority score.

Each attack class has a set of attributes and corresponding values associated with them. A weight value is computed from the associated attributes of each alert. Accordingly, weights can be assigned to each value of an attribute. Thus, for each alert the priority score is the cumulative sum of all weight values on the attributes assigned based on the attribute value.

(d) Alert Correlation: In our method, Granger Causality Test (GCT) [16] is used as a method to correlate alerts. In [18] GCT was used to develop a temporal correlation engine which is a part of a larger correlation framework with a Bayesian network-based as well as a causal discovery theory-based correlation engine. The author's [16] GCT-based temporal correlation engine aims to discover loop dependency pattern and strong temporal pattern. The intuition is that attack steps that do not have well-known patterns or obvious relationships may nonetheless have some temporal correlations in the alert data.

(d.1) Time-Series Analysis: This step aims to (i) identify the nature of the phenomenon represented by the sequence of observations and (ii) predict future values of the time-series variable. It is essential to study the patterns of observed time-series data followed by a formal description. On establishment of this pattern, it can be interpreted as well as integrated with other techniques for extrapolation of future events. Typically, a time series is an ordered finite set of numeric values of a variable of interest along the time axis. Assuming the time interval between consecutively recorded values to be constant, a univariate time series is denoted as $x(k)$, where $k = 0, 1, 2, \ldots, (N - 1)$, and N denotes the number of elements in $x(k)$. Time-series causal analysis discovers causal relationships by analysing the correlation between time-series variable. Granger Causality Test (GCT) is a time series-based statistical analysis method that performs a statistical hypothesis test in order to determine if a time-series variable X correlates with another time-series variable Y.

(d.2) Granger Causality and Granger Causality Test (GCT): The Granger Causality Test (GCT) uses statistical functions to test if lagged information on a time-series variable x provides any statistically significant information about another time-series variable y and if the answer is yes, then it can be concluded that x Granger-causes Y [18]. The caused variable y is modelled by two autoregression models (a) Autoregressive Model (AR Model) and (b) Autoregressive Moving Average Model (ARMA Model). GCT then compares the residuals of the AR Model with the residuals of the ARMA Model. Thus, for two time-series variables y and x with size N, the Autoregressive model of y is defined as [18]:

$$y(k) = \sum_{i=1}^{p} \theta_i (k - i) + e_0(k) \tag{1}$$

And the autoregressive moving average model of y is defined as [18]

$$y(k) = \sum_{i=1}^{p} \alpha_i y(k - i) + \sum_{i=1}^{p} \beta_i x(k - i) + e_1(k) \tag{2}$$

Here, p is a particular lag length, and parameters α_i, β_i and $\theta_i (1 \leq i \leq p)$ are computed in the process of solving the Ordinary Least Square (OLS) problem. The residuals of the AR Model is

$$R_0 = \sum_{k=1}^{T} e_0^2(k) \tag{3}$$

and the residuals of the ARMA Model is

$$R_1 = \sum_{k=1}^{T} e_1^2(k) \qquad (4)$$

Here, T = N -p. The null hypothesis H_0 of GCT is $H_0 : \beta_i = 0, i = 1, 2, \dots, p$ which says that x does not affect y up to a delay of p time units [18]. g is denoted as Granger Causality Index (GCI):

$$g = \frac{(R_0 - R_1)/p}{R_1/(T - 2p - 1)} \sim F(p, T - 2p - 1) \qquad (5)$$

F(a, b) is Fishers F distribution with parameters a and b conducted to verify the validity of the null hypothesis [18]. A g value larger than a critical leads to the conclusion that null hypothesis can be rejected which in turn leads to the conclusion that x Granger-causes y. GCI (g) value represents the strength of the causal relationship. For two GCI values such that $g_1 > g_2$ where g_i, i = 1, 2 denotes the GCI for the input–output pair (x_i, y) and both have passed the F-test, then it can be concluded that it is more likely that $\{x_1(k)\}$ is causally related to y(k) than $\{x_2(k)\}$.

(d.3) Attack Scenario Formulation and Analysis: This component is responsible for providing aggregated information to the analyst in the form of a high-level view of the attack scenario. Attack scenarios are constructed so as to represent the attack strategies from the correlated alert pairs output from the correlated engine. Interpretation of a correlation relationship as an edge of a graph and formulation of complex correlations as a scenario graph makes the analysis of the situation much simpler. An attack scenario can be represented as a correlation graph. [16] defines correlation graph as a directed graph where each edge E_{ij} represents a causal relationship from alert A_i to A_j. A scenario graph consists of alerts with causal relationships. The node corresponding to the causal alert is denoted as cause node whereas the node corresponding to the effected alert is denoted as effect node. Alert A_j is considered to be caused by alert A_i only when $P_{correlation(A_i; A_j)} >$ predefined threshold, t.

The attack scenario formulation unit represents each such causal relationship by directed edges leading to formation of a graph of an attack scenario. Attack scenario graph may also have bidirectional edges that represent mutual causal relationships among alerts. [16] highlights that there are multiple types of relationships among the alerts of an attack scenario. First, there is straightforward causal relationship that is obvious because of the nature of corresponding attacks. Moreover, application of GCT correlation might further lead to the discovery of indirect and loop relationships that might exist among alerts.

4   Experimental Result

We validate our method using DARPA2000 [19]. This dataset has Distributed Denial of Service (DDoS) [20] as the most prominent attack. This attack scenario is carried out over multiple network and audit sessions grouped into five attack phases. The framework has been implemented and evaluated on each of these scenarios separately thus generating four different attack scenario graphs.

Our implementation results in 14/15 clusters with different kinds of alerts grouped into one. The results of aggregation on all four scenarios have been summarized in Table 1. It has been observed that alert fusion component of the aggregation unit has been able to reduce the volume of alerts by almost half. After alert clustering, the volume has been reduced by more than 99% of its original volume. The basic intuition behind our prioritization algorithm is that attributes as well as their corresponding values are of utmost importance in determining the priority of alerts. It has been observed that some attributes provide more significant information over others. Furthermore, DDoS occurs in multiple stages. As such, an attribute value with higher significance in one phase may have low significance in another phase. The algorithm takes into account all these aspects and assigns priority score accordingly.

(a) Alert Correlation: The correlation unit implements the Granger Causality Test (GCT) to all pairs of alerts that have been chosen as a target alert. For this purpose, a simple Granger causality function has been used which is defined as follows: grangertest (A$\sim$B, order = 1).

As in most test statistics, the level of significance is assumed to be 0.05. If the output of the P($>$F) $<$0.05 the null hypothesis is rejected and it can be concluded that B Granger-causes A. The P value signifies the probability of obtaining a sample outcome, given that the value stated in the null hypothesis is true. Figure 2 is an example of implementation of GCT on two hyper-alerts, namely, ICMP_Ping and Sadmind_Ping.

Since the P $<$ 0.05, it can be concluded that ICMP_Ping Granger-causes Sadmind_Ping. The equivalent temporal pattern between the same is shown in Fig. 3a. Similarly, we can compute the Granger causality between Sadmind_Ping and Portmap and the results are shown in Fig. 4.

Table 1 Results of aggregation unit

	LLDOS 1.0		LLDOS 2.0.2	
	DMZ	Inside	Inside	Outside
Packet captures	3,93,354	6,49,787	3,47,987	2,36,753
Elementary alerts	5598	10,397	7790	2655
Aggregated alerts	2616	5097	3793	1330
Hyper-alerts (result of clustering)	15	14	14	14

```
> grangertest(x.x~x.y,order=1)
Granger causality test

Model 1: x.x ~ Lags(x.x, 1:1) + Lags(x.y, 1:1)
Model 2: x.x ~ Lags(x.x, 1:1)
  Res.Df Df      F  Pr(>F)
1     36
2     37 -1 4.1495 0.04905 *
---
Signif. codes:  0 '***' 0.001 '**' 0.01 '*' 0.05 '.' 0.1 ' ' 1
> grangertest(x.y~x.x,order=1)
Granger causality test

Model 1: x.y ~ Lags(x.y, 1:1) + Lags(x.x, 1:1)
Model 2: x.y ~ Lags(x.y, 1:1)
  Res.Df Df      F  Pr(>F)
1     36
2     37 -1 3.8468 0.05761 .
---
Signif. codes:  0 '***' 0.001 '**' 0.01 '*' 0.05 '.' 0.1 ' ' 1
```

Fig. 2 Correlation between ICMP Ping and Sadmind Ping

Since the $P < 0.05$, it can be concluded that Sadmind_Ping Granger-causes Portmap. The equivalent temporal pattern between the same is shown in Fig. 3b. Detailed experimental results have shown that gct is capable of detecting temporal and statistical patterns that have properties such as (i) The time variation must occur in close proximity for the two alerts and (ii) The Time variation must be backed by strong statistical properties. GCT can detect similar alert patterns if they occur in close proximity of each other. In other words, if A and B occur in overlapping time windows, they can be detected. However, if A and B occur in adjacent time windows, they might not be detected. Further, GCT can detect patterns if the alert frequency is high. So, if two alerts A and B show some similar patterns but have low alert frequency, say, 20, there is a probability that they might not be correlated.

(b) Attack Scenario Generation: An attack scenario is generated from the all the correlated alerts. However, in our method, the GCT-based correlation unit is able to detect only few of these correlations, thus making it difficult for us to formulate an attack scenario.

From detailed analysis of the results of the prioritization unit, it has been observed that some pattern exist that can lead to the formulation of the attack scenario graph. In the prioritization algorithm, time was not taken into consideration. Integration of the time as the main deciding factor for priority assignment leads to detection of strong patterns resulting in attack scenario formulation. In the following section, we give an idea on the integration of time into the prioritization algorithm. In the DARPA 2000 dataset, two main observations are as follows: (i) The dataset can be roughly divided into five time windows. (ii) In each of these time windows, each protocol can be assigned a different priority based on their relevance in that phase. For example, in the first phase of DARPA dataset, ICMP has the highest priority followed by UDP and TCP, respectively. As such, priorities are assigned as ICMP = high, UDP = medium and TCP = low. The other attributes are assigned priorities as mentioned in

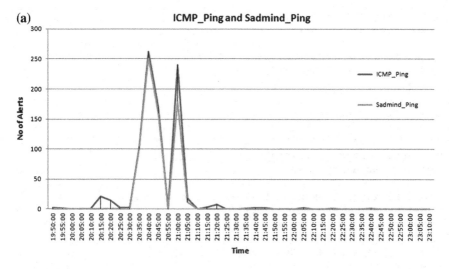

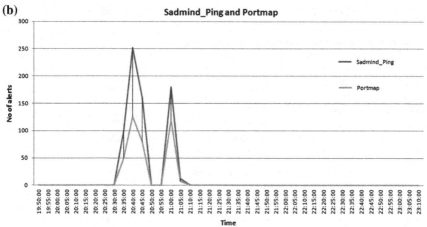

Fig. 3 **a** ICMP_Ping and Sadmind_Ping temporal pattern in LLS_DDOS_1.0-dmz. **b** Sadmind_Ping and Portmap in LLS_DDOS_1.0-dmz

the prioritization algorithm. Table 2 shows the alerts and their corresponding priority scores assigned using the modified prioritization algorithm.

The priority scores exhibit a pattern that partially indicates towards correlation among the alerts. From the priority scores in Table 2, we can conclude that (i) alerts with similar priorities are a part of a single phase of DDOS attack. (ii) Decreasing sequence of alert priorities indicates the sequence of occurrences of the attacks. However, in case of correlation among individual hyper-alerts, only assumptions can be made.

```
> grangertest(x.x~x.y,order=1)
Granger causality test

Model 1: x.x ~ Lags(x.x, 1:1) + Lags(x.y, 1:1)
Model 2: x.x ~ Lags(x.x, 1:1)
  Res.Df Df      F  Pr(>F)
1    37
2    38 -1 3.8733 0.05659 .
---
Signif. codes:  0 '***' 0.001 '**' 0.01 '*' 0.05 '.' 0.1 ' ' 1
> grangertest(x.y~x.x,order=1)
Granger causality test

Model 1: x.y ~ Lags(x.y, 1:1) + Lags(x.x, 1:1)
Model 2: x.y ~ Lags(x.y, 1:1)
  Res.Df Df      F  Pr(>F)
1    37
2    38 -1 4.2724 0.04579 *
---
Signif. codes:  0 '***' 0.001 '**' 0.01 '*' 0.05 '.' 0.1 ' ' 1
```

Fig. 4 Correlation between Sadmind_Ping and Portmap

Table 2 Priority scores assigned to alerts in DARPA2000 1.0-dmz

Alert	Priority score
ICMP Ping	15360
Sadmind Ping	1090
Portmap	420
Sadmind	420
Sadmind overflow	420
Admin	320
Rsh	15
Ftp	15
Telnet	10

5 Conclusion

The proposed MaNaDAC has been able to perform significantly well over DARPA 2000 dataset. Experimental study shows that the preprocessing and aggregation steps of the method are capable of reducing redundant features as well as the amount of alerts. Further, the clustering and the prioritization steps of the method have been successful in forming hyper-alerts and subsequently prioritizing them according to their relevance to the protected network. The Granger Causality Test (GCT) based correlation unit has been found effective in detecting only very strong temporal and statistical correlations. Furthermore, GCT is able to correlate only those patterns that occur within close proximity and ignores the fact that patterns might exist in time windows that are adjacent but further apart.

References

1. Bhuyan, M.H., Bhattacharyya, D.K., Kalita, J.K.: Network Traffic Anomaly Detection and Prevention. Springer (2017)
2. Morin, B., Mé, L., Debar, H., Ducassé, M.: M2D2: A Formal Data Model for IDS Alert Correlation. In: RAID'02 Proceedings of the 5th International Conference on Recent Advances in Intrusion Detection. Springer, pp. 115–137 (2002)
3. Debar, H., Wepsi, A.: Aggregation and correlation of Intrusion Detection Alerts. In: Proceeding RAID '00 Proceedings of the 4th International Symposium on Recent Advances in Intrusion Detection, pp. 85–103. Springer (2001)
4. Valdes, A., Skinner, K.: Probabilistic Alert Correlation. In: Proceeding RAID '00 Proceedings of the 4th International Symposium on Recent Advances in Intrusion Detection, pp. 54–68. Springer (2001)
5. Cuppens, F.: Managing Alerts in a Multi-Intrusion Detection Environment. In: Proceeding ACSAC '01 Proceedings of the 17th Annual Computer Security Applications Conference, pp. 22. IEEE Computer Society (2001)
6. Julisch, K.: Clustering Intrusion Detection Alarms to Support Root Cause Analysis. ACM Trans. Inf. Syst. Secur. (TISSEC) 6(4), 443–471 (2003). https://doi.org/10.1145/950191. 950192
7. Valeur, F., Vigna, G., Kruegel, C., Kemmerer, R.A.: A Comprehensive Approach to Intrusion Detection Alert Correlation. In: IEEE Trans. Dependable Secur. Comput. Arch. 1(3), 146 (2004). https://doi.org/10.1109/TDSC.2004.21
8. Elshoush, H.T., Osman, I.M.: Intrusion Alert Correlation Framework: An Innovative Approach. In: IAENG Transactions on Engineering Technologies, pp. 405–420. Springer (2015)
9. Ning, P., Xu, D.: Adapting Query Optimization Techniques for Efficient Intrusion Alert Correlation. In: North Carolina State University at Raleigh, Raleigh, NC (2002)
10. Cuppens, F., Ortalo, R.: Lambda: A language to model a database for detection of attacks. In: Recent Advances in Intrusion Detection, pp. 197–216. Springer (2000)
11. Cheung, L.U., Fong, S.M.: Modeling Multistep Cyber Attacks for Scenario Recognition. In: Proceedings DARPA Information Survivability Conference and Exposition. IEEE (2003)
12. Michel, C., Mé, L.: ADeLe: An Attack Description Language for Knowledge-Based Intrusion Detection. In: IFIP/Sec '01 Proceedings of the IFIP TC11 Sixteenth Annual Working Conference on Information Security: Trusted Information: The New Decade Challenge, pp. 353–368. Kluwer, B.V. Deventer (2001)
13. Eckmann, S.T., Vigna, G., Kemmerer, R.A.: Statl: An attack language for state-based intrusion detection. J. Comput. Secur. 10(1–2), 71–103 (2000)
14. Nath, B., Bhattacharyya, D.K., Ghosh, A.: Incremental association rule mining: a survey. Wiley Interdisc. Rev.: Data Min. Knowl. Discovery 3(3), 157–169 (2013). https://doi.org/10.1002/dac
15. Cuppens, F., Miége, A.: Alert Correlation in a Cooperative Intrusion Detection Framework. In: Proceeding SP '02 Proceedings of the 2002 IEEE Symposium on Security and Privacy. IEEE Computer Society, pp. 201 (2002)
16. Qin, X., Lee, W.: Statistical Causality analysis of INFOSEC data. In: Recent Advances in Intrusion Detection, pp. 73–93. Springer (2003)
17. Mustapha, Y.B.: Alert correlation towards an efficient response decision support. Ph.D. thesis, Institut National des Telecommunications (2015)

18. Qin, X.: Probabilistic-based framework for INFOSEC alert correlation. Ph.D. thesis, Georgia Institute of Technology (2005)
19. L.L. Massachusetts Institute of Technology. Darpa 2000 (2000). http://www.ll.mit.edu/ideval/data/2000data.html
20. Bhattacharyya, D.K., Kalita, J.K.: DDoS attacks: Evolution, Detection, Prevention, Reaction and Tolerance. CRC Press, Taylor and Francis Group (2016)

Fuzzy Clustering with Ensemble Classification Techniques to Improve the Customer Churn Prediction in Telecommunication Sector

J. Vijaya, E. Sivasankar and S. Gayathri

Abstract Retention of customer improves the profit growth of a marketing firm. Customer retention is the process of identifying and retaining the customers who are about to slide from one brand to another on any dissatisfaction. Customer Correlation Management (CCM) process helps the organization in this retention of customer. A company which performs the retention of customer efficiently is achieving a high market value. Nowadays, customer retention is highly required in the telecommunication systems as the sliding of customers from one provider to the other is increasing. Earlier this customer retention was made using single classification techniques which were less efficient than the hybrid models. Clustering the customers who all belong to similar characteristics into groups and then classifying each group is called the hybrid model. Further to improve the classification in our hybrid model, ensemble techniques are used, which are more efficient. In the first phase, fuzzy-based clustering methods such as Fuzzy C-Means (FCM), Possibility C-Means (PCM), and Possibility Fuzzy C-Means (PFCM) are used for clustering the customers into groups. In the second phase, those clustered groups are partitioned into training and testing data using holdout. In the third phase, the training data are given to ensemble models like bagging, boosting, and Random Subspace (RS) algorithms for building the model. The test data are predicted based on the majority voting, provided by the ensemble techniques. Through this analysis, it is found that the proposed fuzzy clustering with ensemble classification techniques provides more accuracy than single classifier and clustering with base classifier.

Keywords CCM · Churn · Clustering · FCM · PCM · FPCM
Ensemble classification · Bagging · Boosting · Random subspace

J. Vijaya (✉) · E. Sivasankar
NIT Trichy, Trichy, India
e-mail: 406114003@nitt.edu; vijayacsedept@gmail.com

E. Sivasankar
e-mail: sivasankar@nitt.edu

S. Gayathri
Temporary Faculty, NIT Trichy, Trichy, India
e-mail: sgayathri@nitt.edu

© Springer Nature Singapore Pte Ltd. 2019
J. Kalita et al. (eds.), *Recent Developments in Machine Learning and Data Analytics*,
Advances in Intelligent Systems and Computing 740,
https://doi.org/10.1007/978-981-13-1280-9_25

1 Introduction

Due to privatizations throughout the world, telecommunication industry is getting multiplied in the marketing environment. Nowadays, along with voice services telecommunication industry provide data services, online gaming, e-tickets booking, online purchasing, online banking, entertainments, educational services, and many more. Many customers are highly using these provisions that are provided by the telecommunication industries, which help the customer in many ways [1]. Along with the various services provided there are evolving of technology like a shift from 4G to 5G technology also. Because of this, the customers are searching for better service providers and technologies day by day. After identifying a better service provider the customers could easily shift to another service provider even without changing their existing mobile numbers. For example, in India due to the evolving of Jio service provider, churn rate from other service providers was increased enormously [2]. The cost of adding new customers into an organization is six times more than preserving an existing customer. Hence, in order to retain their customer churn prediction becomes a highly wanted technique for every telecommunication service provider [3]. Through the study of many literatures, it is found that machine learning algorithms such as linear regression, logistic regression, Support Vector Machine (SVM) cart, Naive Bayes (NB), K-Nearest Neighbor (KNN), Apriori, K-means, Principle Component Analysis (PCA), bagging, random forest, boosting with Ada-Boost are highly contributing in telecommunication customer churn prediction [4–7]. These machine learning techniques helps not only in the churn prediction of the telecommunication industries but also in various other industries like banking, gaming, social media, online marketing, insurances, restaurants, and so on [8–10]. In this paper, a combination of fuzzy clustering with an ensemble classification techniques-based hybrid churn prediction model is proposed. The data set provided by KDD cup 2009 a French-based telecommunication company on customer information is used for our analysis.

2 Literature Survey

Earlier the customer retention was made using single classification techniques [1] which were less efficient than the hybrid models [4, 5, 8, 11–14]. To cluster the customers who are having the similar characteristics into segments, Hudaib et al. [4] used the clustering methods like K-means, Self-Organizing Map (SOM), and hierarchical clustering algorithms. Then, the clustered segments are fed into the Artificial Neural Networks (MLP-ANN) classifier for hybrid model building and the accuracy is evaluated [4]. Out of 271 attributes in the data set Bose and Chen [5] selected 14 important attribute in which 7 is based on revenue and 7 is based on minute of usage. Then, the selected 14 attributes are fed into five known clustering techniques for segmentation. Further, the segmented clusters are input into the boosted Decision

Tree (DT) classifier for hybrid model building and top-decile lift is evaluated [5]. Rajamohamed and Manokaran [8] used the data set of credit card churn prediction from UCI repository for their hybrid model building which uses improved rough K-means algorithm for clustering and KNN, DT, SVM, NB, and ANN for classification. Hence, they proved their proposed hybrid model to be efficient when compared with a single classifier [8]. Huang and Kechadi [11] built a hybrid model on telecommunication data set which uses proposed weighted K-means algorithm for clustering and first-order inductive learning for classification [11]. Tsai and Lu [12] tried with two hybrid models in which one was the combination of SOM and back-propagation ANN and another by combining a neural with the same neural network and found that the earlier was better than the forth [12]. Two hybrid models were proposed by Hung et al. [13] in which one was a combination of K-mean with DT and another model combining DT with NN and evaluated the result using hit ratio and LIFT in both the models [13]. Vijaya and Sivasankar [14] built a hybrid model on telecommunication data set which uses proposed K-means and K-medoids algorithms for clustering and DT, SVM, NB, KNN, and LDA algorithm for classification.

3   Churn Prediction Models and Methods

Figure 1 depicts the flow diagram of the proposed hybrid model. The data set provided by KDD cup 2009 a French-based telecommunication company on customer information is preprocessed for further processing. In the first phase, fuzzy-based clustering methods such as Fuzzy C-Means (FCM), Possibility C-Means (PCM), and Possibility Fuzzy C-Means (PFCM) are used for clustering the customers into groups. In the second phase, those clustered groups are partitioned into training and testing data using holdout method. In the third phase, the training data are given to ensemble models like bagging, boosting, and random subspace algorithms for building the model. The test data are predicted based on the majority voting, provided by the ensemble techniques. The efficiency is measured using the performance metrics such as accuracy, TPR, and FPR.

3.1   Data Set

The French-based company, orange provides two different sets of data set, namely, small data set and large data set [15]. Here, we have considered the training data of the small data set for our experimentation. The attribute names of the data set are not revealed for maintaining the privacy of the customer. There are 50,000 samples and 230 features in the data set. Among the 50,000 samples, 46328 samples are non-churn samples and 3672 samples are churn samples. Among 230 attributes, 190 are numerical and 40 are string attributes.

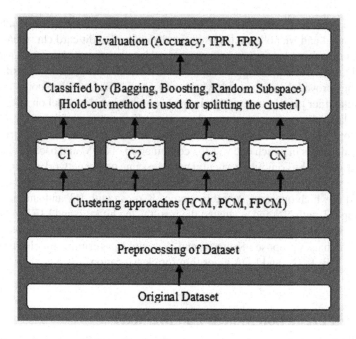

Fig. 1 Fuzzy clustering with ensemble classification customer churn prediction model

3.2 Data Set Preprocessing

In the above considered small data set there are 60% of missing values, which needs effective preprocessing is needed to obtain an effective data set for the experimentation. Preprocessing is carried out using the following procedure. If there is 25% of missing value in an attribute they are totally removed from the data set. In doing this, the data set gets reduced to 67 attributes. If there are further missing values in these 67 attributes they are filled in two different principles. That is if it a numeric value it is filled by means and if it's a string value it is filled by mode. Thus, filled string value had a higher deviation, and hence, it is removed from the data set which reduces the count to 49. Further, the string values are converted into numeric value because the clustering algorithm supports only a numerical value. The data set thus preprocessed are normalized using min-max normalization.

3.3 Clustering Techniques

Clustering is basically an unsupervised learning methodology in the wide area of data mining. Considering any data set, clustering can be used to group data with similar features into a common cluster, in which an efficient clustering method will generate clusters such that the interdependency between the clusters is very less. There are many clustering methods that exist. We have chosen three different fuzzy-based clustering techniques which are discussed below.

3.3.1 FCM

J. C. Dunn in 1973, based on fuzzy technique, propounds a clustering algorithm called fuzzy C-means algorithm [16]. There is a drawback in partition-based clustering algorithms where if an object is closer to more than one cluster formed, it has to be clustered only in one cluster which has minimum distance. To overcome this drawback in fuzzy-based clustering algorithm, even if an object is closer to more than one cluster it can be clustered into all the clusters formed to whichever it is closer. In FCM, entire data set D has N tuples is clustered into K number of clusters, where $K \leq N$ is a user-defined variable. Based on the value of K, K cluster centers are formed randomly. For every tuple in the data set, the fuzzy membership value is calculated using Eq. 1 for every cluster center, where the summation of all the fuzzy membership function should be equal to 1. Now, the clusters are formed by grouping the tuples which are having the maximum fuzzy membership function with the cluster center. Next, new cluster centers are identified for every cluster formed using Eq. 2. These steps are repeated again and again till we obtain a minimum objective function which is defined using Eq. 3.

$$M_{ij} = \frac{1}{\sum_{k=1}^{c} \left(\frac{d_{ij}}{d_{ik}}\right)^{\frac{2}{m-1}}} \tag{1}$$

$$V_j = \frac{\sum_{i=1}^{n} (M_{ij})^m . x_i}{\sum_{i=1}^{n} (M_{ij})^m} \tag{2}$$

$$J(U, V) = \sum_{i=1}^{n} \sum_{j=1}^{c} (M_{ij})^m \left\| x_i - V_j \right\|^2 \tag{3}$$

3.3.2 PCM

To overcome the outliers that are obtained during FCM, possibility C-means algorithm was propounded by R. Krishnapuram in 1996 [16]. Outliers in FCM is found because it depends highly on membership function alone, and hence, another function called typicality matrix is used. FCM also faces the problem of handling large feature data set. In PCM, entire data set D has N tuples is clustered into K number of clusters, where $K \leq N$ is a user-defined variable. Based on the value of K, K cluster centers are formed randomly. For every tuple in the data set, typicality matrix is calculated using Eq. 4 for every cluster center. Now, the clusters are formed by grouping the tuples based on the typicality matrix. Next, new cluster centers are identified for every cluster formed using Eq. 5. These steps are repeated again and again till we obtain a minimum objective function which is defined using Eq. 6.

$$T_{ij} = \frac{1}{(1 + \frac{d_{ji}}{\eta_j})^{\frac{1}{m-1}}} \tag{4}$$

$$V_j = \frac{\sum_{i=1}^{n}(T_{ij})^m . x_i}{\sum_{i=1}^{n}(T_{ij})^m} \tag{5}$$

$$J(U, V) = \sum_{j=1}^{n}\sum_{i=1}^{c}(T_{ij})^m (d_{ji})^2 + \sum_{j=1}^{n}\eta_i \sum_{i=1}^{c}(1 - T_{ij})^m \tag{6}$$

3.3.3 FPCM

Combining the membership function of FCM and typicality matrix function of PCM, Pal and Bezdek in 1997 propounded fuzzy possibility C-means algorithm [16]. For every tuple in the data set, membership matrix is calculated using Eq. 1 and typicality matrix is calculated using Eq. 4 for every cluster center. Now, the clusters are formed by grouping the tuples based on the membership matrix and typicality matrix. Next, new cluster centers are identified for every cluster formed using Eq. 7. These steps are repeated again and again till we obtain a minimum objective function which is defined using Eq. 8.

$$V_j = \frac{\sum_{k=1}^{n}(a.M_{jk}^m + b.T_{jk}^n).x_k}{\sum_{k=1}^{n}(a.M_{jk}^m + b.T_{jk}^n)} \tag{7}$$

$$J(U, V) = \sum_{j=1}^{n}\sum_{i=1}^{c} a * M_{jk}^m + b.T_{jk}^n * \|x_i - V_j\|^2 + \sum_{j=1}^{n}\eta_i \sum_{i=1}^{c}(1 - T_{ij})^m \tag{8}$$

3.4 Ensemble Classification Techniques

To enhance the weak classifiers formed using a single classification technique, ensemble classification technique is introduced. Here, the size of the ensemble and the classification technique used are the important factors. The following are the most important ensemble classification techniques.

3.4.1 Bagging

Bagging is a bootstrap aggregation algorithm propound by Breiman in 1996 [17]. Considering the training data set, N number of bags is formed using sampling with replacement mechanism. Further, each bag is fed into any classifier to build the model where each model produces separate prediction results for every test instance. The majority of the similar output produced by each model will be the final result of each test instance.

3.4.2 Boosting

Boosting is an algorithm that uses the weighted training set [17]. That is it assigned a weight commonly to all the tuples and are given to the classifier to build the model and the performance is evaluated in the model. The weight of the miss predicted tuples is enhanced, and these boosted tuples are again fed into the next classifier for building the model. The previous step is repeated continuously till the miss prediction is reduced.

3.4.3 Random Subspace

Random Subspace (RS) is a feature bagging algorithm propound by Ho in 1998 [17]. Considering the training data set, N number of subsets are formed based on the various combinations of attributes. Further, each subset is fed into any classifier to build the model where each model produces separate prediction results for every test instance. The majority of the similar output produced by each model will be the final result of each test instance.

4 Experiments and Result Analysis

4.1 Performance Measures

The performance of single classification, ensemble classification, and the proposed methods are compared using the performance metrics like accuracy, TPR, and FPR. A good churn prediction model should show high accuracy, high true positive rate, and low false positive rate. Accuracy is the measure of the ratio between accurately predicted churn and non-churn customer with entire customers in the data set. True positive rate is the accurate prediction of churn customers over the given churn customers. False positive rate is the miss prediction of non-churn customer as a churn customer.

Table 1 Performance of single classifiers [14]

Classifiers	DT	KNN	SVM	NB	LDA
Accuracy	87.61	90.91	**92.55**	91.39	89.52
TPR	93.68	98.00	100.0	99.00	97.96
FPR	12.23	02.84	00.00	01.00	00.16

Table 2 Performance of ensemble classifiers

Ensemble classifiers	Bagging	Boosting	Random subspace
Accuracy	92.33	**94.19**	93.32
TPR	97.07	97.73	97.62
FPR	04.29	07.69	06.62

4.2 Setup 1: Performance of Single Classifiers

The preprocessed data set of the experiment consists of 50,000 samples and 49 attributes with one churn attribute. This data set is input into the single classifier model and the performance is evaluated using the accuracy, TPR, and FPR. Holdout method is used to divide the preprocessed data into training part and the test part. The obtained results are tabulated in Table 1 which is already discussed in our earlier paper [14].

4.3 Setup 2: Performance of Ensemble Classifiers

The preprocessed data set of the experiment consists of 50,000 samples and 49 attributes with one churn attribute. This data set is input into the ensemble classifier model and the performance is evaluated using the accuracy, TPR, and FPR. Holdout method is used to divide the preprocessed data into training part and the test part. The obtained results are tabulated in Table 2.

4.4 Setup 3: Performance of Proposed Fuzzy Clustering with Ensemble Techniques

In the proposed hybrid model, fuzzy-based clustering methods such as Fuzzy C-Means (FCM), Possibility C-Means (PCM), and Possibility Fuzzy C-Means (PFCM) are used for grouping the customers into clusters with different C values, where C is the number of clusters to be generated. Those clusters are partitioned into training and testing data using holdout the method. Later, the training data are given to ensem-

Table 3 Performance of ensemble classifiers with FCM using various *C* values

Clustering technique	FCM								
Ensemble techniques	Bagging			Boosting			RS		
Number of clusters	C = 2	C = 4	C = 6	C = 2	C = 4	C = 6	C = 2	C = 4	C = 6
Accuracy	92.09	**93.52**	91.44	95.47	95.38	**96.50**	93.86	93.50	**94.54**
TPR	96.53	97.42	97.06	96.48	100.0	97.53	98.20	99.87	99.91
FPR	04.24	07.40	05.97	02.73	00.00	01.17	06.06	00.97	01.23

Table 4 Performance of ensemble classifiers with PCM using various *C* values

Clustering technique	PCM								
Ensemble techniques	Bagging			Boosting			RS		
Number of clusters	C = 2	C = 4	C = 6	C = 2	C = 4	C = 6	C = 2	C = 4	C = 6
Accuracy	93.14	**93.27**	91.67	96.20	**97.03**	96.68	94.91	94.43	**95.20**
TPR	97.23	97.59	96.12	96.65	97.24	96.88	95.33	94.42	95.11
FPR	07.81	06.60	05.15	13.51	06.66	06.97	16.66	05.26	02.77

ble models like bagging, boosting, and random subspace algorithms for building the model. The test data are predicted based on the majority voting, provided by the ensemble techniques. The efficiency is measured using the performance metrics such as accuracy, TPR, and FPR, and the results are tabulated in Tables 3, 4 and 5. Figure 2 depicts the accuracy comparison between the proposed fuzzy clustering with ensemble classification techniques and single ensemble classification techniques. Figure 2 shows that the Possibility Fuzzy C-Means (PFCM) hybrid with boosting produce maximum accuracy of 97.86%. Table 6 result shows the performance comparison between the proposed churn prediction model and existing techniques.

4.5 Prediction of Various Other Applications

To test the capability of our propound fuzzy clustering with ensemble classification technique, we have considered other predicting applications like bank marketing data set, credit approval data set, heart disease predicting data set, and telecommunication churn prediction data set from UCI repository for our analysis [18]. The results are discussed in Tables 7 and 8, which show that our propound model could predict the other benchmark applications efficiently. First, the collected data sets are input

Table 5 Performance of ensemble classifiers with FPCM using various *C* values

Clustering technique	FPCM								
Ensemble techniques	Bagging			Boosting			RS		
Number of clusters	C = 2	C = 4	C = 6	C = 2	C = 4	C = 6	C = 2	C = 4	C = 6
Accuracy	94.11	**94.20**	93.13	96.95	97.75	**97.86**	**96.68**	96.09	95.61
TPR	99.77	99.78	99.89	96.90	97.74	98.36	97.13	96.53	96.05
FPR	01.12	00.23	00.78	04.68	02.12	09.80	12.19	13.88	15.62

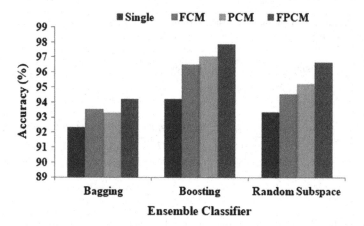

Fig. 2 Accuracy comparison between hybrid ensemble classifier and single ensemble classifier

Table 6 Accuracy comparison with existing techniques

Churn prediction models	Accuracy
Proposed clustering with ensemble (FPCM+Boosting)	**97.86**
Single classifier (SVM) [14]	92.55
Ensemble of classifier (Boosting)	94.17
Clustering with classifier (K-Means+KNN) [14]	94.72
Clustering with classifier (K-Means+BoostedC5.0) [14]	92.84

Table 7 Predicted accuracy of single classifiers versus ensemble classifiers for the collected four benchmark data sets

Data set information			Single classifiers					Ensemble classifiers		
Data sets	# Samples	#Features	DT	KNN	SVM	NB	LDA	Bagging	Boosting	RS
Bank marketing	45211	17	77.62	77.11	79.78	52.63	65.75	78.10	83.49	82.22
Credit approval	690	15	85.54	84.73	85.33	84.80	82.19	86.66	89.22	88.71
Heart disease	303	75	82.01	81.70	83.24	81.88	72.65	84.15	82.70	83.97
Telecom churn	5000	21	91.01	91.94	92.04	92.75	92.23	93.37	94.30	93.90

Table 8 Predicted accuracy of proposed fuzzy clustering with ensemble classifiers for the collected four benchmark data sets

| Data set information | | | Number of clusters C = 2 | | | | | | | | |
| | | | FCM | | | PCM | | | FPCM | | |
Data sets	# Samples	#Features	Bagging	Boosting	RS	Bagging	Boosting	RS	Bagging	Boosting	RS
Bank marketing	45211	17	82.20	83.74	83.72	83.88	83.87	86.34	87.29	**88.75**	88.69
Credit approval	690	15	92.66	92.72	93.43	92.79	93.70	92.15	93.18	94.42	**95.25**
Heart disease	303	75	85.30	85.42	84.24	87.32	87.14	87.32	**88.74**	88.59	88.44
Telecom churn	5000	21	94.46	95.32	96.11	96.79	95.53	94.65	97.53	**98.40**	97.09

into the single classifier model and ensemble classifier model then the performance is evaluated using the accuracy. Holdout method is used to divide the data into training part and the test part. The obtained results are tabulated in Table 7. Next, the collected data sets are input into the proposed hybrid model, fuzzy-based clustering methods such as Fuzzy C-Means (FCM), Possibility C-Means (PCM), and Possibility Fuzzy C-Means (PFCM) with ensemble models like bagging, boosting, and random subspace algorithms. The efficiency is measured using the performance metrics such as accuracy, and the results are tabulated in Table 8.

5 Conclusion

Nowadays, effective churn prediction for an organization has become an essential process to withstand its position in the market. In this paper, we have deployed a hybrid fuzzy clustering with an ensemble classification model for telecommunication firm. Through this experimentation, it can be concluded that (1) the result produced by the proposed hybrid fuzzy clustering with ensemble technique is providing better performance than the single classifier and the ensemble classifier. (2) Among the hybrid fuzzy clustering with ensemble technique, boosting with FPCM produces a better performance compared with the others. (3) FPCM performs better than FCM and PCM because clustering is done efficiently. This proposed hybrid model can be extended for any firm for their churn prediction.

References

1. Huang, B., Kechadi, M.T., Buckley, B.: Customer churn prediction in telecommunications. Expert Syst. Appl. **39**(1), 1414–1425 (2012)
2. Web page reference. https://gadgets.ndtv.com/telecom/opinion/reliance-jio-business-model-how-can-it-make-money-1454531
3. Vafeiadis, T., Diamantaras, K.I., Sarigiannidis, G., Chatzisavvas, K.C.: A comparison of machine learning techniques for customer churn prediction. Simul. Model. Pract. Theory **55**, 1–9 (2015)
4. Hudaib, A., Dannoun, R., Harfoushi, O., Obiedat, R., Faris, H.: Hybrid data mining models for predicting customer churn. Int. J. Commun. Netw. Syst. Sci. **8**(05), 91 (2015)
5. Bose, I., Chen, X.: Hybrid models using unsupervised clustering for prediction of customer churn. J. Organ. Comput. Electron Commer. **19**(2), 133–151 (2009)
6. Xiao, J., Xiao, Y., Huang, A., Liu, D., Wang, S.: Feature-selection-based dynamic transfer ensemble model for customer churn prediction. Knowl. Inf. Syst. **43**(1), 29–51 (2015)
7. Idris, A., Khan, A., Lee, Y.S.: Intelligent churn prediction in telecom: employing mRMR feature selection and RotBoost based ensemble classification. Appl. Intell. **39**(3), 659–672 (2013)
8. Rajamohamed, R., Manokaran, J.: Improved credit card churn prediction based on rough clustering and supervised learning techniques. Clust. Comput. 1–13 (2017)
9. Runge, J., Gao, P., Garcin, F., Faltings, B.: Churn prediction for high-value players in casual social games. In: Computational Intelligence and Games (CIG), pp. 1–8. IEEE, Aug 2014

10. Huigevoort, C., Dijkman, R.: Customer churn prediction for an insurance company (Doctoral dissertation, M. Sc. Thesis, Eindhoven University of Technology, Eindhoven, Netherland) (2015)
11. Huang, Y., Kechadi, T.: An effective hybrid learning system for telecommunication churn prediction. Expert Syst. Appl. **40**(14), 5635–5647 (2013)
12. Tsai, C.F., Lu, Y.H.: Customer churn prediction by hybrid neural networks. Expert Syst. Appl. **36**(10), 12547–12553 (2009)
13. Hung, S.Y., Yen, D.C., Wang, H.Y.: Applying data mining to telecom churn management. Expert Syst. Appl. **31**(3), 515–524 (2006)
14. Vijaya, J., Sivasankar, E. (2018). Improved Churn Prediction Based on Supervised and Unsupervised Hybrid Data Mining System. ICT4SD2016, pp. 485–499. Springer, Singapore
15. http://www.kdd.org/kdd-cup/view/kdd-cup-2009/Data
16. Grover, N.: A study of various fuzzy clustering algorithms. Int. J. Eng. Res. (IJER) **3**(3), 177–181 (2014)
17. Skurichina, M., Duin, R.P.W.: Bagging, boosting and the random subspace method for linear classifiers. Pattern Anal. Appl. **5**(2), 121–135 (2002)
18. http://archive.ics.uci.edu/ml/datasets.html

Interval-Valued Complex Fuzzy Concept Lattice and Its Granular Decomposition

Prem Kumar Singh, Ganeshsree Selvachandran and Ch. Aswani Kumar

Abstract This paper introduces a mathematical model for precise analysis of uncertainty and its fluctuation in the given interval-valued fuzzy attributes. In this regard, a method is introduced for drawing the interval-valued complex lattice and its navigation at user required complex granules with demonstration.

Keywords Complex fuzzy set · Concept lattice · Formal concept analysis
Granular computing

1 Introduction

In the current decade, a problem is addressed while dealing with periodic attributes in which the uncertainty changes. One of the suitable examples is economic data. To represent these data sets, fuzzy sets [1] is extended as complex [2], and interval-valued complex fuzzy set [3]. This paper focuses on graphical structure visualization of interval-valued complex fuzzy attributes using mathematics of Formal Concept Analysis (FCA) [4]. The concept lattice provides super and subconcept hierarchical format among the investigated concepts from the given context [5, 6]. As the demand for this mathematical tool increases, the FCA framework has also been extended

P. K. Singh
Amity Institute of Information Technology, Amity University, Sector 125, Noida 201313,
Uttar Pradesh, India
e-mail: premsingh.csjm@gmail.com

G. Selvachandran
Department of Actuarial Science and Applied Statistics, Faculty of Business and Information
Sciences, UCSI University, Jalan Menara Gadging, 56000 Cheras, Kuala Lumpur, Malaysia
e-mail: ganeshsree86@yahoo.com

Ch. Aswani Kumar (✉)
School of Information Technology and Engineering, VIT University, Vellore 632014,
Tamil Nadu, India
e-mail: cherukuri@acm.org

© Springer Nature Singapore Pte Ltd. 2019
J. Kalita et al. (eds.), *Recent Developments in Machine Learning and Data Analytics*,
Advances in Intelligent Systems and Computing 740,
https://doi.org/10.1007/978-981-13-1280-9_26

from fuzzy [7] to interval-valued fuzzy setting [8–11] as well as bipolar fuzzy setting [12, 13]. In this process, a problem arises while dealing with complex linguistics words and its graphical visualization.

Recently, the mathematical algebra of complex fuzzy set [2] in vague space [14] is studied for graphical analytics as well as paradigm of granulation [15, 16]. The interval-valued complex fuzzy set is studied for adequate analysis of partial ignorance in complex fuzzy attributes [17–20]. Motivated from these recent studies current paper focuses on processing interval-valued complex fuzzy context, and its graphical analytics using the mathematical algebra of concept lattice and granular computing. The first method provides a way to represent the periodic changes in complex linguistics, whereas the second method given multiple ways to decompose the interval-valued complex fuzzy context. The proposed methods are demonstrated for predicting the economic changes of any given country. To accomplish this task, basic preliminary is shown in Sect. 2. The steps of the proposed method are shown in Sect. 3. Section 4 provides demonstration followed by conclusions and references.

2 Preliminaries

Definition 2.1 [2]. A *complex fuzzy set* A defined on a universe of discourse U is characterized by a membership function $\mu_A(x)$ lies within the unit circle of a complex plane and are expressed by $\mu_A(x) = r_A(x)e^{i\omega_A(x)}$, where $i = \sqrt{-1}$, $r_A(x)$ and $\omega_A(x)$ are both real-valued, $r_A(x) \in [0, 1]$ and $\omega_A(x) \in (0, 2\pi]$. A complex fuzzy set A is of the form $A = \{(x, \mu_A(x)): x \in U\} = \{(x, r_A(x)e^{i\omega_A(x)}): x \in U\}$.

Definition 2.2 ([3, 21]). An *interval-valued complex fuzzy set* X' on U is a mapping $X': U \rightarrow \{a|a \in C: |a| \leq 1\} \times \{a|a \in C: |a| \leq 1\}$. For all $x \in U$, the term $\mu_X(x) = [\mu_X^-(x), \mu_X^+(x)]$ is called the complex degree of membership of an element x. Here, $\mu_X^-(x)$ and $\mu_X^+(x)$ refers to the lower and upper bounds of the membership function of x, with $\mu_X^-(x), \mu_X^+(x): U \rightarrow \{a|a \in C: |a| \leq 1\}$, where $\mu_X^-(x) = r^-e^{i\theta^-}$ and $\mu_X^+(x) = r^+e^{i\theta^+}$ such that $0 \leq r^- \leq r^+ \leq 1$ and $0 < \theta^- \leq \theta^+ \leq 2\pi$. The scaling factor $(0, 2\pi]$ is used to map the phase term to confine the performance of the membership functions within the interval $(0, 2\pi]$, and the unit circle.

Example 1 Let $U = \{x_1, x_2, x_3, x_4\}$ be a set of objects then an IV-CFS A' over U can be defined as follows:

$$A' = \left\{ \frac{[0.8, 0.9]e^{i\left[\frac{\pi}{2}, \frac{2\pi}{3}\right]}}{x_1}, \frac{[0.3, 0.4]e^{i\left[\frac{5\pi}{3}, 2\pi\right]}}{x_2}, \frac{[0, 0.05]e^{i\left[\frac{\pi}{3}, \frac{\pi}{2}\right]}}{x_3}, \frac{[0.5, 0.7]e^{i\left[\frac{11\pi}{6}, 2\pi\right]}}{x_4} \right\}.$$

Definition 2.3 ([14, 20]) (Interval-valued complex fuzzy context). It is a triplet $K = (X, Y, R)$, where X is a set of objects, Y is the set of interval-valued complex fuzzy attributes and R is the interval-valued complex fuzzy relations among them.

Definition 2.4 ([7, 8]). (Concept-forming operator). For any L-set of $A \in L^X$ objects and L-set of $B \in L^Y$ attributes, we can define an L-set $A^{\uparrow} \in L^Y$ of attributes and an L-set $B^{\downarrow} \in L^X$ of objects which are defined below:

- $A^{\uparrow}(x) = \wedge_{x \in X} \left(A(x) \to \underset{\sim}{R}(x, y) \right)$

- $B^{\downarrow}(y) = \wedge_{y \in Y} \left(B(y) \to \underset{\sim}{R}(x, y) \right)$

The $A^{\uparrow}(x)$ contains the truth degree of attribute y which is covered by all the objects from A, whereas the values of $B^{\downarrow}(y)$ contains the truth degree of object x that has all the attributes from B. In this way, the ordered pair $(A, B) \in L^X \times L^Y$ is called as a concept if $A^{\uparrow} = B$ and $B^{\downarrow} = A$.

3 Proposed Method

This section contains two methods as given below.

3.1 A Method for Discovery of Interval-Valued Complex Fuzzy Concept

Step 1. Try to represent the IV-CFSs among A and B using the Definitions 2.2.

Step 2. Represent them in form of interval-valued complex fuzzy contexts:

$R(X, Y)$

$$
= \begin{pmatrix}
 & y_1 & y_2 \\
x_1 \left[r_R^-(x_1, y_1), r_R^+(x_1, y_1) \right] e^{i \left[\theta_R^-(x_1, y_1), \theta_R^+(x_1, y_1) \right]} & \left[r_R^-(x_1, y_2), r_R^+(x_1, y_2) \right] e^{i \left[\theta_R^-(x_1, y_2), \theta_R^+(x_1, y_2) \right]} \\
x_2 \left[r_R^-(x_2, y_1), r_R^+(x_2, y_1) \right] e^{i \left[\theta_R^-(x_2, y_1), \theta_R^+(x_2, y_1) \right]} & \left[r_R^-(x_2, y_2), r_R^+(x_2, y_2) \right] e^{i \left[\theta_R^-(x_2, y_2), \theta_R^+(x_2, y_2) \right]}
\end{pmatrix}.
$$

Step 3. The covering objects (or attributes) can be discovered using the properties of Galois connection using the concept-forming operators defined in Definition 2.4.

Step 4. **Compute the union and intersection among concepts as follows:**

(i) The union of A' and B' can be defined as:

$$
\mu_{D'}(x) = \mu_{A' \cup B'}(x) = \sup(\mu_{A'}(x), \mu_{B'}(x))
$$
$$
= \left[\sup(r_{A'}^-, r_{B'}^-), \sup(r_{A'}^+, r_{B'}^+) \right] e^{i \left[\sup(\theta_{A'}^-, \theta_{B'}^-), \sup(\theta_{A'}^+, \theta_{B'}^+) \right]}.
$$

Fig. 1 Interval-valued
complex fuzzy graph

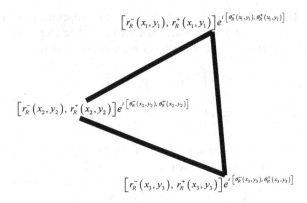

(ii)   The intersection of A' and B' can be defined as:

$$\mu_{E'}(x) = \mu_{A' \cap B'}(x) = \inf(\mu_{A'}(x), \mu_{B'}(x))$$
$$= \left[\inf\left(r_{A'}^-, r_{B'}^-\right), \inf\left(r_{A'}^+, r_{B'}^+\right)\right]e^{i\left[\inf\left(\theta_{A'}^-, \theta_{B'}^-\right), \inf\left(\theta_{A'}^+, \theta_{B'}^+\right)\right]}$$

Step 5.   Visualize the concepts using interval-valued complex fuzzy graph as shown
          in Fig. 1.

Complexity: This method finds interval-valued complex concepts by user
required subsets. It may take $O(2^m)$ time complexity. Same time, connecting these
attributes with objects and attribute set for amplitude and phase term may take
$O(2^m.n^2)$ complexity maximally.

3.2   A Method for Granular Decomposition of Interval-Valued Complex Context

Step 1.   Write the interval-valued complex fuzzy context.
Step 2.   Define complex information granule $\alpha_R(U, V)$ of the form:

$$\alpha_R(U, V) = \left[r_\alpha^-(x, y), r_\alpha^+(x, y)\right]e^{i\left[\theta_\alpha^-(x, y), \theta_\alpha^+(x, y)\right]}.$$

Step 3.   The entries in the interval-valued complex fuzzy matrix can be represented
          as 1 if and only if $\alpha_R(X, Y) \geq R(X, Y)$, otherwise 0.
Step 4.   Determine values for the entries of the matrix as shown below:

$$\left[r_\alpha^-(x, y), r_\alpha^+(x, y)\right]e^{i\left[\theta_\alpha^-(x, y), \theta_\alpha^+(x, y)\right]} \geq \left[r_R^-(x, y), r_R^+(x, y)\right]e^{i\left[\theta_R^-(x, y), \theta_R^+(x, y)\right]}$$

Step 5.   Decide an appropriate granulation level for knowledge processing tasks.

Table 1 An interval-valued complex fuzzy context

$R(a, b)$	y_1	y_2	y_3	y_4
x_1	$[0.4, 0.6]e^{i\left[\frac{5\pi}{3}, 2\pi\right]}$	$[0, 0.1]e^{i\left[\frac{5\pi}{3}, 2\pi\right]}$	$[0.8, 1]e^{i\left[\frac{\pi}{2}, \pi\right]}$	$[0.8, 0.9]e^{i\left[\frac{3\pi}{2}, 2\pi\right]}$
x_2	$[0.2, 0.3]e^{i\left[\frac{5\pi}{3}, 2\pi\right]}$	$[0.6, 0.8]e^{i\left[\frac{2\pi}{3}, \pi\right]}$	$[0.9, 1]e^{i\left[\frac{2\pi}{3}, \pi\right]}$	$[0.6, 0.9]e^{i\left[\frac{3\pi}{2}, 2\pi\right]}$
x_3	$[0.05, 0.15]e^{i\left[\frac{\pi}{2}, \pi\right]}$	$[0, 0.2]e^{i[\pi, 2\pi]}$	$[0.7, 1]e^{i\left[\frac{2\pi}{3}, \pi\right]}$	$[0.8, 1]e^{i\left[\frac{3\pi}{2}, 2\pi\right]}$

Complexity: This method decomposes the interval-valued complex context at user required complex granules for each entries, i.e. $O(m.n)$ of the matrix. In this case, for the amplitude, and phase term it will take overall $O(m^2.n^2)$ complexity.

4 Illustration

The proposed method is demonstrated in this section with their comparative study.

4.1 Illustrative Example for Interval-Valued Complex Fuzzy Concept Lattice

The interval-valued complex fuzzy attributes data set can be found in mostly in stock market or economic data sets as given below:

Example 2 Let an economist want to analyse the effect of government policies, i.e. x_1 = Reform in tax, x_2 = signing with different trade agreements, x_3 = Reforms in the given subsidy on the following parameters y_1 = exchange rate of economy, y_2 = Prices of commodities, y_3 = Exports and y_4 = GDP growth for a given year. The degree of association between all of these parameters and its dependency can be written in a context format as shown in Table 1 [21]. Consider the entry $(R)(x_1, y_1) = [0.4, 0.6]e^{i\left[\frac{5\pi}{3}, 2\pi\right]}$ for the interpretation. It shows that, reform in tax has a 40–60% effect on the exchange rate of economy in 1 year which is the result of normalizing the time lag by 2π (1 year). Subsequently, other complex relation can be interpreted on Table 1. The proposed method shown in Sect. 3.1 provides following complex concepts from the context shown in this table:
1.

$$
\emptyset^{\uparrow} = \left\{ \frac{[0.4, 0.6]e^{i\left[\frac{5\pi}{3}, \frac{\pi}{1}\right]}}{y_1} + \frac{[0.6, 0.5]e^{i\left[\frac{5\pi}{3}, 2\pi\right]}}{y_2} \right.
$$

$$
\left. + \frac{[0.9, 0.1]e^{i\left[\frac{2\pi}{3}, \frac{\pi}{1}\right]}}{y_3} + \frac{[0.8, 0.9]e^{i\left[\frac{3\pi}{2}, 2\pi\right]}}{y_4} \right\}
$$

Discover its covering objects set using down arrow as follows:

$$\left\{ \frac{[0.4,0.6]e^{i\left[\frac{5\pi}{3},\frac{\pi}{1}\right]}}{y_1} + \frac{[0.6,0.5]e^{i\left[\frac{5\pi}{3},2\pi\right]}}{y_2} + \frac{[0.9,0.1]e^{i\left[\frac{2\pi}{3},\frac{\pi}{1}\right]}}{y_3} + \frac{[0.8,0.9]e^{i\left[\frac{3\pi}{2},2\pi\right]}}{y_4} \right\}^{\downarrow}$$

$$= \left\{ \frac{[0,0.1]e^{i\left[\frac{\pi}{2},\pi\right]}}{x_1} + \frac{[0.2,0.3]e^{i\left[\frac{2\pi}{3},\pi\right]}}{x_2} + \frac{[0,0.15]e^{i\left[\frac{\pi}{2},\pi\right]}}{x_3} \right\}$$

Hence, it provides the following extent–intent pair:

Extent: $\left\{ \frac{[0,0.1]e^{i\left[\frac{\pi}{2},\pi\right]}}{x_1} + \frac{[0.2,0.3]e^{i\left[\frac{2\pi}{3},\pi\right]}}{x_2} + \frac{[0,0.15]e^{i\left[\frac{\pi}{2},\pi\right]}}{x_3} \right\}$

Intent: $\left\{ \frac{[0.4,0.6]e^{i\left[\frac{5\pi}{3},\frac{\pi}{1}\right]}}{y_1} + \frac{[0.6,0.5]e^{i\left[\frac{5\pi}{3},2\pi\right]}}{y_2} + \frac{[0.9,0.1]e^{i\left[\frac{2\pi}{3},\frac{\pi}{1}\right]}}{y_3} + \frac{[0.8,0.9]e^{i\left[\frac{3\pi}{2},2\pi\right]}}{y_4} \right\}$

It shows export (x_2) is closed with signing different trade agreements (y_3) in 3–6 months.

Similarly, the following concepts can be generated as given below:

2. Extent: $\left\{ \frac{[1.0,1.0]e^{i[2\pi,2\pi]}}{x_1} + \frac{[0.2,0.3]e^{i\left[\frac{2\pi}{3},\pi\right]}}{x_2} + \frac{[0.05,0.15]e^{i\left[\frac{\pi}{2},\pi\right]}}{x_3} \right\}$

Intent: $\left\{ \frac{[0.4,0.6]e^{i\left[\frac{5\pi}{3},2\pi\right]}}{y_1} + \frac{[0,0.1]e^{i\left[\frac{5\pi}{3},2\pi\right]}}{y_2} + \frac{[0.8,1.0]e^{i\left[\frac{\pi}{2},\pi\right]}}{y_3} + \frac{[0.8,0.9]e^{i\left[\frac{3\pi}{2},2\pi\right]}}{y_4} \right\}$

It shows export (y_3) and GDP growth (y_4) is closely related with tax reforms (x_1) in a given year.

3. Extent: $\left\{ \frac{[0,0.1]e^{i\left[\frac{5\pi}{3},\pi\right]}}{x_1} + \frac{[1.0,1.0]e^{i[2\pi,2\pi]}}{x_2} + \frac{[0.05,0.15]e^{i\left[\frac{\pi}{2},\pi\right]}}{x_3} \right\}$

Intent: $\left\{ \frac{[0.2,0.3]e^{i\left[\frac{5\pi}{3},2\pi\right]}}{y_1} + \frac{[0.6,0.8]e^{i\left[\frac{2\pi}{3},\pi\right]}}{y_2} + \frac{[0.9,1.0]e^{i\left[\frac{2\pi}{3},\pi\right]}}{y_3} + \frac{[0.6,0.9]e^{i\left[\frac{3\pi}{2},2\pi\right]}}{y_4} \right\}$

It shows export (y_3) is highly related with signing different trade agreements (x_2) in the given year.

4. Extent: $\left\{ \frac{[0,0.1]e^{i\left[\frac{\pi}{2},\pi\right]}}{x_1} + \frac{[0.2,0.3]e^{i\left[\frac{5\pi}{3},\pi\right]}}{x_2} + \frac{[1.0,1.0]e^{i[2\pi,2\pi]}}{x_3} \right\}$

Intent: $\left\{ \frac{[0.05,0.15]e^{i\left[\frac{\pi}{2},\pi\right]}}{y_1} + \frac{[0,0.2]e^{i[\pi,2\pi]}}{y_2} + \frac{[0.7,0.1]e^{i\left[\frac{2\pi}{3},\pi\right]}}{y_3} + \frac{[0.8,1.0]e^{i\left[\frac{3\pi}{2},2\pi\right]}}{y_4} \right\}$

It shows GDP growth (y_4) is highly related with subsidiary reforms (x_3) in the given year.

5. Extent: $\left\{ \frac{[1.0,1.0]e^{i[2\pi,2\pi]}}{x_1} + \frac{[1.0,1.0]e^{i[2\pi,2\pi]}}{x_2} + \frac{[0.5,0.15]e^{i\left[\frac{\pi}{2},\pi\right]}}{x_3} \right\}$

Intent: $\left\{ \frac{[0.2,0.3]e^{i\left[\frac{5\pi}{3},2\pi\right]}}{y_1} + \frac{[0,0.1]e^{i\left[\frac{2\pi}{3},\pi\right]}}{y_2} + \frac{[0.8,1.0]e^{i\left[\frac{\pi}{2},\pi\right]}}{y_3} + \frac{[0.6,0.9]e^{i\left[\frac{3\pi}{2},2\pi\right]}}{y_4} \right\}$

6. Extent: $\left\{ \frac{[1.0,1.0]e^{i[2\pi,2\pi]}}{x_1} + \frac{[0.2,0.3]e^{i\left[\frac{2\pi}{3},\pi\right]}}{x_2} + \frac{[1.0,1.0]e^{i[2\pi,2\pi]}}{x_3} \right\}$

Intent: $\left\{ \frac{[0.05,0.15]e^{i\left[\frac{\pi}{2},\pi\right]}}{y_1} + \frac{[0,0.1]e^{i[\pi,2\pi]}}{y_2} + \frac{[0.7,1.0]e^{i\left[\frac{\pi}{2},\pi\right]}}{y_3} + \frac{[0.8,0.9]e^{i\left[\frac{3\pi}{2},2\pi\right]}}{y_4} \right\}$

Fig. 2 Interval-valued
complex lattice generated
from Table 1

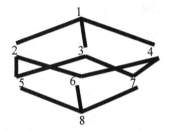

7. Extent: $\left\{ \dfrac{[0,0.1]e^{i\left[\frac{\pi}{2},\pi\right]}}{x_1} + \dfrac{[1.0,1.0]e^{i[2\pi,2\pi]}}{x_2} + \dfrac{[1.0,1.0]e^{i[2\pi,2\pi]}}{x_3} \right\}$

   Intent: $\left\{ \dfrac{[0.05,0.15]e^{i\left[\frac{\pi}{2},\pi\right]}}{y_1} + \dfrac{[0,0.2]e^{i\left[\frac{2\pi}{3},\pi\right]}}{y_2} + \dfrac{[0.7,1.0]e^{i\left[\frac{2\pi}{3},\pi\right]}}{y_3} + \dfrac{[0.6,0.9]e^{i\left[\frac{3\pi}{2},2\pi\right]}}{y_4} \right\}$

8. Extent: $\left\{ \dfrac{[1.0,1.0]e^{i\,[2\pi,2\pi]}}{x_1} + \dfrac{[1.0,1.0]e^{i[2\pi,2\pi]}}{x_2} + \dfrac{[1.0,1.0]e^{i[2\pi,2\pi]}}{x_3} \right\}$

   Intent: $\left\{ \dfrac{[0.05,0.15]e^{i\left[\frac{\pi}{2},\pi\right]}}{y_1} + \dfrac{[0,0.1]e^{i\left[\frac{2\pi}{3},\pi\right]}}{y_2} + \dfrac{[0.7,1.0]e^{i\left[\frac{\pi}{2},\pi\right]}}{y_3} + \dfrac{[0.6,0.9]e^{i\left[\frac{3\pi}{2},2\pi\right]}}{y_4} \right\}$

It shows each of the reforms done by the government highly affects the volume
of export (y_3) in a given year and its GDP growth (y_4).

The interval-valued complex fuzzy lattice shown in Fig. 2 represents that the
policy of government highly affects the volume of export (y_3) and its GDP growth
(y_4) in the given year. In this case, the policy should be precise to increase the growth
of country.

4.2 Granular-Based Decomposition of Interval-Valued Complex Fuzzy Context

Recently, decomposition of interval-context [19–21] is considered as one of the major
issues for researchers to zoom in and zoom out the knowledge. To resolve this issue,
Sect. 3.2 introduced a new mathematical paradigm for refining the contexts with the
condition that given complex-valued relation is accepted if its amplitude and phase
terms are equal to or exceed the threshold value defined by the user of α (indicating
an acceptance of more than $\alpha\%$). Table 2 depicts a context for the complex granules
$[0.7, 1.0]e^{i[2\pi,2\pi]}$ on Table 1. It shows that attribute y_3 and y_4 has maximum effect
on each of the objects at chosen granulation. These results correspond to the results
obtained via the concept lattice shown in Fig. 2. The obtained knowledge echo with
each other. Hence, the applicability of both methods will be decided by the user to
solve the particular problem of given contexts.

Table 2 Decomposed contexts for Table 1 at a granulation of $[0.7, 1.0]e^{i[2\pi, 2\pi]}$

$R(a, b)$	y_1	y_2	y_3	y_4
x_1	0	0	1	1
x_2	0	0	1	1
x_3	0	0	1	1

5 Conclusions

This paper aimed at the descriptive analytics of interval-valued complex contexts using the paradigm of applied abstract algebra and granular computing within $O(2^m.n^2)$ and $O(m^2.n^2)$ time complexity, respectively. It is shown that the knowledge extracted from both of the methods are corroborating with each other. In the near future, analysis will be focused on reducing the time complexity for processing the complex contexts.

References

1. Zadeh, L.A.: Fuzzy sets. Inf. Control **8**, 338–353 (1965)
2. Ramot, D., Milo, R., Friedman, M., Kandel, A.: Complex fuzzy sets. IEEE Trans. Fuzzy Syst. **10**(2), 171–186 (2002)
3. Greenfield, S., Chiclana, F., Dick, S.: Interval-valued complex fuzzy logic. In: IEEE International Conference on Fuzzy Systems (FUZZ-IEEE), pp. 1–6 (2016)
4. Wille, R.: Restructuring lattice theory: an approach based on hierarchies of concepts. In: Rival, I. (ed.) Ordered Sets, pp. 445–470. Reidel, Dordrecht, Boston, MA (1982)
5. Aswani Kumar, C., Srinivas, S.: Concept lattice reduction from fuzzy K-means clustering. Expert Syst. Appl. **37**(3), 2696–2704 (2010)
6. Aswani Kumar, C., Singh, P.K.: Knowledge representation using formal concept analysis: a study on concept generation. In: Global Trends in Knowledge Representation and Computational Intelligence, pp. 306–336. IGI Global Publishers (2014)
7. Burusco, A., Fuentes-Gonzales, R.: The study on interval–valued contexts. Fuzzy Sets Syst. **121**(3), 439–452 (2001)
8. Djouadi, Y., Prade, H.: Interval-valued fuzzy formal concept analysis. Lect. Notes Comput. Sci. **5722**, 592–601 (2010)
9. Singh, P.K., Aswani Kumar, C.: Interval-valued fuzzy graph representation of concept lattice. In: Proceedings of the 12th ISDA, Kochi, India, pp. 604–609 (2012)
10. Zerarga, L., Djouadi, Y.: Interval-valued fuzzy extension of formal concept analysis for information retrieval. In: Huang, T., et al. (ed.) ICONIP 2012, Part 1, LNCS, vol. 7663, pp. 608–615. Springer–Verlag (2012)
11. Zhai, Y., Li, D., Qu, D.: Probability fuzzy attribute implications for interval–valued fuzzy set. Int. J. Database Theor. Appl. **5**, 95–108 (2012)
12. Singh, P.K., Aswani Kumar, C., Li, J.: Knowledge representation using interval-valued fuzzy formal concept lattice. Soft Comput. **20**(4), 1485–1502 (2016)
13. Singh, P.K., Aswani Kumar, C.: Bipolar fuzzy graph representation of concept lattice. Inf. Sci. **288**, 437–448 (2014)
14. Singh, P.K.: Complex vague set based concept lattice. Chaos Solitons Fractals **96**, 145–153 (2017)

15. Zhang, G., Dillon, T.S., Cai, K.-Y., Ma, J., Lu, J.: Operation properties and δ-equalities of complex fuzzy sets. Int. J. Approximate Reasoning **50**, 1227–1249 (2009)
16. Yazdanbakhsh, O., Dick, S.: A systematic review of complex fuzzy sets and logic. Fuzzy Sets Syst. **338**, 1–22 (2018)
17. Kumar, T., Bajaj, R.K.: On complex intuitionistic fuzzy soft sets with distance measures and entropies. J. Math. **2014**, 1–12 (2014)
18. Selvachandran, G., Maji, P.K., Abed, I.E., Salleh, A.R.: Relations between complex vague soft sets. Appl. Soft Comput. **47**, 438–448 (2016)
19. Singh, P.K.: Complex neutrosophic concept lattice and its application for air quality analysis. Chaos Solitons Fractals **109**, 206–213 (2018)
20. Singh, P.K.: Interval-valued neutrosophic graph representation of concept lattice and its(α, β, γ)-decomposition. Arab. J. Sci. Eng. **43**(2), 723–740 (2018). (Springer)
21. Selvachandran, G., Singh, P.K.: Interval-valued complex fuzzy soft set and its application. Int. J. Uncertainty Quantification **8**(2), 101–117 (2018)

SbFP-Growth: A Step to Remove the Bottleneck of FP-Tree

Shafiul Alom Ahmed, Bhabesh Nath and Abhijeet Talukdar

Abstract Many association rule mining techniques have been proposed and most of them are greatly dependent on physical memory. Therefore, the limited amount of physical memory becomes a bottleneck of the existing techniques for large-scale datasets. In this paper, we propose a new approach, (Secondary Storage-Based Frequent Pattern) SbFP-Growth, which is a modification to the FP-Growth algorithm. SbFP-Growth uses secondary storage to store the tree, and therefore, it overcomes the main memory bottleneck at FP-Growth algorithm. By this way, we are able to mine for frequent itemsets from databases of arbitrary size large datasets. Moreover, SbFP-Growth constructs complete tree (SbFP-Tree), i.e., a tree with minimum support count equal to one. It provides the freedom of mining rules for different lower minimum support values without reconstructing the tree from scratch. We can store the SbFP-Tree in secondary storage so that it can be mined later with any minimum support value.

Keywords Association rule · Frequent pattern · FP-Tree · Rule mining
Data mining

1 Introduction

Association rule mining was identified in 1993 [3], to find the informative relationships among attributes/attribute values of the database. The association mining can be explained as follows: Let I = i1, i2, ..., in be a set of items. Let D be a set of transactions, where each transaction T is a set of items, such that $T \subseteq I$. An asso-

S. A. Ahmed (✉) · B. Nath · A. Talukdar
Tezpur University, Tezpur, Assam, India
e-mail: tezu.shafiul@gmail.com

B. Nath
e-mail: bnath@tezu.ernet.in

A. Talukdar
e-mail: talukdar.abhijeet@gmail.com

© Springer Nature Singapore Pte Ltd. 2019
J. Kalita et al. (eds.), *Recent Developments in Machine Learning and Data Analytics*,
Advances in Intelligent Systems and Computing 740,
https://doi.org/10.1007/978-981-13-1280-9_27

ciation rule is an implication of the form $A \Rightarrow B$, where $A \subset I$, $B \subset I$ and $A \cap B$ $= \Phi$. The support of an itemset X in D, denoted sup(X), is the fraction of the total number of transactions in D that contain X. An itemset X is said to be frequent in D if $sup(X) \geq$ minsup (user-defined). The rule $X \Rightarrow Y$ holds in D with confidence conf($X \Rightarrow Y$), if conf($X \Rightarrow Y$)% of the transactions in D that contain X also contain Y. The rules hold which satisfy both minsup and minconf constraints.

Most of the rule mining approaches use some kind of main memory resident data structures to compute the set of frequent itemsets from transactional databases. The limited physical memory issue is studied by Goethals [6], who tested many association rule mining methods on some real-world datasets. Therefore, it is required to find a solution to this problem by either increasing the physical memory to a large extent or by using the secondary storage to store the data structure. As extending the physical memory is not a feasible solution currently, we are going to develop an approach which uses secondary storage (disk).

Here, we present a secondary storage (disk) based algorithm (Secondary Storage-Based Frequent Pattern) SbFP-Growth, which eliminates the drawbacks of FP-Growth [11] algorithm. The algorithm uses a data structure SbFP-Tree which is stored in the secondary storage (disk). The various advantages of this algorithm will be discussed in the latter part.

2 Related Works

Apriori [3] algorithm proposed by R. Agrawal and R. Srikant has received great attention from researchers in the data mining arena. When the support threshold for mining is low and the size of the maximal frequent itemset is large, this approach is not efficient with respect to both space and time because of the sheer large number of candidate itemsets and the multiple database scans. There are many other techniques which are basically Apriori-like or level-wise in nature such as DIC [4], Pincer Search [18], partition algorithm [25], GenMax [7], CLOSET [15, 21, 22, 28, 30], etc., and hence suffer from the same kind of problems. To cope up with this challenge of rule mining, researchers used various data structures which can be realized in [23].

FP-Growth [11] algorithm uses a special kind of prefix-tree called as Frequent Pattern Tree (FP-Tree). This method requires two database scans when mining all frequent itemsets. The FP-Growth compresses the entire database into a smaller data structure (FP-Tree), resulting in the need to only scan the database twice. It avoids the generation of massive numbers of candidate itemsets and generates the frequent itemsets using conditional pattern tree, and hence reduces the search space.

Though the FP-Growth works better compared to other algorithms, it has some drawbacks also. The time taken to construct the FP-Tree is quite large, particularly when the dimensionality is large. With the increase in support count, its performance degrades and at certain instant of time it becomes almost similar to Apriori and also the FP-Tree may not fit into the main memory if the database is large and scattered.

There are many variants of FP-Tree Growth such as Nonordf [24], CFP-tree [26], Improved FP-Tree Growth [11], CT-PRO [27], FP-Growth* [8], and many significant works can be found in [10, 12] also. There exist many algorithms in literature such as PRE-FUFP [17], FUFP [13], CP-tree [9], CanTree [16], IFP-Tree [2], MFS [14, 19], B+ tree, DRFP-Growth [1], prefix-path tree (PP-Tree) [29], etc., tried to mitigate this main memory limitation so that it becomes feasible to mine frequent itemsets in incremental scenario. In this report, we propose an efficient approach, SbFP-Growth, to overcome the limitations of FP-growth.

3   Proposed Algorithm

The proposed secondary storage-based **SbFP-Growth** algorithm is used to efficiently mine frequent patterns from very large databases without being restricted by limited physical memory. The proposed algorithm can be described in two phases, (Secondary Storage-Based Frequent Pattern) SbFP-Tree construction and the pattern mining is performed by invoking SbFP-Growth that utilizes SbFP-Tree.

This algorithm builds a complete FP-Tree with minimum support count of 1. This algorithm uses two structured files for storing the FP-Tree to the disk, one is *NodeTable* and *ChildTable*. NodeTable contains data about the nodes and ChildTable contains the child node information of nodes. The algorithm also uses an AVL-Tree and a linked list, AVL-Tree keeps track of the nodes in physical memory, and liked list keeps the information of the nodes to be unloaded. Every time a node is created it is given a unique node number, which helps us to identify each node. We used AVL-tree as its time complexity is $O(log(n))$ for all insert, delete, and search (for both average case and worst case). This algorithm is explained in the following example.

3.1   Build the SbFP-Tree

In this section, we will illustrate the algorithm by considering the dataset D as an example (Table 1) and the maximum no. of node that can be stored in physical memory is 6.

- Dataset D is scanned to collect the frequency of each item and they are sorted in a Header Table according to their frequency counts. Then, the algorithm constructs the SbFP-Tree and generates the conditional pattern tree to mine frequent itemsets.
- The first transaction TID = 1 is then read and sorted according to the order of items in header table. Nodes in the Fig. 1 are labeled in format **name: node no** (Table 2).
  When the root is created for the FP-Tree, a node with key value 1 is created in the AVL-tree as the node number of root is 1. Similarly, nodes with key values 2 and 3 are created in the AVL-tree when node "a" and "b" are created in the FP-Tree.

Table 1 D:Transaction
dataset

TID	ITEMS
1	{a,b}
2	{b,c,d}
3	{a,c,d,e}
4	{a,d,e}
5	{a,b,c}
6	{a,b,c,d}
7	{a}
8	{a,b,c}
9	{a,b,d}
10	{b,c,e}

Fig. 1 After TID = 1

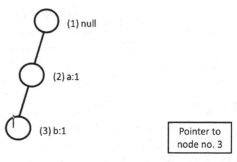

(a) Tree after TID=1   (b) Linked list after TID=1

Table 2 *NodeTable* after
TID = 1

Line no.	Node data
1	Data about node no 1
2	Data about node no 2
3	Data about node no 3

At the last when item "b" is read from the transaction, a node is created for the linked list (Fig. 1b) which points to node "b" with node number 3.

- Now after reading TID = 2, we have the tree as shown in the Fig. 2.
- After reading TID = 3, we have.

  – Item "a" has no child named as "c" therefore we are to create a new node named "c" , but the limit to maximum number of node in physical memory is reached (here 6 nodes). Therefore, we are to free some memory space by unloading some node from memory space (Fig. 2).
  – Before deleting the first node with node number 3 of the linked list, we update its entry in ChildTable file (which contains the child information of the nodes); only if it is a non-leaf node, after that we update its entry in the NodeTable file.

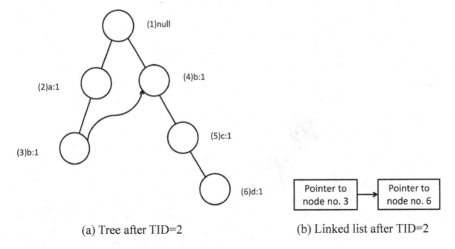

(a) Tree after TID=2

(b) Linked list after TID=2

Fig. 2 After TID = 2

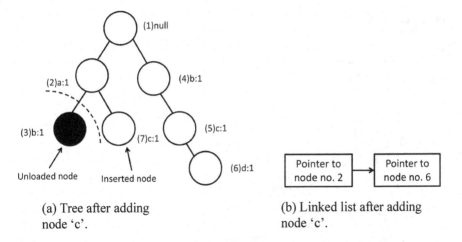

(a) Tree after adding node 'c'.

(b) Linked list after adding node 'c'.

Fig. 3 After handling node "c"

The linked list now points to the parent of the node 3, i.e., 2 (Fig. 3b). The tree after adding node "c" is shown in Fig. 3a.

- Similarly, the node "d" and node "e" are added. The tree after TID = 3 is shown in Fig. 4a.

• Similarly, the transaction TID = 4 is read and items are added to the tree. Tree and linked list after TID = 4 are shown in Fig. 5a and Fig. 5b, respectively.

• Similarly as before, all the transactions in the table.

• The algorithm to build the SbFP-Tree is shown in Algorithm 1.

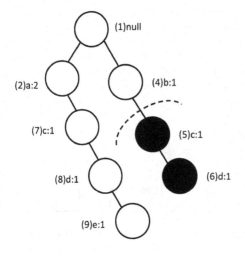

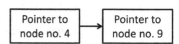

(a) Tree after TID=3.

(b) Linked list after TID=3.

Fig. 4 After TID = 3

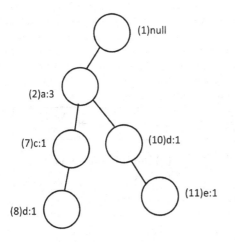

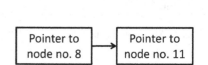

(a) Tree after TID=4.

(b) Linked list after TID=4.

Fig. 5 After TID = 4

3.2   Mining the SbFP-Tree

SbFP-Tree (Algorithm 1) is a complete tree, i.e., with minimum support count equal to one. During the mining process, conditional SbFP-Trees are build from the SbFP-Tree. The conditional SbFP-Trees may also be large enough to not fit in the main memory. Therefore, we are to build the conditional SbFP-Trees in the same way as the SbFP-Tree was built.

Algorithm 1: Build the SbFP-Tree

---

    input : Transactional database(**D**).
    output: SbFP-Tree.

1 **Procedure** *Build_SbFP (D)*
2   Create the file NodeTable and ChildTable, to store the node data and child nodes information ;
3   Create AVL-tree and linked list, which keeps track of the nodes in physical memory and nodes to be removed from physical memory, respectively ;
4   Scan **D** to collect the frequency of each item that appears in the database;
5   Create the root (null) node of the tree;
6   **while** *EOF(D)* **do**
7   *Transaction* ← *Read* transactions in D;
8   Sort Transaction according to frequencies of each node;
9   **Insert_Transaction**(Transaction, root, No. of items in Transaction array, 0(zero));
10   The remaining nodes in the physical memory are written into the files NodeTable and ChildTable, as to completely store the FP Tree into the secondary storage. Thus, the SbFP-Tree is build;
11   **return**;

The approach taken by this algorithm (Algorithm 2) is somewhat similar to the mining process in the FP-Growth; the input to this algorithm is the disk-based SbFP-Tree and minimum support count, and the output will be the complete set of frequent patterns. We first find the conditional pattern bases and then construct the conditional SbFP-Tree. The conditional SbFP-Trees are build the same way the SbFP-Tree was built. SbFP-Growth is recursive in nature, and it terminates under two conditions: first when the conditional SbFP-Tree has only single path and second when SbFP-Tree is null. The conditional SbFP-Trees can be deleted from the secondary storage, when it is no longer needed in the mining process.

4   Experimental Results

In this section, we analyze the performance of the FP-Growth and SbFP-Growth algorithms on different datasets. The datasets are downloaded from the website UCI Machine Learning Repository and are analyzed using market basket analysis.

Algorithm 2: Build the SbFP-Tree (continued)

1: **Procedure Insert_Transaction**(*Transaction*, root, int k, int index);

2: /*'k'is the number of items in the transaction*/

3: /* 'index'is the index of the item in *Transaction* array, to be added to the tree */

4: **if** k==index **then**

5:     **return** ;

6: **end if**

7: Item ← *Transaction*[index];

8: **for** all child nodes of root **do**

9:     **if** *ChildNode.name*== *Item* **then**

10:         Found=1;/*such a node is found*/

11:     **else**

12:         Found=0;

13:     **end if**

14: **end for**

15: **if** Found==1 **then**

16:     **if** ChildNode is present in the memory **then**

17:         /*if node present in AVL-tree*/

18:         Count is incremented, ChildNode.count++;

19:     **else**

20:         **if** nodes in main memory ≥ maximum allowed number of nodes in the memory **then**

21:             Free memory by removing nodes; /*referring linked list*/

22:         **end if**

23:         Create the node(ChildNode) and load the node data from file NodeTable and its child information from ChildTable;

24:         Create a node in AVL-Tree with key value==ChildNode.node_no, and a link to ChildNode;

25:         Count is incremented, ChildNode.count++;

26:     **end if**

27:     **if** Last item of the *Transaction* array **then**

28:         Create a node in linked list, which points to ChildNode;

29:     **end if**

30:     **Insert_Transaction**(Transaction, ChildNode, k, index++);

31: **else**

32:     **if** nodes in main memory ≥ maximum allowed number of nodes in the memory **then**

33:         Free memory by removing nodes; /* referring linked list*/

34:     **end if**

35:     Create a new node,*Node*, with *Node.name* =*Item*;

36:     Node is given a unique node number. (*Node.node_no*= *node_no_t*++);

37:     Data about Node is written to files NodeTable;/* Line number= *Node.node_no*/

38:     Node is connected to the Tree (by connecting it to root);

39:     Create a node in AVL-Tree with key value==Node.node_no, and a link to Node;

40:     **if** Last item of the *Transaction* array **then**

41:         Create a node in linked list, which points to Node;

42:     **end if**

43:     **Insert_Transaction**(Transaction, Node, k, index++);

44: **end if**

4.1 Discretization of Dataset

The rule mining algorithms use either the market basket or transactional databases. But all real-life databases are not of these forms. Most of the real-life datasets contain continuous-valued attributes. To support the rule mining algorithms, the datasets need to be discretized. During this study, following two discretization techniques: **Equal Width Interval Binning** and **Recursive Minimal Entropy Partitioning (RMEP)** [5, 20] are used.

The datasets are converted to market basket using both equal width interval binning and recursive minimal entropy partitioning, and different market basket file sizes are shown in Table 3.

Algorithm 3: MINING THE SbFP-Tree.

---

input: Minimum support count (*minsup*), *SbFP-Tree*.

output: Set of frequent patterns.

1: **Procedure SbFP_Growth**(Tree, α)
2: **if** Tree contains a single path P **then**
3:    set← SubSets(P) /*power set to nodes in the path P */
4:    **for** all elements β of set **do**
5:      Generate pattern $\beta \cup \alpha$ with support count = minimum support count of nodes in β;
6:    **end for**

7: **else**
8:    **for** each ai in the header of Tree **do**
9:      generate pattern β = ai $\cup$ α with support count = ai.support count;
10:      Construct β 's conditional pattern base and then β 's conditional SbFP-Tree, TreeSbFP;
11:      **if** TreeSbFP is not equal to null **then**
12:       SbFP_Growth (SbFP-Tree, β);
13:      **end if**
14:    **end for**
15: **end if**

Table 3 File size of different datasets (market basket)

Dataset name	Market basket file size (in KB)	
	Equal width	Minimal entropy
Connect-4	10580	10580
Nursery	835	835
Mushroom	2016	2016
Adult income	5597	5088
Acute inflammation	2.83	2.83

4.2 Results and Analysis

First of all, we examine the datasets with FP-Tree growth algorithm and found the results in Table 4. Now in the Connect-4 dataset, the number of nodes created in FP-Tree is 238019, and if we consider the size of a node to be 36 bytes, the total size of the FP-Tree for Connect-4 dataset will be $238019 \times 36 = 8568684$ bytes, which is approximately equal to 9 MB. Suppose the physical memory size is only 5 MB, then FP-Growth algorithm will fail for this Connect-4 dataset. Connect-4 dataset is a small dataset with only 129 attributes and 67554 instances, but if the dataset will have 10 millions of attributes and 100 millions of instances, then GBs of physical memory will not be sufficient.

Table 4 No. of nodes generated in different datasets

Dataset name	No. of nodes in RAM (FP-Tree)	
	Equal width	Minimal entropy
Connect-4	238019	238019
Nursery	18867	18867
Mushroom	27068	27068
Adult income	89674	87122
Acute inflammation	40	45

Size of the main memory resident part of the SbFP-Tree can be controlled as per requirement as shown in Step 32 of Algorithm 2, i.e., we can bound the algorithm to use only certain amount of physical memory. Suppose if we allow SbFP-Growth to use 200 nodes as maximum nodes in physical memory, then the algorithm uses only $(200 \times 36 = 7200$ (tree) $+ 200 \times 24 = 4800$ (AVL-tree) $+ 200 \times 8 = 1600$ (linked list)) 13600 bytes of physical memory at max. SbFP-Growth algorithm is tested for Connect-4 dataset, with 200 nodes as the maximum number of nodes in the physical memory; it worked and took less than 14KB of physical memory space. Thus, SbFP-Growth algorithm is able to handle the problem of limited physical memory, faced by FP-Growth algorithm.

Now in Tables 5 and 6, we see the SbFP-Tree file sizes in secondary storage for different datasets (market basket) generated using equal width interval binning and

Table 5 SbFP-Tree file size in secondary storage for datasets in market basket (equal width interval binning)

Dataset Name	SbFP-Tree file size in physical memory (in KB)		
	Minimum nodes (200)	50% of total nodes	100% of total nodes
Connect-4	8347	8042	8042
Nursery	661	561	561
Mushroom	951	889	889
Adult income	3152	2972	2950
Acute inflammation	1.34	1.32	1.30

Table 6 SbFP-Tree file size in secondary storage for datasets in market basket (recursive minimal entropy partitioning)

Dataset name	SbFP-Tree file size in physical memory (in KB)		
	Minimum nodes (200)	50% of total nodes	100% of total nodes
Connect-4	8347	8042	8042
Nursery	661	561	561
Mushroom	951	889	889
Adult income	2980	2852	2840
Acute inflammation	1.45	1.43	1.41

recursive minimal entropy partitioning methods, respectively. File sizes of SbFP-Tree are tested by allowing minimum number of nodes, 50% of total nodes created by the tree and 100% of the total nodes created in physical memory.

From the results represented in Tables 5 and 6, it can be observed that if more than 50% nodes of the actual tree are maintained in main memory, very small deduction of the tree size is there. Similarly, the minimum number of nodes to be maintained in memory should be more than or equal to the length of the longest path of the tree, which is equivalent to the maximal frequent itemset cardinality. In this experiment, a magic figure of 200 nodes is used. The number 200 falls between these two observed lower and upper bounds of number of nodes for all the datasets, after discretizing with both the techniques. However, more experiments need to be carried out to tune this value to some generic one or to make it self-adaptive, based on the total available physical memory of the system.

5 Conclusion

From the results, it can be observed that the size of the SbFP-Tree is smaller than the respective market basket dataset. For example, Connect-4 dataset, market basket file size is 10580 KB (Table 3), and SbFP-Tree file size is 8347 KB (Table 5); therefore, we can say that the SbFP-Tree is compressed form of the dataset, and hence, SbFP-Tree provides compression of the dataset. From the experiments, it is observed that SbFP-Growth algorithm overcomes the limitation of FP-Tree growth algorithm.

References

1. Adnan, M., Alhajj, R.: DRFP-tree: disk-resident frequent pattern tree. Appl. Intell **30**(2) (2009)
2. Adnan, M., Alhajj, R., Barker, K.: Alternative method for incrementally constructing the fp-tree. In: Proceedings of IEEE International Conference on Intelligent Systems, UK, Sept 2006
3. Agrawal, R., Imielinski, T., Swami, A.: Tmining association rules between sets of items in large databases. ACM SIGMOD Int. Conf. Manag. Data **22**, 207–216 (1993)
4. Brin, S., Motwani, R., Ullman, J., Tsur, S.: Dynamic itemset counting and implication rules for market basket data. In: Proceedings of the 1997 ACM SIGMOD International Conference on Management of Data, vol. 26(2), pp. 255–264 (1997)
5. Fayyad, U., Irani, K.: Multi-interval discretization of continuous-valued attributes for classification learning. In:Proceedings of the 13th International Joint Conference on Artificial Intelligence, pp. 1022–1029 (1993)
6. Goethals, B.: Memory issues in frequent itemset mining. In: Proceedings of ACM Symposium on Applied Computing, pp. 530–534 (2004)
7. Gouda, K., Zaki, M.J.: Genmax: an efficient algorithm for mining maximal frequent itemsets. Data Min. Knowl. Discov. **11**(3), 223–242 (2005)
8. Grahne, G., Zhu, J.: Efficiently using prefix-trees in mining frequent itemsets. In: FIMI, vol. 90 (2003)
9. Hamedanian, M., Nadimi, M., Naderi, M.: An efficient prefix tree for incremental frequent pattern mining. Int. J. Inf. **3**(2) (2013)

10. Han, J., Cheng, H., Xin, D., Yan, X.: Frequent pattern mining: current status and future directions. Data Min. Knowl. Discov. **15**(1), 55–86 (2007)
11. Han, J., Pei, J., Yin, Y.: Mining frequent patterns without candidate generation:a frequent-pattern tree approach. In: Proceedings of ACMSIGMOD, Dallas, TX, pp. 1–12 (2000)
12. Han, J., Pei, J., Yin, Y., Mao, R.: Mining frequent patterns without candidate generation: a frequent-pattern tree approach. Data Min. Knowl. Discov. **8**(1), 53–87 (2004)
13. Hong, T.P., Lin, C.W., Wu, Y.L.: Incrementally fast updated frequent pattern trees. Expert Syst. Appl. **34**(4), 2424–2435 (2008)
14. Kao, B., Zhang, M., Yip, C.L., Cheung, D.W., Fayyad, U.: Efficient algorithms for mining and incremental update of maximal frequent sequences. Data Min. Knowl. Discov. **10**(2), 87–116 (2005)
15. Kum, H.C., Chang, J.H., Wang, W.: Sequential pattern mining in multi-databases via multiple alignment. Data Min. Knowl. Discov. **12**(2–3), 151–180 (2006)
16. Leung, C.K.S., Hoque, T., Khan, Q.I.: CanTree: a tree structure for efficient incremental mining of frequent patterns. In: Proceedings of IEEE ICDM, pp. 274–281 (2005)
17. Lin, C.W., Hong, T.P., Lu, W.H.: The Pre-FUFP algorithm for incremental mining. Expert Syst. Appl. **36**(5), 9498–9505 (2009)
18. Lin, D.I., Kedem, Z.M.: Pincer-search: a new algorithm for discovering the maximum frequent set. In: Advances in Database TechnologyEDBT'98, pp. 103–119. Springer (1998)
19. Liu, G., Lu, H., Lou, W., Xu, Y., Yu, J.X.: Efficient mining of frequent patterns using ascending frequency ordered prefix-tree. Data Min. Knowl. Discov. **9**(2), 249–274 (2004)
20. Liu, H., Hussain, F., Tan, C.L., Dash, M.: Discretization: an enabling technique. Data Min. Knowl. Discov. **6**(4), 393–423 (2002)
21. Park, J.S., Chen, M.S., Yu, P.S.: Using a hash-based method with transaction trimming for mining association rules. IEEE Trans. Knowl. Data Eng. **9**(5), 813–825 (1997)
22. Pei, J., Han, J., Mao, R., et al.: CLOSET: An efficient algorithm for mining frequent closed itemsets. In: ACM SIGMOD Workshop on Research Issues in Data Mining and Knowledge Discovery, vol. 4, pp. 21–30 (2000)
23. Pietracaprina, A., Riondato, M., Upfal, E., Vandin, F.: Mining top-K frequent itemsets through progressive sampling. Data Min. Knowl. Discov. **21**(2), 310–326 (2010)
24. Racz, B.: Nonordfp: an FP-growth variation without rebuilding the fp-tree. In: Proceedings of IEEE ICDM Workshop on Frequent Itemset Mining Implementations (2004)
25. Savasere, A., Omiecinski, E.R., Navathe, S.B.: An Efficient Algorithm for Mining Association Rules in Large Databases (1995)
26. Schlegel, B., Gemulla, R., Lehner, W.: Memory-efficient frequent-itemset mining. In: Proceedings of the 14th International Conference on Extending Database Technology, pp. 461–472. ACM (2011)
27. Sucahyo, Y.G., Gopalan, R.P.: CT-PRO: a bottom-up non recursive frequent itemset mining algorithm using compressed fp-tree data structure. FIMI **4**, 212–223 (2004)
28. Wang, E.T., Chen, A.L.: A novel hash-based approach for mining frequent itemsets over data streams requiring less memory space. Data Min. Knowl. Discov. **19**(1), 132–172 (2009)
29. Xu, Y., Yu, J., Liu, G., Lu, H.: From path tree to frequent patterns: a framework for mining frequent patterns. In: Proceedings of 2002 IEEE International Conference on Data Mining, 2002. ICDM 2003, pp. 514–521 (2002)
30. Zaki, M.J.: Scalable algorithms for association mining. IEEE Trans. Knowl. Data Eng. **12**(3), 372–390 (2000)

Combining Multilevel Contexts of Superpixel Using Convolutional Neural Networks to Perform Natural Scene Labeling

Aritra Das, Swarnendu Ghosh, Ritesh Sarkhel, Sandipan Choudhuri, Nibaran Das and Mita Nasipuri

Abstract Modern deep learning algorithms have triggered various image segmentation approaches. However, most of them deal with pixel-based segmentation. Superpixels, on the other hand, provide a certain degree of contextual information while reducing computation cost. In our approach, we have performed superpixel-level semantic segmentation considering three various levels as neighbors for semantic contexts. Furthermore, we have enlisted a number of ensemble approaches like max-voting and weighted average. We have also used the Dempster–Shafer theory of uncertainty to analyze confusion among various classes. Our method has proved to be superior to a number of different modern approaches on the same dataset.

Keywords Scene segmentation · Superpixel · Convolutional neural network
Dempster–Shafer theory

A. Das · S. Ghosh · N. Das (✉) · M. Nasipuri
Jadavpur University, Kolkata 700032, West Bengal, India
e-mail: nibarandas@jadavpuruniversity.in

A. Das
e-mail: dasaritra93@gmail.com

S. Ghosh
e-mail: swarbir@gmail.com

M. Nasipuri
e-mail: mitanasipuri@jadavpuruniversity.in

R. Sarkhel
Ohio State University, Columbus, OH 43210, USA
e-mail: sarkhelritesh@gmail.com

S. Choudhuri
Arizona State University, Tempe, AZ 85281, USA
e-mail: sandipanchoudhuri90@gmail.com

© Springer Nature Singapore Pte Ltd. 2019
J. Kalita et al. (eds.), *Recent Developments in Machine Learning and Data Analytics*,
Advances in Intelligent Systems and Computing 740,
https://doi.org/10.1007/978-981-13-1280-9_28

1 Introduction

Deep learning has brought a new era in machine learning. Being able to learn more complex features from images, problems such as classification, localization, and segmentation have seen remarkable progress especially for natural images. Previously, most significant research in the domain of natural image processing was performed using some sort of pattern recognition over pixels [3, 5, 9]. The problem that has been dealt in this paper is semantic image segmentation. Image segmentation goes beyond tasks like object recognition or localization. In this problem, we are mainly interested in precise segments which semantically separate one object from another. While pixel level algorithms [8, 10, 12] provide very fine level segmentation, superpixels [18] provide much lesser computational complexity while not compromising performance. Superpixels refer to small patches of adjacent similar pixels grouped together. We have used these superpixels for our algorithms thus providing real-time performance. Convolutional neural networks (CNNs) have showed tremendous performance in the field of natural image processing as well as segmentation. In our approach we have implemented multiple convolutional neural networks to obtain results. Any classification problem can be associated with uncertainty in the decision process. We have used some ensemble methods as well as Dempster–Shafer Theory to handle such uncertainty. The next section will give a brief review of some related works. Section 3 will explain the methodologies. Sections 4 and 5 will cover the experimentations and discussions regarding obtained results.

2 Related Works

Segmentation algorithms also gained momentum with the onset of deep learning. In 2015, Ross Girshick in his paper Faster-RCNN [13] outlined the quickest way of detecting multiple regions in an image. However, his proposed architecture does not segment the whole image but can find where the objects are in the image. In SegNet [2], the idea of convolution and deconvolution have been used together to generated segmented regions. Farabet et al. [6] showed how superpixel-level classification may be performed by using CNNs. Previously superpixels were only used to generate a scene parsing tree rather than considering them for the actual segmentation. Our approach, on the other hand, trains the CNN directly on the superpixel patches. While a variety of superpixels have been seen in the field of image segmentation [18], our choice is the SLIC [1], for its speed and boundary adherence. Uncertainty is a common challenge in machine learning problems. In image segmentation we have seen the use of Dempster–Shafer Theory [4, 14, 16] for elimination of such uncertainties as well. For our current work, the ICCV09 Dataset [7] was used.

3 Methodologies

In this work, three different CNNs were first used to classify the superpixel patch along with multiple levels of context, whose scores were later ensembled using the three proposed approachs. The overall workflow is clearly demonstrated in Fig. 1.

3.1 Superpixel-Based Segmentation (Module 1)

Pixel level classification is a tedious process primarily due to two factors. First, even a small image contains quite a high number of pixels, and second, the information content of a pixel is very limited to consider classification into various segments. By using superpixels, we capture much more information than a single pixel and number of superpixels in an image is much lesser than the number of pixels. First, each image was divided into superpixels by using SLIC [1]. To keep uniformity in the sizes of superpixels across images of various sizes of images, the *minimum object resolution* (*MOR*) was fixed. The *number of superpixels NOS* of an image I is given by

$$NOS(I) = \frac{Height(I) * Width(I)}{MOR} \tag{1}$$

A patch of superpixel shows significant amount of texture information with respect to just pixels. However, for semantic segmentation we also need to consider the context in which this superpixel occurs. So each superpixel was augmented with its neighbors to create a larger patch for the CNN to extract features from. For our experiments, we have considered the first, second, and third neighbor of each superpixel for its classification. Three different CNNs were trained for each of this neighbor category. It can be clearly seen in Fig. 2 Each CNN for classifying the superpixels consisted of two layers. The first layer has 32, 5×5 convolution kernels followed by

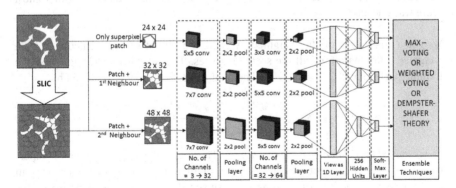

Fig. 1 Overall flowchart

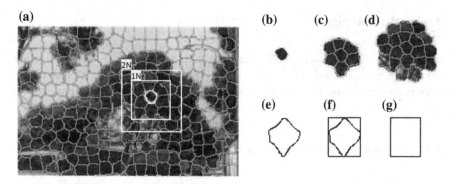

Fig. 2 Superpixels neighbor. **a** Image with selected superpixel (in red); **b** single superpixel patch, noted as 0N; **c** superpixel patch with first neighbor, noted as 1N; **d** superpixel patch with 2n d neighbor, noted as 2N; **e** single superpixel patch or patch with neighbors cropped out from image; **f** minimal covering bounding box; and **g** regular size cropped out patch fed into the CNNs

a standard 2×2 pooling. The second layer consists of 64, 3×3 convolution kernels followed by a standard 2×2 pooling. This is followed by a fully connected layer with 256 hidden units and a softmax output layer.

3.2 Ensemble Strategy

Each of three CNN outputs an eight-dimensional softmax distribution. These are ensembled using three different methods, namely, max-voting, combination of mass function with the help of Dempster–Shafer theory of uncertainty and weighted sum techniques.

Max-Voting: This techniques takes the three predictions from three CNNs and chooses the winner on the basis of votes. In case of a tie, the prediction with highest score was chosen.

Weighted Average: For the weighted average, the output score is calculated by a weighted combination of all the softmax scores. The weight is determined by the training performance of each CNN. The final score S_i for patch i is given by

$$S_i = \frac{S_i^1 * r_1 + S_i^2 * r_2 + S_i^3 * r_3}{r_1 + r_2 + r_3} \qquad (2)$$

Dempster–Shafer Theory of Uncertainty: There are a certain number of superpixels for which the network gives poor predictions. The uncertainty rises in training because of lots of reasons like skewed datasets, similar superpixels for different classes, wrong ground truth level annotations. To deal with such uncertainties Dempster–Shafer [50] theory of evidence is taken into account. Unlike normal classification which uses a probability distributions across the number of classes, Dempster–Shafer theory of uncertainty deal with the masses and beliefs which are distributions across the all the possible combination of the classes. Henceforth, mass value of a certain combination is defined by $m_j, where j \in 2^C and C = no.of classes$. So we designed an approach to simulate mass distribution by using the confusion matrix obtained during training. In theory of evidence, $mass(A) <= Prob(A) <= Bel(A)$. The power set $\Phi(C)$ is written as

$$\Phi(C) = \{m_1, m_2, \ldots, m_C, m_{11}, m_{12}, \ldots, m_{1C}, \ldots, m_{12\ldots C}\} \tag{3}$$

So this difference in the between the mass value m_i and probability $p_i \forall i \in C$ is defined in terms of the confusion (misclassification) related to that class.

$$m_i = p_i - p_i * \left(\frac{\sum_{i \neq j} miss(i, j)}{number\ of\ samples\ in\ class\ i} \right) \quad \forall j \in C \tag{4}$$

The computation of mass values for other elements of the power set such as $m_{ij}, m_{ijk} \ldots$ is more complicated. The confusion matrix provides us with information regarding misclassification among two classes as well. Higher combination of classes we not considered henceforth because they needlessly increase the computation while not providing significant information. In other words while considering the predicted class of a patch we are giving consideration to one more class which has a high probability of confusion with chosen class. So mass values such as $m_{ijk}, m_{ijkl}, \ldots$ are ignored. If we remember from Eq. 4 the probability of each class was deducted by a certain amount to obtain corresponding mass values. If all these deductions are accumulated and redistributed among other members of the power set as their mass values then the requisite of a mass distribution is satisfied which is given by $\sum_{i \in 2^C} m_i = 1$. Let the accumulated deductions be defined as D.

$$D = \sum_{i \in C} p_i * \left(\frac{\sum_{i \neq j} miss(i, j)}{number\ of\ samples\ in\ class\ i} \right) \tag{5}$$

The mass values of higher order members of the power set with a cardinality of 2 are given by

$$m_{jk} = \frac{miss(j, k) + miss(k, j)}{\sum_{p,q \in C, p \neq q} (miss(p, q) + miss(p, q))} * D \tag{6}$$

After computing the mass distribution for each of the three CNNs, we combine them to find the final mass distribution using the Dempster–Shafer rule of combination of evidence as described in Sect. 2.2.1 in the works of Sentz et al. [15].

Table 1 Classification performance of individual CNNs. $nCk = n$ number of $k \times k$ convolution kernels, $mP = Max - pooling$ with $m \times m$ window and $stride = m$, $FCp = Fully\ connected\ layer$ with p units, qN refers to input patch along with qth-level neighbors

Patch	Input size	Network architecture	Test classification accuracy
0N	24×24	32C5-2P-64C3-2P-FC256	72.32
1N	32×32	32C7-2P-64C5-2P-FC256	72.45
2N	48×48	32C7-2P-64C5-2P-FC256	72.24

4 Experimentations

The first part of our experimentation trains and tabulates the performance of the three individual CNNs. Optimum size for the raw superpixel patch and first and second neighbor images were chosen as 24×24, 32×32, and 48×48, respectively, based on validation performance. The architecture and the corresponding performance is given in Table 1. The second phase records the result of the various ensemble methods. The ICCV 09 Database was used for the experimentation. It contains 715 images with ground truths showing eight semantically segmented classes. The dataset was split into 500 training, 72 validation and 143 test samples. The minimum object size considered for generating the superpixels was approximately 20×20. The total number of superpixels was 291,911.

5 Results and Analysis

In Fig. 3, we can see some segmented examples as generated by our approach. In the next subsections, we shall look into the performance of the individual CNNs and how they improved upon using ensemble techniques.

5.1 Individual CNNs

Each individual CNN was trained over 500 images. The optimal architecture was selected according to their performance on the validation dataset. The final test accuracy along with the optimum configurations is shown in Table 1.

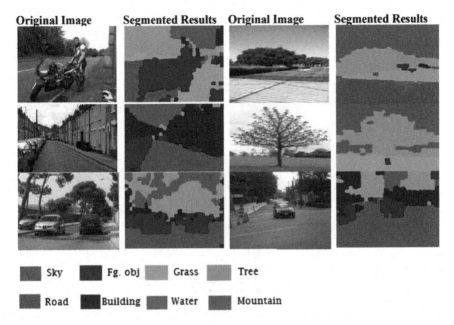

Sky	Fg. obj	Grass	Tree
Road	Building	Water	Mountain

Fig. 3 Segmentation results for our proposed approach

It can be seen that all the individual CNNs perform almost at the same level. Thus, it may seem that the choice of different neighbors is ineffective. However, if we ensemble the softmax outputs of these three CNNs we see a different story.

5.2 Ensemble Methods

We chose three ensemble strategies to deal with disagreement among the individual CNNs along with Dempster–Shafer theory to remove uncertainty in the obtained results. Table 2 shows the performance of the three ensemble strategies across all classes.

It can be seen that for some classes, Dempster–Shafer wins, whereas for other classes the weighted average is ahead. The poor performance in the mountain category was due to the fact that segments with mountains were quite scarce throughout the dataset.

Finally, in Table 3 we can see how our approach performs against some fantastic works on this database.

Moreover, the testing time has been calculated to be in the range of 25–30 ms on a GTX 1080 GPU, thus ensuring successful real-time implementation.

Table 2 Performance of ensemble approaches with respect to various classes

Type	Test accuracy								Avg. Acc.
	Sky	Tree	Grass	Ground	Building	Mountain	Water	Object	
Dempster–Shafer	**88.07**	76.74	**80.29**	**86.59**	**77.8**	2.56	59.41	59.18	77.07
Max-voting	84.05	73.37	73.65	80.80	69.90	4.15	63.90	60.79	72.88
Weighted average	87.75	**77.17**	79.69	85.43	75.85	**4.16**	**64.69**	**63.24**	**77.14**

Table 3 Our approach compared with other approaches on the ICCV09 dataset

Approaches	Methodology	Accuracy
Baseline (ICCV 09) [7]	Pixel CRF	74.3
Gould et al. (ICCV 09) [7]	Region-based energy	76.4
Munoz et al. (ECCV 10) [11]	Probabilistic model	76.9
Farabet et al. (PAMI 13) [6]	CNN + superpixel	74.56
Tighe et al. (ECCV 10) [17]	Features + superpixel	76.3
Our approach	**Superpixel + CNN + ensemble**	**77.14**

6 Conclusion

We have implemented a novel approach for superpixel-level segmentation and boosted its performance by various ensemble methods and uncertainty handling. Our approach shows a fast method for creating decent segments. When compared with other methods that were applied in this dataset it showed its strength. In the future, it is possible to extend this work to video segmentations. Overall, we believe that speed the algorithm combined with a relatively small sized CNN our approach shows promising results.

Acknowledgements This work is partially supported by the project entitled Development of Knowledge Graph from Images Using Deep Learning sponsored by SERB (Government of India, order no. SB/S3/EECE/054/2016) (dated 25/11/2016), and carried out at the Centre for Microprocessor Application for Training Education and Research, CSE Department, Jadavpur University.

References

1. Achanta, R., Shaji, A., Smith, K., Lucchi, A., Fua, P., Süsstrunk, S.: Slic superpixels compared to state-of-the-art superpixel methods. IEEE Trans. Pattern Anal. Mach. Intell. **34**(11), 2274–2282 (2012)
2. Badrinarayanan, V., Kendall, A., Cipolla, R.: Segnet: a deep convolutional encoder-decoder architecture for image segmentation. arXiv:1511.00561 (2015)
3. Belongie, S., Malik, J., Puzicha, J.: Shape matching and object recognition using shape contexts. IEEE Trans. Pattern Anal. Mach. Intell. **24**(4), 509–522 (2002)
4. Bendjebbour, A., Delignon, Y., Fouque, L., Samson, V., Pieczynski, W.: Multisensor image segmentation using Dempster-Shafer fusion in Markov fields context. IEEE Trans. Geosci. Remote Sens. **39**(8), 1789–1798 (2001)
5. Campbell, R.J., Flynn, P.J.: A survey of free-form object representation and recognition techniques. Comput. Vis. Image Underst. **81**(2), 166–210 (2001)
6. Farabet, C., Couprie, C., Najman, L., LeCun, Y.: Learning hierarchical features for scene labeling. IEEE Trans. Pattern Anal. Mach. Intell. **35**(8), 1915–1929 (2013)
7. Gould, S., Fulton, R., Koller, D.: Decomposing a scene into geometric and semantically consistent regions. In: 2009 IEEE 12th International Conference on Computer Vision, pp. 1–8. IEEE (2009)

 8. Ilea, D.E., Whelan, P.F.: Image segmentation based on the integration of colour-texture descriptorsa review. Pattern Recognit. **44**(10), 2479–2501 (2011)
 9. Liu, Y., Zhang, D., Lu, G., Ma, W.Y.: A survey of content-based image retrieval with high-level semantics. Pattern Recognit. **40**(1), 262–282 (2007)
10. Luccheseyz, L., Mitray, S.: Color image segmentation: a state-of-the-art survey. In: Proc. Indian Natl. Sci. Acad. (INSA-A) **67**(2), 207–221 (2001)
11. Munoz, D., Bagnell, J.A., Hebert, M.: Stacked hierarchical labeling. In: European Conference on Computer Vision, pp. 57–70. Springer (2010)
12. Pal, N.R., Pal, S.K.: A review on image segmentation techniques. Pattern Recognit. **26**(9), 1277–1294 (1993)
13. Ren, S., He, K., Girshick, R., Sun, J.: Faster r-cnn: towards real-time object detection with region proposal networks. In: Advances in Neural Information Processing Systems, pp. 91–99 (2015)
14. Rombaut, M., Zhu, Y.M.: Study of Dempster-Shafer theory for image segmentation applications. Image Vis. Comput. **20**(1), 15–23 (2002)
15. Sentz, K., Ferson, S., et al.: Combination of evidence in Dempster-Shafer theory, vol. 4015. Citeseer (2002)
16. Shafer, G., et al.: A mathematical theory of evidence, vol. 1. Princeton university Press, Princeton (1976)
17. Tighe, J., Lazebnik, S.: Superparsing: scalable nonparametric image parsing with superpixels. Comput. Vis.-ECCV **2010**, 352–365 (2010)
18. Wang, C., Chen, J., Li, W.: Review on superpixel segmentation algorithms. Appl. Res. Comput. **31**(1), 6–12 (2014)

A Fast Algorithm for Automatic Segmentation of Pancreas Histological Images for Glucose Intolerance Identification

Tathagata Bandyopadhyay, Shyamali Mitra, Sreetama Mitra, Nibaran Das, Luis Miguel Rato and Mrinal Kanti Naskar

Abstract This paper describes a novel fast algorithm for automatic segmentation of islets of Langerhans and β-cell region from pancreas histological images for automatic identification of glucose intolerance. Here, LUV colour space and connected component analysis are used on 134 images among which 56 are of normal and rest 78 are of prediabetic type. The paper also talks about a supervised learning approach for classifying the images based on their morphological features. In the present work, we have introduced a modern classifier weighted ELM (Extreme Learning Machine) for prediabetes identification. Performances of weighted ELM are comparable with all the present-day's robust classifiers such as Support Vector Machines (SVM), Multilayer Perceptron (MLP), etc. We have also compared the result with traditional ELM and observed better performance in the present skewed dataset with substantial improvement in training time.

T. Bandyopadhyay · S. Mitra
School of Computer Engineering, KIIT University, Bhubaneswar, India
e-mail: tathagatabanerjee15@rocketmail.com

S. Mitra
e-mail: msreetama10@gmail.com

S. Mitra · M. K. Naskar
Department of Electronics and Telecommunication Engineering,
Jadavpur University, Kolkata 700032, West Bengal, India
e-mail: cshyamali.mitraa@gmail.com

M. K. Naskar
e-mail: mknaskar@etce.jdvu.ac.in

L. M. Rato
Department of Informatics, University of Evora, Evora, Portugal
e-mail: dlmr@di.uevora.pt

N. Das (✉)
Department of Computer Science and Engineering, Jadavpur University, Kolkata 700032,
West Bengal, India
e-mail: nibaran@gmail.com

© Springer Nature Singapore Pte Ltd. 2019
J. Kalita et al. (eds.), *Recent Developments in Machine Learning and Data Analytics*,
Advances in Intelligent Systems and Computing 740,
https://doi.org/10.1007/978-981-13-1280-9_29

307

Keywords Automatic segmentation · Histological image · Islets of Langerhans β-cell · Diabetes · Computerized diagnostic system · Extreme learning machine

1 Introduction

In present days, diabetes is prevalent among most of the people in the world [1]. If not controlled at the right time, it can lead to multi-organ failure and eventually death. Genetic factors coupled with environmental influences, rising living standards and lifestyle changes are the causes of diabetes. Mortality due to diabetes and its potential complications are enormous. Therefore, it is commonly known as silent killer. So proper identification and treatment at early stage are very important. Generally, diabetes refers to body's inability to produce insulin hormone with consequent abnormal metabolism of carbohydrate and an elevated level of glucose in blood. Glucose-intolerant condition also refers to increased level of glucose in blood but here the amount of glucose is less than that of diabetic condition [2]. In other words, it can be said that glucose intolerant is a precondition of diabetes, and thus by detecting this we can detect the subjects prone to diabetes. In glucose-intolerant rats, an autoimmune cell destruction happens in pancreas [3], which leads to morphological changes in the pancreas histological images and thus opens the way of image-based computerized prediabetic diagnostic system.

Computerized and image-based diabetes diagnostic system has its foundation on three pillars: (a) Segmentation of islets of Langerhans area and insulin-positive β-cell area [4], (b) Feature extraction from those segmented regions and (c) Classification of the images based on extracted features. Segmentation refers to the process of separating out the islets of Langerhans and β-cell region from the microscopic slide image. Rato et al. described a manual segmentation method using ImageJ [5]. Although manual segmentation is very precise, it is tiresome and time-taking process and due to its requirement of human intervention, it incurs dependency on expertise and experience level of the analyst. This motivated the researchers to develop and use fully automated segmentation algorithms to solve the said segmentation problem. In [6–8], different colour-based segmentation approaches were proposed for automated separation and quantification of insulin-stained β-cell regions, respectively. Colour-based segmentation approach works very well in segmenting the β-cell region as it significantly differs in colour stain from the surrounding regions. But, segmentation of the islets of Langerhans region is much more challenging due to its colour similarity with the background and thus only colour-based segmentation method is not capable enough in this case. Chen et al. [9] proposed connected component analysis and active contour model for automatically segment out the islets area. Berclaz et al. also used active contour model but with level set method for initialization of the contour [10]. Active contour model does not work well in case of noisy images [11] and also this algorithm is computationally complex. Bandyopadhyay et al. in [3] proposed wavelet decomposition-based segmentation methodology, but its huge processing time per image limits its applicability in real life. In the present work, we have introduced a huge improvement with respect to time over the approach [3] by using LUV colour space transform instead of wavelet decomposition used there.

In this paper, we propose a much simpler and fast segmentation method for both the β-cell and islets of Langerhans regions, based on colour space transform and connected component analysis. After segmentation, we extracted morphological features and used classifier like Extreme Learning Machine (ELM), Support Vector Machine (SVM), etc., to classify the images as normal and prediabetic (i.e. it is a binary classification problem). Input RGB image is transformed to LUV colour space for segmenting out the β-cell regions. Again, colour stretching and thresholding are used to identify the islets of Langerhans regions. Finally, connected component analysis is done to segment out the islets including the β-cell regions. Based on the algorithm used for β-cell segmentation, we differentiate in methods, namely, method A and method B which will be discussed in details in the image segmentation subsection of the proposed methodology section. Totally, six features are used in various combinations and will be discussed in details in the feature extraction subsection of the proposed methodology section. For classification, our focus is on the relatively new classifier ELM. As our dataset is imbalanced, we have used weighted regularized ELM. Results are compared with basic ELM and wildly used classifiers like SVM and MLP.

This paper is organized in five sections including this introduction as the Sect. 1. Experimental setups like dataset description, hardware and software configuration, etc. are described in Sect. 2. Section 3 sheds light on the proposed approach. Results and discussion are there in Sect. 4 Finally, we conclude in Sect. 5.

2 Experimental Setup

In this work, an image dataset of pancreas histological images of Wistar rats has been used. The dataset is collected from the Department of Biology, University of Evora, Portugal and it contains 134 colour (RGB) pancreas histological images of Wistar rats, among which 56 are of normal rats and rest 78 are of prediabetic or diabetic-prone rats. Each of the images is of size 1280×960 pixels with both horizontal and vertical resolutions 72 dpi and bit depth 24. Matlab 2014Ra is used for image segmentation, feature extraction, ELM and WELM. For SVM and MLP, Weka 3.8 is used.

2.1 Proposed Methodology

As mentioned earlier, our approach consists of three phases which are (i) image segmentation, (ii) feature extraction and (iii) classification. Among these, we have mostly contributed in first two phases.

2.2 Image Segmentation

Here, in our study, there are three regions of interest which are (i) the region which covers total islets of Langerhans including insulin-positive β-cell areas (roi1), (ii) the region which includes only the insulin-positive β-cell areas (roi2) and (iii) region containing the islets of Langerhans excluding the insulin-positive β-cell areas (roi3). Segmentation process of roi1, roi2 and roi3 is discussed below:

ROI 1 Segmentation

In this study, we have used colour space transform and connected component analysis to do this segmentation. First, the original RGB image (I) is transformed to LUV colour space using Matlab inbuilt function. LUV transformation is done because it can clearly distinguish the bluish background and the reddish brown β-cell region. Then, LUV image is then converted into binary. Hole filling and opening are done to remove spurious areas. The resulting image is named I_1. Then, complement of I_1 is taken and it is named I_1'.

The blue shade in both the regions in the original image (I) (islets of Langerhans and background) makes the task of separation difficult using colour segmentation. So, RGB stretching, HSV stretching and sharpening are done on image I to enhance the fine shade difference between the regions. Resulting image is called I_2. Top hat

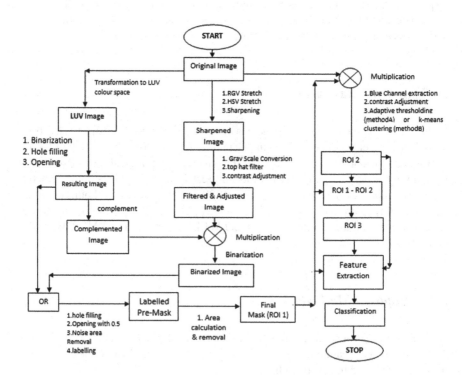

Fig. 1 Flowchart of the proposed method

filtering is done on the grayscale-converted image of I_2 to remove an approximate background. Adjusting contrast of the resulting image, we obtain image I_3.

Image I_3 is multiplied with image I_1' to mask out the β-cell region and concentrate only on the surrounding islets of Langerhans region. Resulting image is converted into binary to obtain I_4. Logical OR operation is done on I_1 and I_4 to merge both the regions, namely, islets of Langerhans region and β-cell region. After hole filling and opening (done to remove thin connection) and noise area (very small regions) removal, resulting image is labelled to name each of its connected components with numbers. Thus we obtain labelled premask ($I_{premask}$). Now, final mask or ROI 1 is generated keeping the largest connected area in $I_{premask}$. This is done under the assumption, which is made based on the images of our data set, that this region of interest will occupy larger area than the different noise areas in the background (Fig. 1).

ROI 2 and ROI 3 Segmentation
ROI 2 refers to only the insulin-positive reddish brown β-cell region excluding the rest of islets of Langerhans both inside and outside. Though LUV colour space transform approximately identifies the β-cell regions, it fails to exclude non-β-cell portions of islets inside the β-cell portions. To address this issue, we have used two different approaches, namely, (i) Method A and (ii) Method B. Method A uses median filtering-based adaptive thresholding to separate out the β-cells from the inside non-β-cells, whereas Method B uses k-means clustering to cluster the different colour shades, thus making the separation task very easy just by using the cluster indices.

IROI1 is multiplied with original image I. This is followed by blue channel extraction, contrast adjustment. Then, the resulting image is processed with Method A and Method B as discussed above to do the segmentation. We report the results for both Method A and Method B in the results section.

After segmentation of ROI 1 and ROI 2, ROI 3 can be easily obtained by subtracting ROI 2 from ROI 1.

2.3 Feature Extraction

We have used six features each of length one in different combinations. The features are (i) β-cell area to islets area ratio (F_1), (ii) α-cell area to islets area ratio (F_2), (iii) perimeter per area (ppa) ratio (F_3), chi-squared distances of the histograms of red (iv), and green (v) and blue (vi) channel between β-cell and α-cell regions and are denoted by F_4, F_5 and F_6, respectively. Three combinations of the features, namely, C_1, C_2 and C_3 are used with taking $\{F_1, F_2, F_3\}$, $\{F_4, F_5, F_6\}$ and $\{F_1, F_2, F_3, F_4, F_5, F_6\}$, respectively.

To extract the features F_1, F_2 and F_3 first, areas (number of ON pixel in the region) of ROI 1, ROI 2 and ROI 3 have been calculated and these are named as A_1, A_2 and A_3, respectively. Also, the perimeter of ROI 1 is calculated and its value is denoted by P. Now features are calculated as per the following formulae:

$$F_1 = \frac{A_3}{A_1}, \quad F_2 = \frac{A_2}{A_1}, \quad F_3 = \frac{P}{A_1}$$

where F_1, F_2 and F_3 are three feature values.

Calculation of the last three features is much more straightforward. For all three (Red, Green and Blue) channels of the β-cell and α-cell regions, histograms are calculated. Then, channel-wise chi-squared distance is calculated between the histograms of these two regions and thus F_4, F_5 and F_6 features are formed.

2.4 Classification

Classification refers to the task of categorizing the data in some categories or classes. In this study, our task is to classify the images into two classes, namely, normal and prediabetic which corresponds to normal and diabetic-prone rats, respectively. To address this binary classification problem, we have used SVM, MLP, ELM and regularized ELM [12] classifiers. For MLP and SVM, we have used Weka 3.8 tool. MLP is used with default Weka 3.8 configuration. For SVM, we have used nu-SVC with linear kernel keeping another parameter as Weka 3.8 default. ELM and regularized ELM (WELM) are done in Matlab. For ELM, we have empirically used 11 hidden nodes and for WELM we have empirically used N × N version with 300 hidden nodes and with a C value equals to 227. For each of the classifiers, 80% of the data is used for training and rest 20% is used for testing.

3 Results and Discussion

As our present work involves two major stages, namely, segmentation and classification, our results are also of twofolds. Figure 2 shows some segmented images along with its original one to compare our segmentation performance with the previous works. It clearly shows that our segmentation result is closely in line with the ground truth produced by Rato et al. in [5].

As stated earlier, we have used three different types of feature combinations on the segmented images using method A and B by the proposed algorithm. To compare the performance, we have used four different classifiers, namely, SVM, MLP, ELM and WELM. The results are shown in Tables 1 and 2. It is observed from Table 1 that weighted ELM performs better than other methods in terms of accuracy for method A. From Table 2, we can see that SVM for feature combination C_2 outperforms other methods and achieved a maximum recognition accuracy of 85.1852%, whereas ELM requires least training time for all feature types and classifiers. Figure 3 shows the comparison between method A and method B for different classifiers in terms of ROC area.

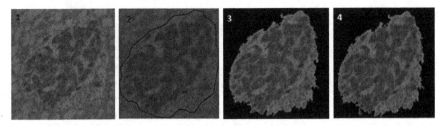

Fig. 2 Comparison of segmentation result with previous methods (1) original image, (2) manual segmentation by Rato et al. [5], (3) automatic segmentation by Bandyopadhyay et al. [3], and (4) automatic segmentation by the proposed method

Table 1 Classification results for method A

Classifier	Feature type	Accuracy (%)	TPR	FPR	F-measure	ROC area	Training time (s)
SVM	C_1	74.0741	0.741	0.462	0.734	0.639	0.07
MLP		**77.7778**	**0.778**	**0.078**	**0.792**	0.836	0.05
ELM		76.8519	0.769	0.169	0.782	**0.845**	**0.000547**
WELM		74.0741	0.741	0.186	0.756	**0.845**	0.00182
SVM	C_2	40.7407	0.407	0.579	0.441	0.414	0.08
MLP		**77.7778**	**0.778**	0.635	**0.709**	**0.614**	0.05
ELM		72.4815	0.725	**0.572**	0.697	0.564	**0.000964**
WELM		69.2593	0.693	0.600	0.670	0.465	0.001813
SVM	C_3	77.7778	0.778	0.171	0.790	0.804	0.11
MLP		70.3704	0.704	0.197	0.722	**0.907**	0.08
ELM		75.5556	0.756	0.297	0.764	0.819	**0.000467**
WELM		**81.1111**	**0.811**	**0.161**	**0.820**	0.879	0.001755

The bold indicates the maximum in respective areas

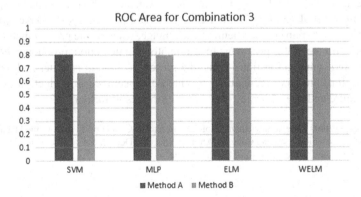

Fig. 3 ROC area comparison for best feature combination #3

Table 2 Classification results for method B

Classifier	Feature type	Accuracy (%)	TPR	FPR	F-measure	ROC area	Training time (s)
SVM	C_1	48.1481	0.481	**0.274**	0.490	0.604	0.05
MLP		**74.0741**	**0.741**	0.741	0.630	0.764	0.05
ELM		73.1482	0.731	0.365	**0.739**	0.808	**0.000429**
WELM		66.0000	0.660	0.398	0.677	**0.859**	**0.001757**
SVM	C_2	**85.1852**	**0.852**	0.423	0.829	0.714	0.09
MLP		81.4815	0.815	0.343	0.810	0.800	0.05
ELM		84.2593	0.843	**0.244**	**0.842**	**0.892**	**0.000539**
WELM		78.6667	0.787	0.338	0.787	0.797	**0.001754**
SVM	C_3	70.3704	0.704	0.382	0.714	0.661	0.06
MLP		**81.4815**	**0.815**	0.343	0.810	0.800	0.08
ELM		81.3704	0.814	**0.250**	**0.818**	**0.850**	**0.000454**
WELM		74.4074	0.744	0.278	0.756	0.848	**0.001770**

The bold indicates the maximum in respective areas

4 Conclusion

In the present work, we have introduced a fast automatic segmentation of islets of Langerhans and β-cell region from pancreas histological images. The automatic accurate segmentation helps to improvise the performance of identification of glucose intolerance to distinguish normal and prediabetic images. The segmentation is challenging due to having minute morphological differences between those images of normal and prediabetic cases. The present method is an improvement over the previously reported method in terms of both accuracy and time. The achieved recognition accuracy is 0.85% higher than the accuracy achieved by Bandyopadhyay et al. and Kakimoto et al. [3, 4]. Also, our earlier method in [3, 4] used to take approximately 5400 s (90 min) for segmenting each image, whereas the present method just takes approximately 30 s per image for automated segmentation, producing **99.44%** reduction in segmentation time. Thus, the present technique is better and helps to save time for automatic recognition of glucose intolerance. Introduction of more robust feature and deep learning may improve the result further.

Acknowledgements The authors thank Professor Fernando Capela e Silva, from the Department of Biology and Ana R. Costa and Célia M. Antunes, from the Department of Chemistry, University of Évora, Portugal, for the dataset used in this article.

References

1. World Health Organization (WHO).: Disease Incidence, Prevalence and Disability (2004)
2. Wondermom, N.: HealthBoards Message (2003). http://www.healthboards.com/boards/diabet es/39426-what-differance-between-glucose-intolerant-diabetes.html. Accessed 30 Oct 2017
3. Bandyopadhyay, T., Mitra, S., Mitra, S., et al.: Analysis of pancreas histological images for glucose intolerance identification using wavelet decomposition. In: Satapathy, S.C., Bhateja, V., Udgata, S.K., Pattnaik, P.K. (eds.) Proceedings of the 5th International Conference on Frontiers in Intelligent Computing: Theory and Applications : FICTA 2016, vol. 1, pp 653–661. Springer Singapore, Singapore (2017)
4. Kakimoto, T., Kimata, H., Iwasaki, S., et al.: Automated recognition of pancreatic islets in Zucker diabetic fatty rats treated with exendin-4. J. Endocrinol. **216**, 1–24 (2012)
5. Rato, L.M., e Silva, F.C., Costa, A.R., Antunes, C.M.: Analysis of pancreas histological images for glucose intolerance identification using imagej—preliminary results. In: 4th Eccomas Thematic Conference on Computational Vision and Medical Image Processing (VipIMAGE), pp 319–322. CRC Press (2013)
6. Rojo, M.G., Bueno, G., Slodkowska, J.: Review of imaging solutions for integrated quantitative immunohistochemistry in the pathology daily practice. Folia Histochem. Cytobiol. **47**, 349–354 (2009)
7. Prasad, K., Prabhu, G.K.: Image analysis tools for evaluation of microscopic views of immunohistochemically stained specimen in medical research—a review. J. Med. Syst. **36**, 2621–2631 (2012)
8. Isse, K., Lesniak, A., Grama, K., et al.: Digital transplantation pathology: combining whole slide imaging, multiplex staining and automated image analysis. Am. J. Transplant. **12**, 27–37 (2012)
9. Chen, H., Martin, B., Cai, H., et al.: Pancreas++: automated quantification of pancreatic islet cells in microscopy images. Front. Physiol. **3**, 482 (2013)
10. Berclaz, C., Goulley, J., Villiger, M., et al.: Diabetes imaging—quantitative assessment of islets of Langerhans distribution in murine pancreas using extended-focus optical coherence microscopy. Biomed. Opt. Expr. **3**, 1365–1380 (2012)
11. Aswathy, M.A., Jagannath, M.: Detection of breast cancer on digital histopathology images: present status and future possibilities. Inf. Med. Unlocked **8**, 74–79 (2017)
12. Zong, W., Huang, G.B., Chen, Y.: Weighted extreme learning machine for imbalance learning. Neurocomputing **101**, 229–242 (2013)

Elephant Herding Algorithm for Clustering

Karanjekar Pranay Jaiprakash and Satyasai Jagannath Nanda

Abstract Elephant Herding Optimization (EHO) is a nature-inspired algorithm reported by Wang et al in 2015. The algorithm has a mixed nature of swarm intelligence (behaviour of elephant living in groups) and evolutionary algorithm (reproduction to create baby elephant). It has both exploitation (clan updating operator) and exploration (separating operator) capability to make it a potential algorithm for optimization. In this chapter, the EHO has been suitably formulated to perform clustering task by minimizing intra-cluster distance as cost function. Simulation is demonstrated on cluster analysis of three synthetic and six benchmark datasets. Comparative analysis with RCGA, PSO, and K-means algorithm demonstrate superior percentage accuracy of EHO in the form of box plots. It is also observed that computational time of EHO is higher than K-means but lower than PSO and RCGA.

Keywords Elephant herding optimization · Clustering · Intra-cluster distance

1 Introduction

Clustering is a method of categorizing elements of a dataset into groups in such a manner that element belongs to a group (one cluster) have similar characteristics than other groups (other clusters). It finds several applications in diversified areas including data mining, pattern recognition, image segmentation, analyzing genetic information in bioinformatics, voice mining, text mining, web cluster engines, weather report analysis, etc. The survey papers [1, 2] highlight the developments in clustering techniques, their validation and many more potential applications in the past 50 years.

K. P. Jaiprakash · S. J. Nanda (✉)
Department of Electronics and Communication Engineering, Malaviya National Institute of Technology, Jaipur 302017, Rajasthan, India
e-mail: nanda.satyasai@gmail.com

K. P. Jaiprakash
e-mail: 2016pwc5338@mnit.ac.in

© Springer Nature Singapore Pte Ltd. 2019
J. Kalita et al. (eds.), *Recent Developments in Machine Learning and Data Analytics*,
Advances in Intelligent Systems and Computing 740,
https://doi.org/10.1007/978-981-13-1280-9_30

The nature-inspired metaheuristic algorithms have been enormously employed to perform the cluster analysis in the past two decades [3]. With these algorithms, the clustering is formulated as an optimization problem and the objective becomes minimization/maximization of a fitness function, which partitions the dataset into groups. The techniques have the advantage to perform random search in a given space and ability to avoid local minima in case of multimodal fitness function. Some of these prominent algorithms for clustering include Real-Coded Genetic Algorithm (RCGA) [4], Particle Swarm Optimization (PSO) [5], Improved Differential Evolution (IDE) [6] and Social Spider Optimization [7].

Wang et al. [8] proposed a new Elephant Herding Optimization (EHO) algorithm considering the natural behaviour of elephants living in groups and their process of reproduction for next-generation baby elephants. In any metaheuristic algorithm, a balance between exploitation and exploration of search space using search agents is essential to obtain the global optimum. In EHO, the clan update performs the task of exploitation and the separating operator is responsible for exploration. In [8], Wang et al. applied the EHO on optimization of 15 benchmark functions and reported superior performance over GA, DE and Biogeography-Based Optimization (BBO). Deb et al. [9] reported superior optimization performance of EHO over PSO, Bat algorithm, Grey Wolf Optimization, Firefly algorithm on 9 benchmark functions for 10, 20 and 30 dimensionalities. Wang et al. [10] applied the EHO to solve optimal control of a nonlinear-stirred tank reactor and to solve dynamic economic load dispatch problem. Recently, Meena et al. [11] proposed an improved EHO for multi-objective DER accommodation in distribution systems.

Inspired by the performance of the EHO algorithm over the benchmark meta-heuristic algorithms in this chapter, the concept is suitably modified to solve the task of clustering for benchmark datasets. The sum of intra-cluster distance between the centroid to the data points is considered as the objective function for minimization. This fitness function leads to the formulation of compact clusters. The validation of clusters is carried out by observing the scatter plot of points in each cluster after grouping. Also, the box plot of percentage of accuracy achieved over ten runs signifies the performance of each algorithm.

2 EHO for Clustering

2.1 Background of EHO Algorithm

Elephants live in the form of groups called herds, which contain 10–100 elephants including male and female elephants with their calf. The complete herd is led by the matriarch, which is the eldest and largest female elephant in the group. The herding behaviour is shown in Fig. 1. Male elephants in a group used to come and go away from the group in search of another food source by enlarging the search range, while female elephants search more in a specific region only and take care of the calf.

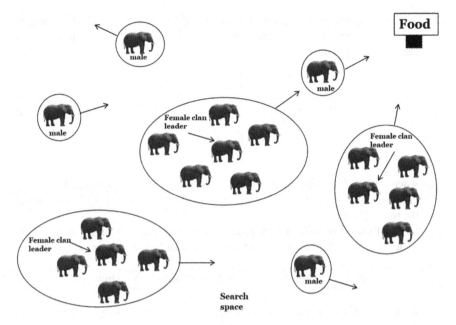

Fig. 1 Behaviour of elephants herding nature

So, it is assumed that male elephants go for exploration and female elephants focus on exploitation. When a baby elephant is born it inherits the properties from the fittest female elephant. So, based on this, the updated solution inherits the location information from fittest solution in the group. In the food search process, if the male elephants find a better location, the whole group will slowly move to that location and after that female elephants do a local search of that region.

2.2 Proposed EHO-Based Clustering Algorithm

Consider a dataset $P_{N \times D}$ given by

$$P_{N \times D} = \begin{bmatrix} p_{11} & \cdots & p_{1D} \\ \vdots & \ddots & \vdots \\ p_{N1} & \cdots & p_{ND} \end{bmatrix} \tag{1}$$

The objective is to group this dataset into K number of clusters. The stepwise EHO algorithm is given here as follows:

(1) **Initialization of cluster centre**: Initial population of elephants which represents the cluster centre is selected from the dataset. For K clusters, the initial cluster centroids are given by

$$
C_{M \times K} =
\begin{bmatrix}
c_{11} c_{12} & \cdots & c_{1K} \\
c_{21} c_{22} & & c_{2K} \\
\vdots & \ddots & \vdots \\
c_{M1} c_{M2} & \cdots & c_{MK}
\end{bmatrix}
\tag{2}
$$

where M is the total population of elephants. Any cluster centre $c_{ij} = \{p_{r1}, p_{r2}, \ldots, p_{rD}\}$ with $i \in [1, M], j \in [1, K]$. r is a random number in the range $[1, N]$.

(2) **Evaluation of fitness function**: The intra-cluster distance from the cluster centre c_{ij} to the data point is used as the fitness function for minimization

$$
D = \sum Dmin\_c1 + \sum Dmin\_c2 + \cdots + \sum Dmin\_cK
\tag{3}
$$

where $Dmin\_c1$ represents minimum Euclidean distance between centre c_{11} to those data points they belong to cluster c_{11}.

(3) **Clan updating operator**: The entire population of elephants $C_{M \times K}$ is divided into T clans. Each elephant in clan L_i with $i \in [1, T]$ is updated, except the worst fit elephant of that clan. The update equation for elephant j in clan L_i is given by

$$
c_{new, L_{i,j}} = c_{L_{i,j}} + \alpha * \left(c_{best, L_i} - c_{L_{i,j}} \right) * rand,
\tag{4}
$$

where rand is a random number in range $[0, 1]$. The α is a scale factor taken in the range $[0, 1]$. If the $C_{L_{i,j}} = C_{best, L_i}$ then update according to equation

$$
c_{new, L_{i,j}} = \beta * c_{center, L_i}
\tag{5}
$$

$$
c_{center, L_i} = \frac{1}{n_{L_i}} * \sum_{j=1}^{n_{L_i}} c_{L_{i,j}}
\tag{6}
$$

where n_{L_i} is the number of elephants in clan L_i. These updated elephants form a female elephants group, which carries out the exploitation operation.

(4) **Separating operator**: All the worst elephants in each clan are eliminated and replaced by new baby elephants as search agents. The update equation for elephant j in clan L_i is

$$
c_{worst, L_i} = \{p_{r1}, p_{r2}, \ldots, p_{rD}\}
\tag{7}
$$

where r is a random number in the range $[1, N]$. These updated elephants form a male elephant group which leads to exploration operation.

(5) **Repetition till convergence**: The steps 2–4 are repeated for a fixed number of iterations till convergence is achieved and the final clusters with their centroid values are reported.

3 System Simulation

In order to evaluate the performance of EHO, in this section, simulation study is carried out in MATLAB 2015B environment in an Intel i5 7th generation laptop with 4 GB RAM and 1.50 GHz clock frequency with Windows 10, 64-bit operating system.

3.1 Datasets Used for Analysis

The simulation study is carried out on three synthetic datasets taken from [5] and six real-life datasets taken from UCI machine learning repository [12]. The details about the number of samples, features and number of clusters are given in Table 1.

3.2 Comparative Algorithms

The clustering performance of the proposed EHO is compared with that achieved by benchmark algorithms like K-means, RCGA, PSO. All the four algorithms are allowed to run in identical environments. The parameter settings are given in Table 2.

Table 1 Benchmark datasets used for analysis

S. No.	Name of dataset	No. of instances (N)	No. of attributes (d)	No. of clusters (k)
1	Data 3_2	76	2	3
2	Data 6_2	300	2	6
3	Data 10_2	500	2	10
4	IRIS	150	4	3
5	Wine	178	14	3
6	Thyroid	215	5	2
7	Glass	214	9	7
8	Breast Cancer	286	9	2
9	Yeast	1484	8	10

Table 2 Parameter setting for the metaheuristic algorithms used for simulation

Parameters	K-means	RCGA	PSO	EHO
No. of Gen.	100	100	100	100
Population size	30	30	30	30
Constants	–	$Pc = 0.7$ $Pm = 0.3$	$C1 = 2.1$ $C2 = 2.1$ $W = 0.4$	Parent kept for next gen. = 2 $\alpha = 0.5$ $\beta = 0.1$ Clan num. $T = 5$

3.3 Performance Evaluation

Performance evaluation of the clustering algorithms is carried out as follows:

(1) Cluster Visualization using Scatter Plot: The final number of points obtained in each cluster are plotted using the 'Scatter Plot' of MATLAB.
(2) Clustering Accuracy with Box Plot: Accuracy is evaluated by taking the ratio of correctly clustered points to the total points in the given dataset. As the algorithms are heuristic in nature, therefore, 10 independent executions of each algorithm are carried out and the percentage of accuracy obtained is presented graphically with 'Box Plot' of MATLAB.
(3) Computational time: The average run time obtained by each algorithm (mean and standard deviation values for ten independent runs) are reported.

4 Result and Discussion

The scatter plots obtained with the proposed EHO algorithm for the nine datasets are reported in Fig. 2. It is observed that in the three synthetic datasets the obtained clusters are non-overlapping in nature, whereas in the real datasets they are overlapping in nature. The box plots (representing the overall percentage of accuracy by the algorithms) obtained for the nine datasets are plotted in Fig. 3. It is observed that except two datasets Thyroid and Yeast in the rest of seven datasets clustering accuracy of the proposed EHO algorithm is superior. In Thyroid and Yeast, the performance of PSO is better. The computational time obtained by all the four algorithms is reported in Table 3. It is observed that the computational time of K-means algorithm is least followed by EHO, PSO and real-coded GA.

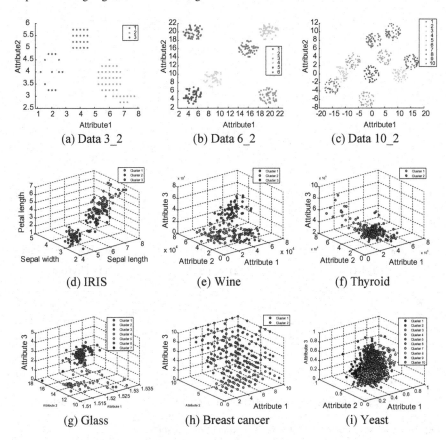

Fig. 2 Scatter plots of obtained clusters using the proposed EHO on nine datasets

Table 3 Average computational time of the metaheuristic algorithms used for analysis

S. No.	Dataset	K-means	Real-coded GA	PSO	EHO
1	Data 3_2	1.06	3.09	1.74	1.82
2	Data 6_2	1.31	3.34	2.18	1.92
3	Data 10_2	1.28	3.53	2.39	2.24
4	IRIS	1.13	2.90	3.89	1.66
5	Wine	0.71	3.92	2.80	2.32
6	Thyroid	0.85	4.12	2.47	2.06
7	Glass	0.84	5.12	3.14	1.62
8	Breast Cancer	1.16	4.34	2.74	2.45
9	Yeast	2.18	5.68	4.23	3.94

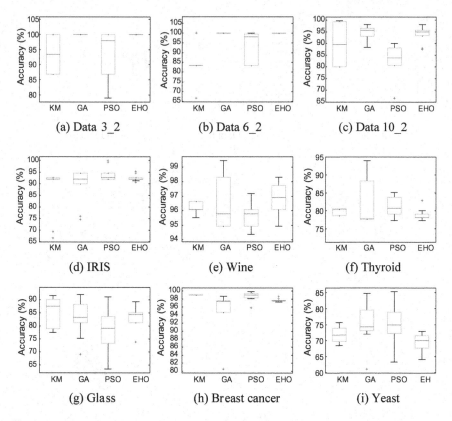

Fig. 3 Box plots of obtained clusters with the proposed EHO on nine datasets

5 Conclusion

In this chapter, the recently proposed metaheuristic algorithm EHO has been suitably modified to perform clustering. The performance of the proposed method is demonstrated on three synthetic and six benchmark datasets. Simulation study reported that the proposed EHO algorithm provides superior accuracy in seven datasets and only in case of Thyroid and Yeast the accuracy of PSO is better. The computational complexity of the K-means is lower followed by EHO, PSO and real-coded GA. Therefore, the EHO algorithm in near future can suitably be applied to solve real-life clustering problems of science and engineering.

References

1. Jain, A.K., Murty, M.N., Flynn, P.J.: Data clustering: a review. ACM Comput. Surv. (CSUR) **31**, 264–323 (1999)
2. Jain, A.K.: Data clustering: 50 years beyond K-means. Pattern Recogn. Lett. **31**, 651–666 (2010)
3. Nanda, S.J., Panda, G.: A survey on nature inspired metaheuristic algorithms for partitional clustering. Swarm Evol. Comput. **16**, 1–18 (2014)
4. Maulik, U., Bandyopadhyay, S.: Genetic algorithm-based clustering technique. Pattern Recogn. **33**, 1455–1465 (2000)
5. Nanda, S.J., Panda, G.: Automatic clustering algorithm based on multi-objective immunized PSO to classify actions of 3D human models. Eng. Appl. Artif. Intell. **26**, 1429–1441 (2013)
6. Das, S., Abraham, A., Konar, A.: Automatic clustering using an improved differential evolution algorithm. IEEE Trans. Syst. Man Cybern. Part A Syst. Hum. **38**, 218–237 (2008)
7. Shukla, U.P., Nanda, S.J.: Parallel social spider clustering algorithm for high dimensional datasets. Eng. Appl. Artif. Intell. **56**, 75–90 (2016)
8. Wang, G.G., Deb, S., Coelho, L.D.S.: Elephant herding optimization. In: 2015 3rd International Symposium on IEEE Computational and Business Intelligence (ISCBI), pp. 1–5 (2015)
9. Deb, S., Fong, S., Tian, Z.: Elephant search algorithm for optimization problems. In: 2015 Tenth International Conference on IEEE Digital Information Management (ICDIM), pp. 249–255 (2015)
10. Wang, G.G., Deb, S., Gao, X.Z., Coelho, L.D.S.: A new metaheuristic optimisation algorithm motivated by elephant herding behaviour. Int. J. Bio-Inspired Comput. **8**, 394–409 (2016)
11. Meena, N.K., Parashar, S., Swarnkar, A., Gupta, N., Niazi, K.R.: Improved Elephant Herding Optimization for Multiobjective DER Accommodation in Distribution Systems. Accepted in IEEE Transactions on Industrial Informatics (2017)
12. UCI Machine Learning Repository. Available online: http://archive.ics.uci.edu/ml/index.php

Fusion-Based Underwater Image Enhancement by Weight Map Techniques

Pooja Honnutagi, V. D. Mytri and Y. S. Lalitha

Abstract Nowadays, underwater vehicles are set with visual sensors. But, it is very likely that underwater images can be captured using optic cameras having low quality due to light conditions in real-life appliances. This is helpful to apply image enhancement techniques to enhance visual quality of the imagery as well as improve interpretability and visibility. We establish a novel strategy that efficiently enhances the visibility of underwater images or videos. The research method is buildup on the fusion approach that takes a series of inputs imitative from the initial image or video. Almost, the proposed fusion-based method aims to give a final picture that overcomes the insufficiency existing in degraded input images by employing several weight maps that differentiate the regions characterized by poor visibility. The widespread experiments exhibit usefulness of our result thus the visible range of underwater images or video is considerably increased by improving both sight difference and color appearance. Our proposed methods are luminance, chromatic, salient weight map, and Laplacian pyramid fusion techniques which enhances a good quality for underwater image.

Keywords Luminance · Chromatic · Salient weight map and Laplacian pyramid fusion technique

P. Honnutagi (✉)
Department of Computer Science Engineering, VTU, Belagavi, Karnataka, India
e-mail: poojahonnutagi@gmail.com

V. D. Mytri
Deperpment of Electronics and Communication Engineering,
AIET, Kalaburagi, Karnataka, India
e-mail: vdmytri.2008@rediffmail.com

Y. S. Lalitha
Deperpment of Electronics and Communication Engineering,
NMIT, Bangalore, Karnataka, India
e-mail: patil_lalitha12@yahoo.com

© Springer Nature Singapore Pte Ltd. 2019
J. Kalita et al. (eds.), *Recent Developments in Machine Learning and Data Analytics*,
Advances in Intelligent Systems and Computing 740,
https://doi.org/10.1007/978-981-13-1280-9_31

1 Introduction

Underwater image is a significant topic in the research field. The most recent problem of capturing image in underwater is mainly due to attenuation of light [1]. Underwater imaging is a big challenging appropriate to the physical properties existing in such surroundings. Dissimilar or familiar pictures, underwater pictures suffer from poor visibility due to reduction of broadcasted light. Then light can be attenuated exponentially by the distance and intensity due to scattering and absorption effects [2]. The absorption considerably decreases the light energy while scattering modifies the light direction. These images typically exhibit low contrast and color distortion [3]. The random attenuation of light is a major problem of underwater appearance while the portions of the light back scattered by the medium with sight significantly degrade the visual contrast. Virtually, objects in sea water, at the distance of more than 10 m can approximately identical though the colors faded because their attribute wavelength can cut according to depth of water.

The amount of images attained under sea pictures are low quality because of the precise broadcast property of brightness in sea water, thus image enhancement is essential to permit valuable analysis for the operators [4]. The main significant cause of the underwater picture degradation is due to the transmission of property of the light in seawater like light changes direction and light disappears. There is another problem which is related to the depth [5], due to underwater optics, red color disappears at deepness of about 3 m and orange color start thinning later, yellow color can lost at deepness of about 10 m and green continues at next depths, so lastly at 25 m only blue color remains, thus image classically show bluish. The drawback in marine snow such as snowflakes suspended in Deep Ocean. Marine snow can produce bright artifacts as a result underwater images suffers from low visibility, nonuniform lighting,low contrast, diminished colors and blurring of image characteristics.

2 Literature Survey

Serikawa et al. [1] propose a new technology to improve the underwater images with dehazing. Color change and scattering are the two main mistakes of the distortion for the underwater imaging. Scattering can be caused by great incomplete particle like turbid irrigate which encloses abundant elements. The color change or distortion of color corresponds to the changing degrees of the attenuation through the light travelling in the water by dissimilar wavelengths, and rendering ambient underwater mediums dominated through the bluish tone. The key assistances proposed the underwater replica to recompense the reduction difference along the broadcast path, and proposed a fast joint trigonometric filtering dehazing technique. The improved images are characterized by noise level reduction, enhanced exposedness of dark areas, enhanced global contrast as the finest edges and details are improved significantly. The proposed system is comparable to higher quality than the previous methods by considering most recent image processing systems.

Fu et al. [2] introduce a novel retinex scheme-based enhancing for underwater single image. Illumination and reflectance are improved by decomposition of underwater image to deal with the underexposure and fuzz. A novel variation retinex method is implemented and optimization technique is initiated to construct the decomposition. An easy and yet efficient post-processing can adopt to enhance degraded images after decomposition. The experimental results exhibit that the improved images has the assets of the color alteration, intensity, naturalness maintenance and fine sharpness. In future, it can show that, new method improves other sorts of degraded image problems.

Yang et al. [6] propose the underwater color image quality evaluation(UCIQE) metric. The method extracts the main applicable CIELab space statistical characteristics that can be a representative for the underwater image degradations are color cast, noise and blurring by attenuation, floating elements and lightning. The experimental results specify that the planned metric have fast processing instance, which can make it appropriate for the real time image processing. It can be competent to effectively forecast the relative destortion through same scene and the variation between the results of improvement. This can also give enhanced correlation by subjective assessment. This system assures in terms of mutually computational competence and sensible dependability for the real-time applications and generally significant a meaningful structural replica to recognize efficient underwater color image quality evaluation for dissimilar applications.

Galdran et al. [3] proposed a novel system for restoration of underwater imagery that undertakes both color corruption and loss of visibility. They extends Dark Channel system; adjusted to the system, these images were degraded. They give a common method to place artificially illumine regions inside an underwater prospect, and propose a system to grip these areas suitably, avoids color artifacts visibility because of incorrect estimation of thes depth within them. Concerning the results, they have projected to achieve a twofold assessment. Initial, to determine the enhancement of contrast image, they resort to the metric anticipated into compute visual enhancement on the imagery degraded through fog. Next is to estimate the value of improved colors, they proposed to employ 3 basic indicators that can adopt properly to difficult of measuring the improvement of underwater image colors after the image restoration. Images restored through the algorithm has been evaluated by 5 various methods. The experimental results demonstrate that the method attained desirable quality image, through a visibility improvement similar or enhanced than the other latest methods. Because of the color recovery, they retrieve natural color, constantly position among the greatest locations between the various images, despite of the dissimilar water conditions.

Ancuti et al. [7] describe the new scheme to improve the underwater images and videos considered on the principles of fusion; the scheme derives necessary weight measures and inputs by the degraded type of image. To overcome drawbacks of underwater medium they describe two inputs that characterize the color corrected also contrast improved version of the earlier underwater image or frame, however

the 4 weight maps desired to improve the visual of distant objects degraded appropriate to average absorption and scattering. The approach represents the single image
method that cannot entail specific hardware or familiarity concerning that the underwater medium or prospect structure. The fusion structure supports chronological
coherence among adjacent structures with performing an efficient edge protecting
noise level reduction scheme. The improved videos and images are considered from
the reduced noise stage, better visibility of dark areas and improved global difference
though the fine detail and edges can be improved considerably. In calculation, the
effectiveness of the enhancing algorithms verified for some challenging applications.

Aysun et al. [8] describe the underwater images are low aspects appropriate to
the partial assortment of blurring and low contrast. The image enhancement was an
imperative task for the underwater image. They presented new enhancement method
for underwater image. The better image is constructing with summing IMF's of RGB
channels through a best weight set attained by this technique. The improved image
is acquired by the proposed method when exposed to give best visual concert than
conservative enhancement schemes like contrast stretching and alternative techniques
accessible in the novel. The improved image gives a more visibility, interpretability
and is best in terms of clarity and color. The poor contrast mistakes commonly
encounter in underwater image can relatively resolve by proposed method. In future
the contrast of images can be improved much more.

Lu et al. [9] effectively designs the new image enhancement algorithms for underwater optical image processing. They proposed the easy prior on the basis of variation
in attenuation between R color channels, can motivated to approximate transmission
map. Another involvement is to recompenset the transmission with guided trigonometric bilateral filter, which not only has benefit of the noise removing and edge preserving but also speedup the computational difficulties. The proposed αACE-based
system of underwater picture color correction system can colorful the underwater
destorted image fine than the state-of-the-art algorithms, also by small computation period. The proposed system can appropriate for the real-time measurement
in Underwater Mining Systems (UMS). This system also includes several mistakes
likecalculation of the probable occurrence of an artificial light source cannot be
allowing for the feature estimation and also inappropriate for the underwater image
quantity.

3 Methodology

Figure 1 represents a proposed block diagram of the underwater image fusion. Initially, we take two input underwater images and apply preprocessing step by resizing
the image. To enhance the image, we apply the enhancement techniques such as luminance weight map, chromatic weight map, and salient weight map. Then, Laplacian
pyramid fusion technique is applied to fuse both images to get the enhanced image.

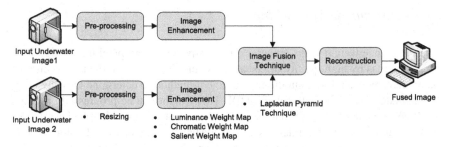

Fig. 1 Proposed system for underwater image fusion

3.1 Preprocessing System

In image processing input image, like photograph; the output image might be set of characteristics or an image or parameters which can related to image. All image-processing algorithms are consider 2D image, then apply standard preprocessing methods to it. Image scaling is a process of resizing an image in graphics. Scaling can be nontrivial procedure that involves the trade-off between smoothness, sharpness, and efficiency. The size of an image might be enlarged or reduced with bitmap graphics, thus pixels form a more visible image. This image is "soft" when pixels were sharped or averaged. The trade-off might be in signal processing for re-rendering of the image, which might be noticeable otherwise, edge skipping and lower edge rate occurs in computer animation.

3.2 Image Enhancement

The global contrast enhancement methods were limited to dealing with hazy sights remarked earlier authors. Because of this, the optical density of haze differences across an image and involves all values are differently at each pixel. The drawbacks of contrast enhancement metods such as histogram equalization, gamma correction, and white balance is due to the constantly performs the identical operation across the complete image. To overcome this drawback, we are introducing three weight maps techniques. These maps were designed in per-pixel fashion to describe the spatial relations of degraded areas. The Proposed weight maps balance the contribution of each input and make sure that area is more contrast or saliency from derived input and receive privileged values.

3.3 Luminance Weight Map

The luminance weight maps processes the prospect of every pixel then assign high ideals to areas by high-quality of visual range and little ideals to the rest. While the hazy image consist low saturation, an efficient method to determine these assets is to calculate loss of the colorfulness [10]. This weight can be progressed based on RGB color channel data. So that we can make use of glowing identified assets that more saturated colors capitulates superior ideals in single or two of color channel [11]. Then this map is just computed (for every input L^k, by k index the imitative input) as the difference (for each pixel place) between R, G and B color channels with L as of input, which is given by Eq. (1),

$$W_L^k = \sqrt{1/3(R^k - L^k)^2 + (G^k - L^k)^2 + (B^k + L^k)^2} \tag{1}$$

While the L luminance can be calculated by averaging a R,G,B channels, this yield difference of superior values meant for saturated pixels supposed to be element of primary haze-free areas. Otherwise, haze generates colorlessness and poor intensity, then compution allocates minute values (decreasing the involvement of those positions to yield) for the hazy area and depreciated areas (for example, in second derivative input refer to areas that contain loss of light and then encompass the dark appearance). The WL map can be a simple, yet effectual recognition of such areas [12].

Algorithm1: Luminance weight Map
Input: *preprocessed image*
Output: *Normalized compressed image*

Step 1: Load the preprocessed image
Step 2: Get the luminance for pixel x from log-norm luminance component.
Step 3: Get the luminance for pixel y form restored luminance component.
Step 4: Calculate scale factor by division.
Step 5: Multiply RGB value by scale factor
Step 6: Repeat till last pixel for step 7
Step 7: Normalized compressed image
End Algorithm

3.4   Chromatic Weight Map

The chromatic weight map manages saturation gain in output image [11]. This can enthused with the reality that common human desired image is considered through the high stage of dispersion. Thus, the color can be intrinsic indicator of image value; often alike color improvement strategy is also executed in quality improvement. To attain this weight map, for every pixel distance among its saturation value and most of saturation series is calculated by Eq. (2).

$$W_c^k(x) = \exp\left(\frac{-(S^k(x) - S_{\max}^k)^2}{2\sigma^2}\right) \tag{2}$$

where k represents the indexes of derived efforts, the default value of standard deviation is $\sigma = 0.3$ and $S_{\max}$ represents the maximum value of saturation range that establishes with color space dispersion. Therefore, the minute values can be allocated to the pixels by reduced saturation though the majority of saturated pixels get more values. This map will ensure that first saturated areas are better described in the final output.

3.5   Saliency Weight Map

The saliency weight map identifies amount of conspicuousness with neighborhood areas. This perceptual value calculate the assured object or person stands from rest of an image, or close by areas. In common, the saliency no matter what the motivation behind it seeks to estimates the contrast image areas which are relative to their environment (on the basis various image characteristics like orientation, intensity or color) [11]. This approach is inspired by biological thought of center-surround contrast. This weight at the pixel position (x, y) of input I_k can be depicted in Eq. (3).

$$W_s^k(x) = \left\| I_k^{\omega hc}(x) - I_k^\mu \right\| \tag{3}$$

where I_k^μ depicts the arithmetic mean value of input I_k (the constant rate during the complete process is calculates only once) while $I_k^{\omega hc}$ depicts the blur version of same input that aspires to take out high frequency like noise. $I_k^{\omega hc}$ depicts attain using 5×5 separable binomial kernel with high-frequency cut-off rate $\omega hc = \pi/2.75$. For minute kernels, the binomial kernel was good estimate of its Gaussian counterpart, but it consists advantage that can be calculated efficiently. The blurred version of image $I_k^{\omega hc}$ and arithmetic mean I_k^μ are calculated, therefore saliency can attain in pixel fashion. Moreover, this is possible to produce maps with the well-defined boundaries as well uniformly highlighted salient areas [5].

This characteristic of saliency map prevents introducing unnecessary artifacts in resultant image. This image yield by fusion method, since neighboring comparable values are assigned equally on saliency map. In addition, this map empha-sizes large areas and estimates the uniform values for complete salient areas. As a result the effect of its gain is to improve the global as well local area of contrast appearance (i.e., corresponds to the more scale). Processing the large and dissimilar set of degraded image, then observed that these three events are uniformly important. First measures the maximum impact on visibility. The resulted weights W^k attained by multiplication of the process of weight maps WLk, W_C^k W_S^k. This constrains that calculation at each pixel place x of regularized weight map to identical one.

3.6 Laplacian Pyramid Fusion Technique

Image fusion is commonly used technology to increase the visual interpretation of images in different applications like medical diagnosis, enhanced vision system, military, surveillance and robotics. In this paper, Laplacian Pyramid Fusion tech-nique is used for underwater images.

The system of Laplacian pyramid construction and reconstruction is demonstrated in Fig. 2. The image at the 0th stage Z_0 of the size $M \times N$ can be reduced (downsam-pling) to attain next stage Z_1 of the size $0.5M \times 0.5N$ where equally spatial density values are decreased [13]. In the same way, Z_2 as a reduced version of Z_1 and so on. Then stage to stage image will perform using function of reduce R. Then reduction function R can be represented as Eq. (4).

$$Z_k = R(Zk - 1) \tag{4}$$

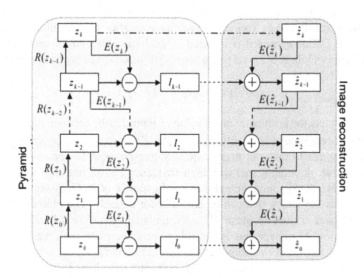

Fig. 2 Laplacian pyramid construction

The reduced reverse function can expand the function of E (upsampling). Its cause is used to enlarge the image of $M \times N$ size to image of $2M \times 2N$ size. The expand function E is defined by Eq. (5).

$$\hat{Z}_k = E(\hat{Z}_{k+1}) \tag{5}$$

Construction of the pyramid can be written as Eqs. (6) and (7):

$$Z_{k+1} = R(Z_k) \tag{6}$$

$$l_k = Z_k - E(Z_{k+1}) \tag{7}$$

where $l_0, l_1, \ldots l_{k-1}$ represents the Laplacian image pyramids that enclose the band-pass filter images and maintaining these files to exploit on rebuilding procedure and xx represents the coarser stage image. Then, k level of the image pyramid can be demonstrated as $P_k \rightarrow \{Z_k, l_0, l_1, \ldots l_{k-1}$. The coarser level is shown in Eq. (8).

$$\hat{Z}_k = Z_k \tag{8}$$

Because there is no extra decomposition beyond this level shown by Eq. (9):

$$Z_{k-1} = l_{k-1} + E(\hat{Z}_k) \tag{9}$$

Let us consider that two images (I_1 and I_2) are used to fuse.

Pyramid construction is done by keeping the error records for each image. Signify the assembled k levels of the Laplacian image pyramid for 1st image as given in Eq. (10).

$$P_k^1 \rightarrow Z_k^1, l_0^1, l_1^1, \ldots, l_{k-1}^1 \tag{10}$$

and similarly for second image is given by Eq. (11).

$$P_k^2 \rightarrow Z_k^2, l_0^2, l_1^2, \ldots, l_{k-1}^2 \tag{11}$$

Thus, the image rule fusion is shown in Eq. (12)
At kth level,

$$Z_k^f = \frac{\hat{Z}_k^1 + \hat{Z}_k^2}{2} \tag{12}$$

For $k - 1$ to 0 levels, it is defined by Eq. (13) and Eq. (14):

$$Z_{k-1}^f = l_{k-1}^f + E(Z_k^f) \tag{13}$$

Fig. 3 Flow chart for
Laplacian pyramid

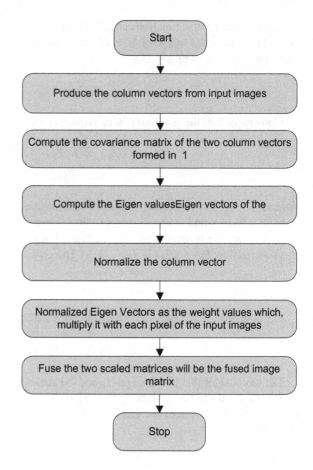

$$l_{k-1}^f = \begin{cases} l_{k-1}^1 |l_{k-1}^1| \ge |l_{k-1}^2| \\ l_{k-2}^2 |l_{k-1}^1| < |l_{k-1}^2| \end{cases} \qquad (14)$$

Then, the amount assessment is done on corresponding pixels. The pyramid $I_f = f Z_0$ is the fused image. The Laplacian Pyramid implements the "pattern selective" method for fusion; so that the composite image can construct not a pixel at a time as show in Fig. 3.

The essential design is to execute pyramid decomposition on every source images, then incorporate all these decomposition to appear as a composite image and finally rebuild the fused image with executing inverse pyramids transform. Once, the arrangement procedure choose the component pattern as the source also copies it to composite pyramid, for discarding the fewer patterns. Then, this method averages a sources pattern. This will decrease the noise then offers constancy; the source pictures enclose the similar pattern data.

Table 1 Existing and proposed model performance comparison table

S. No.	Author	Methods	PSNR	MSE	Entropy
1	Qiao et al. [14]	CLAHE-WT	13.39	49.21	6.68
2	Ahmed Shahrizan et al. [15]	RGB-HSV	14.23	26.6	7.70
3	Ansar et al. [4]	PCA and DWT	17.85	32.63	6.80
4	Proposed model	Laplacian pyramid fusion	18.01	26	8.12

4 Experimental Results

The fusion is commonly used technology to improve the sight interpretation of images in several applications such as medical diagnosis, enhanced vision system, military, surveillance, and robotics. It has been widely used in many fields such as object identification, classification, and change detection. In our research work, we are considering the underwater images or video to fuse it. Figure 4 represents the illustration of underwater image fusion, in which Fig. 4a, b represents input images and Fig. 4c depicts fused image. The Mean Square Error (MSE), the Peak Signal-to-Noise ratio (PSNR), and entropy are [14] used for performance analysis. The quality of processed image is analyzed using Eqs. (15) and (16):

$$\text{MSE} = \frac{1}{MN} \left\{ \sum_{i=0}^{M-1} \sum_{j=0}^{N-1} [f(i, j) - g(i, j)]^2 \right\} \tag{15}$$

$$\text{PSNR} = 10 \log_{10} \left\{ \frac{[\max f(i, j)]^2}{\text{MSE}} \right\} \tag{16}$$

Entropy indicates the abundance of information that measures the image information content, which is given in Eq. (17).

$$H = \sum_{i=1}^{255} p_i * \log_2 p_i \tag{17}$$

The respective performance values are compared with the existing techniques. Existing and proposed functional parameters are presented in Table 1.

Similarly, the respective performance analysis graph is depicted in Fig. 5. The comparison table and performance graph presents that designed system gives better quality of fused image. Further, the proposed system accuracy can also be improved using advanced computer vision algorithm.

(a)   **(b)**   **(c)**

Fig. 4 **a** First input image. **b** Second input image. **c** Fused output

Fig. 5 Comparison graph
for PSNR ratio

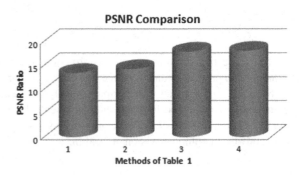

5 Conclusion

Underwater images might be low quality because of limited light range, blurring, and low contrast. Image enhancement is a significant task for underwater images. In this paper, novel enhancement method is presented for underwater images. The improved image can obtain from proposed methods is to provide good illustration performance than the conventional enhancement techniques. The enhanced underwater images give more visible, interpretability, and good color clarity. The low contrast problem usually encounters in underwater images is resolved by the proposed system. Furthermore, the contrasts of images are improved.

References

1. Serikawa, S., Lu, H.: Underwater image dehazing using joint trilateral filter. Elsevier **40**(1), 41–50 (2014)
2. Fu, X., Zhuang, P., Huang, Y., Liao, Y., Zhang, X.P., Ding, X.: A retinex-based enhancing approach for single underwater image, pp. 4572–4576. IEEE (2014)
3. Galdran, A., Pardo, D., Picón, A., Alvarez-Gila, A.: Automatic red-channel underwater image restoration. Elsevier **26**, 132–145 (2015)
4. Ansar, M.K., VR, V.K.: Performance evaluation of image fusion algorithms for underwater images-a study based on PCA and DWT. Int. J. Image Graphics Signal Process **6**(12), 65 (2014)
5. Ancuti, C.O., Ancuti, C., De Vleeschouwer, C., Bekaert, P.: Color balance and fusion for underwater image enhancement. IEEE (2017)
6. Yang, M., Sowmya, A.: An underwater color image quality evaluation metric. IEEE **24**(12), 6062–6071 (2015)
7. Ancuti, C., Ancuti, C.O., Haber, T. Bekaert, P.: Enhancing underwater images and videos by fusion, pp. 81–88. IEEE (2012)
8. Çelebi, A.T., Ertürk, S.: Visual enhancement of underwater images using empirical mode decomposition. Expert Syst. Appl.**39**(1), 800–805 (2012)
9. Lu, H., Li, Y., Serikawa, S.: Underwater image enhancement using guided trigonometric bilateral filter and fast automatic color correction. IEEE (2014)
10. Ancuti, C.O., Ancuti, C., Haber, T., Bekaert, P.: Fusion-based restoration of the underwater images, pp. 1557–1560. IEEE (2011)
11. Drews, P.L.J., Nascimento, E.R., Botelho, S.S., Campos, M.F.M.: Underwater depth estimation and image restoration based on single images. IEEE **36**(2), 24–35 (2016)
12. Choi, B.S., Kim, D.C., Kyung, W.J., Ha, Y.H.: Multi-spectral flash imaging under low-light condition using optimization with weight map. IEEE **1**, 33–39 (2014)
13. Sahu, A., Bhateja, V., Krishn, A.: Medical image fusion with laplacian pyramids, pp. 448–453. IEEE (2017)
14. Qiao, X., Bao, J., Zhang, H., Zeng, L., Li, D.: Underwater image quality enhancement of sea cucumbers based on improved histogram equalization and wavelet transform. Elsevier **4**(3), 206–213 (2017)
15. Ghani, A.S.A., Isa, N.A.M.: Underwater image quality enhancement through rayleigh-stretching and averaging image planes. Int. J. Naval Archit. Ocean Eng. **6**(4), 840–866 (2014)
16. Fang, S., Deng, R., Cao, Y., Fang, C.: Effective single underwater image enhancement by fusion. J. Comput. **8**(4) (2013)
17. Ancuti, C.O., Ancuti, C., De Vleeschouwer, C., Bovik, A.C.: Single-scale fusion: an effective approach to merging images. IEEE(2016)

Analysis and Prediction of City-Scale Transportation System Using XGBOOST Technique

Sai Prabanjan Kumar Kalvapalli and Mala Chelliah

Abstract Recent research has been carried out in predicting the taxi rides that can be shared with different customers with similar pickup point also called as taxi pooling by taking into consideration parameters such as pickup latitude mean waiting time and distance between the pickup points. The influence/impact of weather is vital in deciding the taxis that needed to be present in an area for successful pooling transportation. This paper proposes a methodology incorporating the impact of weather in taxi pooling using XGBOOST technique. The proposed method is tested with New York City taxi dataset and New York City weather dataset. The simulation results show that there is a significant improvement in analysis and prediction of delays after incorporating weather details.

1 Introduction

Pooling of taxi rides is an evolving and also a very high level of complexity involving the process that is proved to be an NP-Complete algorithm and only optimization can be provided to it as a solution an not a best one [1].

Transpiration algorithms, in general, use spatial–temporal characteristics of various ride requests to find taxis for various customers with similar pickup requests that have been limited to not more 200 m in distance and not more than 5 min in delay. Taxi Ride Sharing is an evolving and also a highly evolved form of understanding the urban transportation which uses spatial temporal characteristics of various ride requests to find taxis for various the whole process of pooling can be broadly explained as joining a taxi ride of at least two separate ride requests that are having pickup requests being now away from more than 200 m and waiting time for the taxi to reach both of them being not more than 5 min and the end goal of the optimization

S. P. K. Kalvapalli · M. Chelliah (✉)
National Institute of Technology Trichy, Trichy, India
e-mail: cmala123@gmail.com

S. P. K. Kalvapalli
e-mail: saiprabanjan@gmail.com

© Springer Nature Singapore Pte Ltd. 2019
J. Kalita et al. (eds.), *Recent Developments in Machine Learning and Data Analytics*,
Advances in Intelligent Systems and Computing 740,
https://doi.org/10.1007/978-981-13-1280-9_32

being both saving money for customers and taxi companies to economize the use of their resources [2].

As an ecological point of view, greenhouse gas emissions can also lowered. With an empirical evaluation, the approach of combining at least two rides that are less than 200 m apart and with request times being less than 5 min over a total request of five million trajectories almost of forty-eight percent of the rides are open for merging saving three million kilometers of travel distance, two fifty thousand liters of petroleum, and reducing the greenhouse gases [3]. Taking weather data into consideration is a complicated calculation to do, as prediction is a costly operation and it becomes more and more divergent from accuracy as the no. of parameters that are being taken to the estimation increases [4].

Instead the paper takes a much simpler approach of taking the general promise of how the normal taxi fares are calculated. Taxi fares in NYC follow a fixed approach for calculating various charges like opening charges (Which means base fare that gets added to the journey in spite of its total distance), surcharges (are calculated for providing taxi services in odd timings like night 8 to early morning 6 o'clock and a special charge called as an additional unit (which is used for calculating waiting time, when the taxi is waiting is waiting for more than sixty seconds, also when the taxi is moving slower than six miles per hour) [5]. Additional unit charge is traditionally calculated as the price for one-fifth of a mile.

All of these values should be combined in fractional value of their total measure of distance and time while calculating the final fare for each unit of distance that can be either miles or kilometers. The official website for New York City Taxi and Limousine Commission, 2016, mentions that the taxi fare can be broken down into following independent components, like, Total Travel Fare = Fixed Charge + Distance-Based Charge + Delay Charge, where final charge represents the fare that is added as base price, and the time of the day which is used for surcharge calculations and delay charges will be calculated whenever the taxi is in a stationary position for more than sixty seconds and is traveling at the speed of six miles per hour, where, the fuel consumption is more than the distance that is actually being covered by the taxi [6].

Unless or until any changes happen to the taxi travel by the events like intersection or weather conditions along the trip. The total charge should be equal to the summation of fixed charge and distance charge. As an ideal case calculation, a two-mile trip in the city is taken and it should attract 2.5 dollar as fixed charge and 2.5 * 0.5 dollar as fare for the metered distance and the total fare finalizes to seven and half dollars [7]. This is the fare that is ideal and any fare more than should be considered as surcharge. Any amount that is calculated more than that should be considered the delay related surcharge [8].

Based on the variations like the times of the day and weather conditions comparisons actually define the surcharges that are added at the end of each trip. And the only way to find the base price on which the comparisons can be made is by finding the most ideal trips possible in the day that has varying surcharges and the ideal traffic conditions are from morning five to six under the clear sky [6, 9].

Even though it seems to be logical, the time period is again the assumption that all the factors leading to the recurrent congestion stay normal. With the delays from the congestion and clear weather are considered to be of contributing to the additional price paid is negligible, the only surcharge that will be added is going to come if any is from the intersections alone [10]. Some research papers calculated that the rate of delay signal to be 1.57 times the normal period. These calculations can be considered for the New York City (NYC) taxi dataset and the weather dataset [11].

For the more comprehensive analysis of delays on New York City (NYC) metropolis readers can refer Ashish Lal 2017, in which delays in the taxi distribution system in respect to congestion using regression analysis in three area airports Liberty, John F. Kennedy, LaGuardia. Range of environmental and operational factors have also been taken into account which explains that adverse impact has largest impact on the expected delay which is followed by congestion during the metroplex peak. Congestion due to demand hotspots like Manhattan metropolis area, airport has been taken special care [12].

Recurrent delay and nonrecurrent delay is the theme that is discussed in the M. Anil et al., 2017. The delays that are occurring because of weather is calculated to be leading to crashes and additional charges by rate of ten percent out of all possible reasons. The paper takes only four categories out the available twelve categories for the sake of simplification [13].

By reducing the clear, partly, mostly cloudy, overcast, fog, haze, light rain and light freezing rain all under one category being just clear which also reduces the complexity of the dataset. Whereas light rain, light freezing rain will be considered as light rain. Only heavy rain and rain are given a special coverage.

2   Related Work

Researchers employ techniques like standard deviation (SDTT), Average Travel Time (ATT), and Coefficient of Variation (CoV) for each hourly period throughout the week and various weather conditions. Afterwards, a classification and regression trees methodology is used to determine the temporal patterns are discussed with respect to the findings and assumptions of Value of Time (VoT), and Value of Reliability (VoR). Traditional peak hours are not most of the cases where the congestion are arisen, but the congestion is actually dependent on inter-period heterogeneity in terms of ATT and SDTT. In [5], For the more comprehensive analysis of delays on New York City (NYC) metropolis readers can refer Ashish Lal 2017, in which delays in the taxi distribution system in respect to congestion using regression analysis in three area airports Liberty, John F. Kennedy, LaGuardia. Range of environmental and operational factors have also been taken into account which explains that adverse impact has largest impact on the expected delay which is followed by congestion during the metropolis peak. Congestion due to demand hotspots like Manhattan metropolis area, airport has been taken special care. Recurrent delay and nonrecurrent delay is the theme that is discussed in [7]. The delays that are occurring because of weather is calculated to be leading to crashes and additional charges by rate of ten percent out of all possible reasons. The paper takes only four categories out the available twelve categories for the sake of simplification. By reducing the clear, partly cloudy, scattered clouds, mostly cloudy, overcast, fog, haze, light rain and light freezing rain all under one category being just clear which also reduces the complexity of the dataset Whereas light rain, light freezing rain will be considered as light rain. Only heavy rain and rain are given a special coverage [8].

2.1 Train Versus Test Overlap

The train dataset and the test dataset should be discrete and also cover similar range of values like time and geographical area which needs to be compared before applying XGBOOST algorithm on them. As the feature selection is a very important part of the XGBOOST algorithm comparing the temporal and spatial properties of the train and the test data is a very important part of the whole process. This step can be formally called as a consistency check and it needs to be performed like an exploration.

And the preferences have been taken such that the examination of the training data is done first before going to the test dataset so that the analysis is blindfolded to the patterns of the data before going to the forecasting functions. The Figs. 3 and 4 represent that relevant comparison plots that prove that indeed the test and the train datasets cover the same range of values.

2.2 Data Formatting

Data Formatting: Here selected features will be converted into integer columns, as the classifiers in the utmost cases cannot deal with categorical values.

3 Results and Analysis

This figure represents the changes in mean duration of a taxi drive in result to the temperature changes. The line represents the pattern of the curve with respect to increase of temperature the duration increases substantially.

Most of the work that has been done in the area of taxi pooling has taken consideration of only rainfall statistics but not the fog analysis which is the most determining factor in the four months in the city in the winter season (Figs. 1 and 2).

This figure represents the changes in mean duration of a taxi drive in result to the temperature changes. The line represents the pattern of the curve with respect to increase of temperature the duration increases substantially.

Humidity also plays a key role in defining the travel time delay because of the city New York lies near to the most tropical part of the United States of America it experiences almost all seasons. In 2012 the New York City has been hit by one

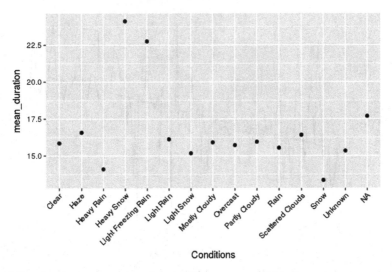

Fig. 1 Complete feature comparison for rainfall impact on travel duration

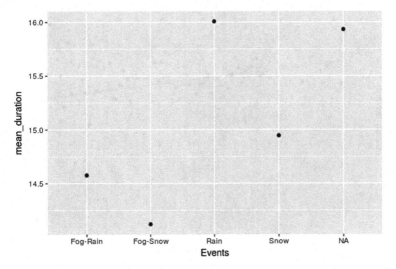

Fig. 2 Comparing Fog and Rainfall impact on travel duration

of the most powerful cyclones in the history of United States of America <Sandy Cyclone> referring the significance of the analysis made here.

Here, the XGBOOST algorithm is used for analyzing the feature that is having the most amount of effect on the travel of the taxi. And temperature stands as the biggest influencer (Figs. 3 and 4).

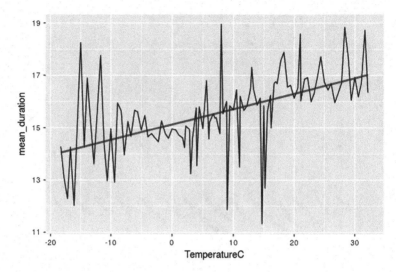

Fig. 3 Representing temperature and its significant impact on travel duration

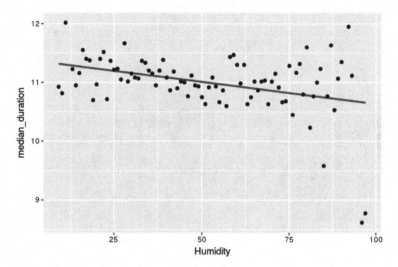

Fig. 4 Humidity representation

4 Conclusion

Any City scale transportation system needs to provide details related to delay, cost involved, traffic congestion, rainfall, snow depth considering real-time data. Weather irregularities play a major role in deciding the delay in city-scale transportation. Most of the work that has been done till now, in ride sharing of taxis didn't take weather data into consideration, either to make sure a good amount of taxis are present in key congestion areas or for calculating the travel time for a certain taxi drive in any given time of the day. This paper proposed a methodology using XGBOOST technique for the analysis and prediction of delays in city scale transportation system, considering the impact of weather and compared the results taken without weather conditions. It is seen from the results that the prediction accuracy is improved with weather conditions (Figs. 5 and 6).

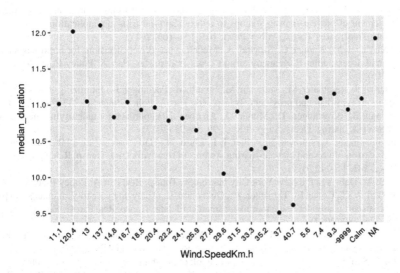

Fig. 5 Representing wind speed and its effects on travel duration

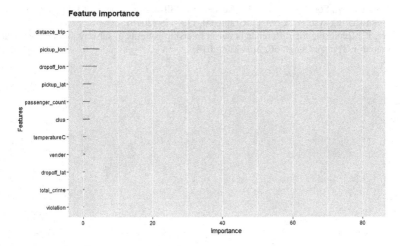

Fig. 6 XGBOOST feature comparision

References

1. Dimitriou, L., Kousta, O., Nikolaou, P.: A discrete-time nonlinear optimal control mechanism for monitoring dynamic signalized urban traffic networks. IFAC-PapersOnLine **49**(3), 19–24 (2016)
2. Dimitriou, L., Kourti, E., Christodoulou, C., Gkania, V.: Dynamic estimation of optimal dispatching locations for taxi services in mega-cities based on detailed gps information. IFAC-PapersOnLine **49**(3), 197–202 (2016)
3. Dimitriou, L., Kourti, E., Christodoulou, C., Gkania, V.: An optimization approach for the placement of bicycle-sharing stations to reduce short car trips: an application to the city of seoul. Transp. Res. Part A: Policy Pract. **105**, 154–166 (2017)
4. Anil Yazici, M., Kamga, C., Singhal, A.: Modeling taxi drivers decisions for improving airport ground access: John F. Kennedy airport case. Transp. Res. Part A: Policy Pract. **91**, 48–60 (2016)
5. Asamer, J., Reinthaler, M., Ruthmair, M., Straub, M., Puchinger, J.: Optimizing charging station locations for urban taxi providers. Transp. Res. Part A: Policy Pract. **85**, 233–246 (2016)
6. Barann, B., Beverungen, D., Müller, O.: An open-data approach for quantifying the potential of taxi ridesharing. Decis. Support Syst. **99**, 86–95 (2017)
7. Faghih-Imani, A., Anowar, S., Miller, E.J., Eluru, N.: Hail a cab or ride a bike? a travel time comparison of taxi and bicycle-sharing systems in New York City. Transp. Res. Part A: Policy Pract. **101**, 11–21 (2017)
8. Lall, A.: Delays in the New York City metroplex. Transp. Res. Part A: Policy Pract. (2017)
9. Briand, A.-S., Côme, E., Trépanier, M., Oukhellou, L.: Analyzing year-to-year changes in public transport passenger behaviour using smart card data. Transp. Res. Part C: Emerg. Technol. **79**, 274–289 (2017)
10. Caulfield, B., O'Mahony, M., Brazil, W., Weldon, P.: Examining usage patterns of a bike-sharing scheme in a medium sized city. Transp. Res. Part A: Policy Pract. **100**, 152–161 (2017)
11. Li, T., Sun, H., Jianjun, W., Ge, Y.: Optimal toll of new highway in the equilibrium framework of heterogeneous households' residential location choice. Transp. Res. Part A: Policy Pract. **105**, 123–137 (2017)
12. Oliveira, G.N., Sotomayor, J.L., Torchelsen, R.P., Silva, C.T., Comba, J.L.D.: Visual analysis of bike-sharing systems. Comput. Graph. **60**, 119–129 (2016)
13. Kim, K.: Exploring the difference between ridership patterns of subway and taxi: case study in seoul. J. Transp. Geogr. **66**, 213–223 (2018)

An Application of Ensemble Random Forest Classifier for Detecting Financial Statement Manipulation of Indian Listed Companies

Hiral Patel, Satyen Parikh, Amit Patel and Abhishek Parikh

Abstract A rising incident of financial frauds in recent time has increased the risk of investor and other stakeholders. Hiding of financial losses through fraud or manipulation in reporting, hence, resulted in the erosion of considerable wealth of their stakeholders. In fact, a number of global companies like WorldCom, Xerox, and Enron and a number of Indian companies such as Satyam, Kingfisher, and Deccan Chronicle had committed fraud in financial statement by manipulation. Hence, it is imperative to create an efficient and effective framework for detection of financial fraud. This can be helpful to regulators, investors, governments, and auditors as preventive steps in avoiding any possible financial fraud cases. In this context, increasing number of researchers these days have started focusing on developing systems, models, and practices to detect fraud in early stage to avoid any attrition of investor's wealth and to reduce the risk of financing. In the current study, the researcher has attempted to explore the various 42 modeling techniques to detect Fraudulent Financial Statements (FFS). To perform the experiment, researcher has chosen 86 FFS and 92 non-Fraudulent Financial Statements (non-FFS) of manufacturing firms. The data were taken from Bombay Stock Exchange for the dimension of 2008–2011. Auditor's report is considered for classification of FFS and non-FFS companies. T-test was applied to 31 important financial ratios, and 10 significant variables were taken into consideration for data mining techniques. 86 FFS and 92 non-FFS during 2008–2017 were taken for testing dataset. Researcher has trained the model using datasets. Then, the trained model was applied to the testing dataset for the accuracy check. Random forest gives the best accuracy. Here, modified random forest model was developed with improved accuracy.

Keywords Fraudulent company · BSE · Random forest · Financial ratio
Predictive model

H. Patel (✉) · S. Parikh · A. Patel · A. Parikh
Ganpat University, Ganpat Vidyanagar, India
e-mail: hiral.patel@ganpatuniversity.ac.in

© Springer Nature Singapore Pte Ltd. 2019
J. Kalita et al. (eds.), *Recent Developments in Machine Learning and Data Analytics*,
Advances in Intelligent Systems and Computing 740,
https://doi.org/10.1007/978-981-13-1280-9_33

1 Introduction

"Prevention is better than cure", this proverb is very well known in health care. However, simple meaning is "It is better to spending ounce on preventing some bad occurring to prevent wastage thousand pounds for damage repair." In fact, this proverb is true not only for the health but also equally true for financial wealth. Fraud can hurt financial wealth of country. Fraud is an act which is done deliberately and intentionally to deprive wealth in any form by deceits. When fraud is taking place, most of the time it affects more than one human being. In fact, fraud at corporate level affects all the stakeholders simultaneously. In corporate, most of the financial frauds are related to asset misappropriation. These types of fraud can fall under the improper asset valuation or concealment of liabilities or fictitious revenues or improper disclosures. These practices lead to serious hurt to the financial wealth of institutions across countries.

It is always better to develop some mechanism to indicate warning signal before any financial fraud in corporate than doing damage repair after scam takes place. Stakeholders do not know any formal system to detect and respond to frauds. It is obvious that none of the models or systems can predict fraud with 100% accuracy, but stakeholders can take some preventive steps if the model helps in giving warning signals about possible frauds. The billions of dollars were assumed to be vanished due to fraud by corporate management annually in the United States [19]. Association of Certified Fraud Examiners in India has noted financial statement fraud in 10% cases with median loss of USD 9.75 lacks [18].

Creative manipulation of financial statement is commonly used practices. Auditing became a very essential tool. In India, any possible fraud in financial statement should be detected by certified auditors. Contrarily, auditors are overburden. Hence, standard auditing procedures are not enough to identify financial frauds. Therefore, some trained model or system can be used as alternatives for the auditors to simplify auditing. It also helps in detecting any possible fraud related to warning signal at prima facie [1]. In this context, if some computer-based model or system is developed, then it can be very helpful to stakeholders for fraud detections.

One of the model development methods is data mining. It is a process of searching insights that may be statistically reliable, actionable, and previously unknown from the available historical datasets [6, 11]. Thus, in current research, 42 different (all possible methods using Weka software) data mining algorithms were applied for training the model by feeding the data and finally were tested for the accuracy. It was found that the random forest method provides best accuracy. In the current study, improved algorithm was developed and proposed using Python computing software for financial fraud detection.

We would like to draw your attention to the fact that it is not possible to modify a paper in any way, once it has been published. This applies to both the printed book and the online version of the publication. Every detail, including the order of the names of the authors, should be checked before the paper is sent to the Volume Editors.

2 Review of Literature

In literature, many researchers have made attempt to define fraud. Fraud is defined as "Intentional act to obtain an unjust advantage" [4] or "A scheme designed to deceive" [20]. However, if the act is not intentional it may not be considered as fraud but considered as an error [10]. Hence, to detect a fraud case is difficult without enough evidence of intentional act by any mechanically defined rule. It is obvious that financial reporting errors that occurred by intentionally or unintentionally create wealth damage to their stakeholders.

In same context, most researchers considered an error in the financial reporting as fraud to develop the model for fraud detection. Eining et al. [5] developed a model to be used as an expert system which can enhance the performance of auditors. In 1997, Green and Choi [8] obtained a Neural Network (NN) model for classifying fraud and non-fraud financial statements. The researchers used five financial ratios resulting from the financial statements as input variables. Fanning and Cogger [7] again tried a neural network for detecting FFS. They analyzed comparison of the performance of their model with logistic regression as well as linear and quadratic function of discriminate analysis and argued that the model developed by them is more efficient in comparison to past models. In 2000, Bell and Carcello [2] built and checked a logistic model for the estimation of fraudulent financial reporting with data samples of 77 FFS and 305 non-FFS. They observed that the significant red flags that differentiate FFS from non-FFS are (1) aggressive attitude of management, (2) a weak internal control environment, (3) difficulty in auditing transaction, and (4) management pressure on meeting earning goals.

Ravisankar [16] worked upon applying intelligent techniques to find fraud in the financial statement of Chinese companies. Researcher had selected 18 dimensions initially, and then for dimension reduction, t-statistic was applied and 10 significant variables are used as input. Researcher observed that PNN was the best performer among all the above methods. Zhou and Kapoor [21] tested four data mining methods, namely, Bayesian networks, regression, neural network, and decision trees.

In past, no literature was found for financial fraud detection that used all available algorithms for model development. Above all, it was not tried in past to improve any available algorithm by modifying the default criteria in the software based on available algorithm. So, in the current study, author tried and successfully achieved more than 5% improvement in accuracy through Python programing language.

3 Research Methodology

3.1 Data

Totally, 31 variables were selected for the development of model based on literature; researcher has decided to refer the listed companies on BSE and/or NSE from time line of 2008 to 2017, as the listed companies represent the total population under consideration. 10 years' time line is the true representation of whole economic cycle, and hence data from time line give an appropriate representation of FFS and non-FFS companies. The researcher has referred the annual auditor's report for the randomly selected 178 companies from manufacturing sector only. Reason behind selecting manufacturing industries is its necessity of huge capital expenditure requirement that leads to possible financial trouble. This leads to creative accounting practices in industry and possible fraud reporting.

Based on auditor's opinion, researcher has classified the companies into two categories, i.e., companies with FFS and companies with non-FFS. Researcher has used auditor's unmodified (Give True and Fair View) versus modified (all others; qualified, declaimer, and adverse) opinion to classify all companies as non-FFS and FFS, respectively. Researcher has selected 86 FFS and 92 non-FFS companies to train the model.

3.2 Process

Step-by-step process with reasoning in bracket for development of decision model is illustrated in Fig. 1. In this figure, reasons for data selection, tool selection, and methodology used are explained.

3.3 Variables Under Consideration

Financial ratios from income statement, balance sheet, and composite were taken for the study based on literature. The data contained 31 financial ratios for each company as Gross Profit Ratio (GPR), Return on Assets (ROA), Return on Sales (ROS), Return on Stockholders' Equity (ROSE), Net Profit Ratio (NPR), Operating Margin Ratio (OMR), Current Ratio (CR), Acid Test Ratio (ATR), Cash Ratio (CashR), Net Working Capital (NWC), Receivable Turnover (RTDebtor), Days Sales Outstanding (DS), Inventory Turnover (ITInventory), Days Inventory Outstanding (DIOS), Days Payable Outstanding (DPOS), Operating Cycle (OC), Cash Conversion Cycle (CCC), Total Asset Turnover Ratio (TATR), Debt Ratio (DR), Equity Ratio (ER), Debt–Equity Ratio (DER), Times Interest Earned (TIE), Earnings per Share (EPS), Price–Earnings Ratio (PE), Dividend Payout Ratio (DPR), Dividend Yield Ratio (DYR), Book Value per Share (BVP), Inventory to sales (IOS), Inventory to Total Assets (IOTA), Taxes to Sales (TS), and Account Payable Turnover (APT).

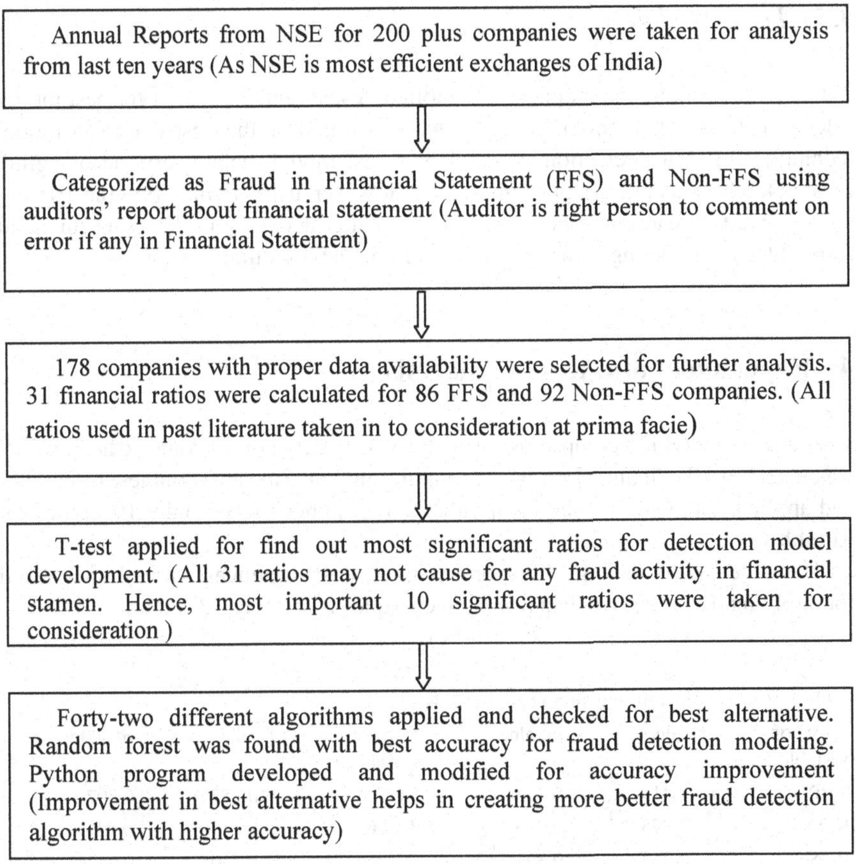

Annual Reports from NSE for 200 plus companies were taken for analysis from last ten years (As NSE is most efficient exchanges of India)

⇩

Categorized as Fraud in Financial Statement (FFS) and Non-FFS using auditors' report about financial statement (Auditor is right person to comment on error if any in Financial Statement)

⇩

178 companies with proper data availability were selected for further analysis. 31 financial ratios were calculated for 86 FFS and 92 Non-FFS companies. (All ratios used in past literature taken in to consideration at prima facie)

⇩

T-test applied for find out most significant ratios for detection model development. (All 31 ratios may not cause for any fraud activity in financial stamen. Hence, most important 10 significant ratios were taken for consideration)

⇩

Forty-two different algorithms applied and checked for best alternative. Random forest was found with best accuracy for fraud detection modeling. Python program developed and modified for accuracy improvement (Improvement in best alternative helps in creating more better fraud detection algorithm with higher accuracy)

Fig. 1 Process for development of detection model

3.4 T-statistic

For identifying significant and important financial ratios that can help in classification of FFS and non-FFS companies, t-test was applied using dummy variable of 0 for FFS and 1 for non-FFS companies. Furthermore, totally 10 financial variables were found significant for identifying FFS companies. Hence, these 10 financial variables namely ROA, ROSE, NWC, DIOS, DPOS, TATR, ER, EPS, BVP, and APT were used for model development using data mining techniques.

3.5 Data Modeling

Totally, 42 modeling techniques were applied, tested, and compared for best model selection. In result, random forest model was found as the best detection model technique for fraud detection. In random forest model, default criteria of "gini" created the best accuracy algorithm model. However, using Python program, it may be possible to create more better criteria for better accuracy in detecting financial fraud through modeling. This was covered in the next section of analysis.

4 Experiment and Result Analysis

The dataset of total 178 companies particularly 92 having non-FFS and 86 having FFS was used with 31 financial ratios. For taking only significant variables, researcher had applied t-statistic. As shown in Table 1, researcher found totally 10 significant variables.

Then, the dataset having 178 records with these 10 significant variables was used for applying 42 predictive data mining techniques using Weka 3.8 tool.

Table 1 t-test for identifying significant variables

Name of variable	t-value	p-value	Name of variable	t-value	p-value
GPR	−0.915	0.363	CCC	0.257	0.797
ROA	2.375	0.019*	TATR	2.256	0.025*
ROS	−0.863	0.391	DR	−1.150	0.252
ROSE	2.736	0.007*	ER	2.820	0.006*
NPR	1.668	0.099	DER	0.924	0.357
OMR	−0.971	0.334	TIE	1.575	0.119
CR	1.291	0.198	EPS	2.746	0.007*
ATR	1.370	0.172	PE	−0.982	0.327
CashR	1.507	0.135	DPR	1.224	0.222
NWC	−2.206	0.030*	DYR	1.463	0.145
RTDebtor	0.018	0.985	BVP	2.901	0.004*
DS	−0.018	0.986	IOS	−1.433	0.156
ITInventory	−0.911	0.365	IOTA	−1.031	0.306
DIOS	−4.722	0.000*	TS	−0.816	0.415
DPOS	−2.443	0.016*	APT	3.411	0.001*
OC	−0.154	0.877			

*Significant at 5%

4.1 Data Mining

Data mining is a statistical technique to find hidden and interesting pattern from the database or data warehouse. It is categorized into two broad categories: descriptive data mining technique and predictive data mining technique. Descriptive data mining technique finds patterns from the large database that is human interpretable. Clustering, association rule discovery, and sequential pattern discovery are descriptive data mining techniques. Predictive data mining technique finds group of an attribute with the help of independent variables under consideration [9]. For this study, researcher has selected predictive data mining and applied 42 predictive data mining algorithms in Weka 3.8 tool in order to test the performance of each model with the respect of accuracy, area under the receiver operating characteristic, and area under precision–recall curve as shown in Table 2.

As shown in Table 2, researcher concluded that random forest is giving best result in terms of all performance measures conducted in the study. With random forest, we are getting 81.46% accuracy, 0.89 ROC, and 0.89 PRC, and these are highest values among all 42 applied algorithms. Random forest is an ensemble learning algorithm, so it is very natural to get highest accuracy because it predicts not from single predictor's output, instead it makes more predictors from which it can make decision. Researcher has still tried to improve the accuracy of a random forest algorithm by setting the different hyperparameters in Python that Weka 3.8 tool does not allow to set. Researcher has explored his study by trying different values of hyperparameters and tested the random classifier model to boost the predictive capacity.

4.2 Ensemble Learning

An ensemble classifier creates multiple predictors for the same problem and makes predictions by aggregating the predictions resulting from the predictors. The multiple predictors or classifiers are called weak learners. To boost predictive capacity, ensemble algorithm combines the outputs of weak learners to get one strong learner. Previous research has shown that an ensemble is often more accurate than any of the single classifier in the ensemble [12, 13, 15]. Multiple base learners can be generated by different strategies like all base learners may use different algorithms to predict the problem or they may use the same learning algorithm having different datasets and different hyperparameters. Two popular methods for ensemble procedure are bagging [3] and boosting [17]. Using re-sampling techniques each Classifier obtains different training sets. Concept of bootstrap is used in bagging method which means bootstrap aggregating. For example, with m individual classifier from N size, training datasets are generated, and then N size of m different data training sets are gener-

Table 2 Comparison of different modeling techniques

Sr. No.	Algorithm name	Accuracy (%)	Incorrectly classified (%)	ROC area	PRC area
1	BayesNet	77.53	22.47	0.8462	0.847
2	NaiveBayes	70.79	29.21	0.7955	0.77
3	NaiveBayesMultiNominalText	51.69	48.31	0.4775	0.489
4	NaiveBayesUpdatable	70.79	29.21	0.7955	0.77
5	Logistic Regression	71.35	28.65	0.7678	0.743
6	MultilayerPerceptron	75.84	24.16	0.8014	0.765
7	SGD	75.84	24.16	0.7561	0.696
8	SGDText	51.69	48.31	0.5	0.501
9	SimpleLogistic	70.79	29.21	0.7681	0.746
10	SMO	70.22	29.77	0.6964	0.643
11	VotedPerceptron	65.73	34.27	0.7123	0.683
12	IBK	73.03	26.97	0.7499	0.695
13	LWL	65.73	34.27	0.718	0.703
14	AdaBoostM1	76.97	23.03	0.8339	0.828
15	Bagging	74.16	25.84	0.8381	0.831
16	ClassificationViaRegression	75.84	24.16	0.8	0.775
17	CVParameterSelection	51.68	48.31	0.4775	0.489
18	FilteredClassifier	74.72	25.28	0.7296	0.676
19	IterativeClassifierOptimizer	76.97	23.03	0.8657	0.867
20	LogitBoost	78.65	21.35	0.87	0.871
21	MultiClassClassifier	71.35	28.65	0.7678	0.743
22	MultiClassClassifierUpdatable	72.78	27.21	0.7561	0.663
23	MultiScheme	51.69	48.31	0.4775	0.489
24	RandomCommittee	79.78	20.22	0.8818	0.876
25	RandomizablefilteredClassifier	73.6	26.4	0.7534	0.697
26	RandomSubSpace	79.78	20.22	0.8469	0.834
27	Staking	51.69	48.31	0.4775	0.489
28	Vote	53.25	46.75	0.4775	0.499
29	WeigtedInstancesHandlerWrapper	51.69	48.31	0.4775	0.489
30	InputMappedClassifier	51.69	48.31	0.4775	0.489
31	DecisionTable	76.97	23.03	0.8198	0.806
32	Jrip	70.22	29.28	0.6757	0.644
33	OneR	64.61	35.39	0.6466	0.595
34	PART	72.47	27.53	0.7516	0.71
35	ZeroR	51.69	48.31	0.4775	0.489
36	DecisionStump	76.97	23.03	0.6602	0.806

(continued)

Table 2 (continued)

Sr. No.	Algorithm name	Accuracy (%)	Incorrectly classified (%)	ROC area	PRC area
37	HoeffdingTree	69.66	30.34	0.7975	0.77
38	J48	70.79	29.21	0.7058	0.661
39	LMT	72.78	27.21	0.7707	0.739
40	**RandomForest**	**81.46**	**18.54**	**0.8926**	**0.897**
41	RandomTree	73.03	26.97	0.7331	0.672
42	REPTree	73.6	26.4	0.7708	0.747

ated by sampling with replacements in datasets. Bagging method generates multiple classifiers that are independent. Also, it has the highest voting ensemble compared to all other multiple classifiers. In fact, result of bagging method is better than other and helpful for reducing issues related to overfittings.

4.3   Random Forest

Random forest is a bagging of de-correlated decision trees. Random forest uses decision tree as a base classifier. Random forest generates multiple decision trees; the randomization is presented in two ways: (1) random sampling of data for bootstrap samples as it is done in bagging and (2) random selection of input features for generating individual base decision trees [14].

We tried to boost the accuracy of predicting of random forest classifier by two ways: one of them is to create multiple decision trees by random training set and random feature selection and the other one is by setting the hyperparameters of the random forest classifier in Python. The code for making random forest classifier by setting the hyperparameters is as follows:

```
In [10]:  import pandas
          import numpy
          from sklearn import datasets, metrics
          from sklearn.ensemble import RandomForestClassifier

          data1=pandas.read_csv('D:/datasheet10.csv',header=None)
          data1=data1.as_matrix() # or data=data.values
          print(data1)
          trainingset=data1[list(range(0,178,2)),0:10]   #iris.data is numpy array
          trainingsettarget=data1[list(range(0,178,2)),10]
          testset=data1[list(range(1,178,2)),0:10]
          testsettarget=data1[list(range(1,178,2)),10]
          clf=RandomForestClassifier(n_estimators=30, criterion='entropy', max_depth=None, min_samples_split=2,
                          min_samples_leaf=1, min_weight_fraction_leaf=0.0, max_features='auto',
                          max_leaf_nodes=None, bootstrap=True, oob_score=False, n_jobs=1, random_state=None,
                          verbose=0, warm_start=False, class_weight=None) #creating instance of a classifier #default criteria
          clf.fit(trainingset, trainingsettarget)
          predictions = clf.predict(testset) #predict using the learnt classifier

          print("############### Predictions #################")
          print(predictions)
          print("###########################################")

          print("Accuracy = {0}".format(metrics.accuracy_score(testsettarget, predictions, normalize=True)))

          print(metrics.classification_report(testsettarget, predictions))

          print(metrics.confusion_matrix(testsettarget, predictions))
```

5 Conclusion

For optimizing the performance of random forest, we did number of experiments by setting different values of two hyperparameters with Scikit learn in Python. We observed that assessing a model only on training data has led to the problem of overfitting. It means that our model gives a very good result with training dataset but not well at unknown dataset. To overcome this problem, we have split our data into two partitions: training dataset and testing dataset. So now our model gives a better result than before. We used two hyperparameters for our experiment: n_estimators and criterion, and evaluated the accuracy. The results obtained are as follows.

n_estimators	Criterion	Accuracy (%)
10	entropy	85
10	gini	80
12	entropy	86
12	gini	83
15	entropy	87
15	gini	86
22	entropy	89
22	gini	87
28	entropy	90
28	gini	88
31	entropy	91
31	gini	89
35	entropy	89
35	gini	88
38	entropy	87
38	gini	87

In our experiment, we built 16 random forests by random subsampling and by setting two hyperparameters and keeping the values of other hyperparameters as default. n_estimators is the number of trees generated by random forest. We tried different values for both the parameters and observed that up to some threshold point the accuracy has been increased but after that point, if the value of n_estimators is increased, then the accuracy has been decreased. So, it is very important to find out the threshold value, so that maximum accuracy can be achieved. Our second hyperparameter is criteria, which are splitting criteria. By default, the splitting criterion is "gini impurity" and the other option is information gain (entropy). We have changed from the default criteria to "entropy" and as a result we are able to achieve 2% of more accuracy than we were achieving by using gini index. It has been noticed that we got the accuracy in the range from 85 to 91% of a strong learner when the criterion set was "entropy" and the average accuracy, in that case, was 88% and we got the accuracy in the range from 80 to 86% in case of "gini" criteria. Also in the "gini" criteria we got average 86% accuracy. When we averaged the output of all 16 learners, we got 89% accuracy which was 6% higher than that we were getting in Weka 3.8 tool. As in Weka we are not able to set some of the hyperparameters that we could do in Python, accuracy can be improved in the latter case. In this experimental research, we tried to make many weak decision tree learners by random subsampling of training set as well as variables to make one strong random forest predictor. In future, it can also be done that instead of taking the same kind of base learners it is possible to make different kinds of base learners from varied background in order to make strong decision by random forest algorithm.

References

1. Ata, H.A., Seyrek, I.H.: The use of data mining techniques in detecting fraudulent financial statements: an application on manufacturing firms. Suleyman Demirel Univ. İktisadi İdari Biliml. Fak. Derg. **14**(2) (2009)
2. Bell, T.B., Carcello, J.V.: A decision aid for assessing the likelihood of fraudulent financial reporting. Audit. J. Pract. Theory **19**(1), 169–184 (2000)
3. Breiman, L.: Bagging Predictors. Technical report No 421 (1994)
4. Buller, D.B., Burgoon, J.K.: Interpersonal deception theory. Commun. Theory **6**(3) (1996)
5. Eining, M.M., Jones, D.R., Loebbecke, J.K.: Reliance on decision aids: an examination of auditors' assessment of management fraud. Audit. J. Pract. Theory **16**(2), 1–19 (1997)
6. Elkan, C.: Magical thinking in data mining: lessons from CoIL challenge 2000. In: Proceeding of KDD-2001, pp. 426–431 (2001)
7. Fanning, K., Cogger, K.: Neural network detection of management fraud using published financial data. Int. J. Syst. Acc. Financ. Manag. **7**(1), 21–24 (1998)
8. Green, B.P., Choi, J.H.: Assessing the risk of management fraud through neural network technology. Auditing **16**(1), 14 (1997)
9. Patel, H., Parikh, S.: Fraudulent financial statements: detection modeling using data mining. Int. J. Latest Trends Eng. Technol. **9**(3) (2018)
10. International Federation of Accountants, International Standard on Auditing 240 (2002): The auditor's responsibility to consider fraud and error in an audit of financial statements
11. Kirkos, E., Spathis, C., Manolopoulos, Y.: Data mining techniques for the detection of fraudulent financial statements. Expert Syst. Appl. **32**(4), 995–1003 (2007)
12. Kosorok, M., Ma, S.: Marginal asymptotics for the large p small n paradigm: with applications to microarray data. Ann. Stat. **35**, 1456–1486 (2007)
13. Krogh, A., Vedelsby, J.: Neural network ensembles, cross validation, and active learning. Adv. Neural Inf. Process. Syst. **7**, 231–238 (1995)
14. Kulkarni, V.Y., Sinha, P.K.: Random forest classifiers: a survey and future research directions. Int. J. Adv. Comput. **36**(1), 1144–1153 (2013)
15. Opitz, D., Maclin, R.: Popular ensemble methods: an empirical study. J. Artif. Intell. **11**, 169–198 (1999)
16. Ravisankar, P., Ravi, V., Rao, G.R., Bose, I.: Detection of financial statement fraud and feature selection using data mining techniques. Decis. Support Syst. **50**(2), 491–500 (2011)
17. Schapire, R.E.: The boosting approach to machine learning an overview. In: Nonlinear Estimation and Classification. Springer, Berlin (2003)
18. Thorton, G.: Financial and corporate fraud. ASSOCHAM report (2016)

19. Wells, J.T.: Occupational Fraud and Abuse. Obsidian Publishing Company, Nottingham (1997)
20. Wallace, W.A.: Auditing. South-Western College Publishing, Cincinnati, OH (1995)
21. Zhou, W., Kapoor, G.: Detecting evolutionary financial statement fraud. Decis. Support Syst. **50**(3), 570–575 (2011)

Selfish Controlled Scheme in Opportunistic Mobile Communication Network

Moirangthem Tiken Singh and Surajit Borkotokey

Abstract The opportunistic mobile communication network is a social communication network, where every node depends on the other in transferring data. Thus, every node makes a potential router. However, due to the lack of resource, nodes tend to become selfish by dropping the messages relayed through them. In this paper, an algorithm is proposed to control such selfish nodes from overtaking the network. The proposed algorithm is designed for an uncertain stochastic network environment, where the popular strategy like Tit for Tat (TFT) fails. It is further compared with Generous popular approaches such as Generous Tit for Tat (GTFT) and Generous Zero-determinant (GenZD). Furthermore, the simulation study has shown that it outperforms the GTFT and GenZD while controlling the free rider selfish nodes.

Keywords Selfish · Uncertain network environment · Tit for Tat
Zero- determinant strategy

1 Introduction

An Opportunistic Mobile Communication Network (OMCN) [4] is a multi-hop ad hoc network with no end-to-end communication path. Each node can store and forward messages. This makes every node a potential router. If a source node cannot find any relay node during a transmission, the source delivers messages directly to the destination without using any relay. In this structure, every node must work for the welfare of the society. However, as every node is endowed with limited resource, some nodes become selfish [8]. It results in another interesting phenomenon, where the selfish nodes start exploiting the selfless nodes. So each node faces the dilemma

M. T. Singh (✉) · S. Borkotokey
School of Science and Engineering, Dibrugarh University, Dibrugarh, Assam, India
e-mail: tiken.m@dibru.ac.in

S. Borkotokey
Department of Mathematics, Dibrugarh University, Dibrugarh, Assam 786004, India
e-mail: surajitbor@yahoo.com

© Springer Nature Singapore Pte Ltd. 2019
J. Kalita et al. (eds.), *Recent Developments in Machine Learning and Data Analytics*,
Advances in Intelligent Systems and Computing 740,
https://doi.org/10.1007/978-981-13-1280-9_34

whether to forward or drop a message. The model of the interaction behavior of nodes in the OMCN is, thus, a generalized version of the multiplayer social dilemma game, where any node can randomly interact with any other node with the strategies of forwarding (F) or dropping (D) the incoming relayed messages. Several techniques have been proposed to deal with the selfish-free riders in OMCN. All the methods are either based on reciprocity or on conditional cooperation, see, for example [2, 7, 11–13]. All these works include a collection of information, especially strategies of the neighbors which are to be used later for controlling the selfish nodes. Nodes in [7] use Generous Tit for Tat (GTFT) [9] for controlling the selfish nodes in the opportunistic network. The disadvantage in this model emerges when GTFT nodes end up in generous cooperation if a node is indifferent to previous decisions or is seemingly choosing at random. Moreover, it is shown that Generous Zero-determinant (GenZD) [9] outperforms GTFT. Therefore, in the present paper, we compare our model with GenZD and GTFT. All these studies found in the literature assume that the communication network signals between the nodes are perfect, i.e., each node assumes an error-free communication network in interpreting the strategies of the others. In any case, a necessary element of associations, in reality, is that decisions cannot be executed without error; we call it noise. This is because mobile nodes do not know whether the ensured action is an error or a deliberate choice. Based on the above discussions, in this work, we define a model to control the free riding selfish nodes by adapting and modifying a stable evolutionary variant of Zero-determinant strategy [5] for the noisy environment.

2 Literature Survey

Research on the message delivery system in opportunistic mobile communication that relies upon cooperation by intermediate nodes has been wide and exhaustive. In all these studies, it is considered that due to the limitations in the resource, the intermediate nodes do not always want to help the other nodes. Each node in the network tries to fulfill its own goal by using the other nodes' resources. Different cooperation-enforcing schemes have been proposed to promote cooperation between nodes in such systems. Credit- based, reputation-based and game-theory-based [2, 7, 11, 13] approaches foster collaboration between nodes. In credit-based approaches [7, 11], the equilibrium state is analyzed where virtual currency or pricing acts as the credit to increase the interest for participation among the nodes. SMART [13] uses a non-game strategy that utilizes multilayered credit coin (virtual currency) scheme to provide security against cheating. However, in all these cases, maintaining of credits increases the extra complexity of the network. In reputation-based approaches [12], every node monitors the reputation for neighboring nodes, separates the selfish nodes, and finally punishes them. In doing so, they propose the protocol for forwarding data and the mechanism for monitoring the behavior. DISCUSS [2] uses evolutionary game theory to control the selfish-free riding nodes. In DISCUSS, every node tests the expected payoff of its neighbors and compares it with their expected payoffs.

According to the comparison, each node either keeps its strategy or changes to the neighbor's strategy.

3 Materials and Methods

Nodes in this work use indirect reciprocity for controlling the selfish riders as used in [10]. This is because here in this paper a node keeps track of the other node's behavior and reactions. Then, we establish the reputation system based on the history of the node's actions. Thus, we recorded 1 and 0, respectively, to the forwarder and the dropper.

To study the behavior of interaction of N nodes in a network, we first scale down it to the interaction of two self-centric mobile nodes. It is possible because at any instant of time only two mobile nodes can interact. So, an isolated pair of nodes may play the message forwarding game. For example, consider three nodes X, Y, and Z of which node X wants to send a message to node Z through Y. Node Y can decide either to forward (F) or to drop (D) the incoming message from X. If Y forwards the message, then X gains utility of r units. Since Y has consumed its resources, it loses b utility units. If node Y sends data to another node K through node X, X consumes its resources, it loses b utility units while Y gains r units. Table 1 is the payoff matrix for the scenario which is indeed the payoff matrix of the prisoner's dilemma. To simplify the payoff matrix for analysis, we do a transformation to the payoff matrix given in Table 1 to the payoff matrix given in Table 2. It is done by using the following formula $y = \frac{x+r}{2r-b}$, where x is the entry in Table 1, and y is the respective entry in Table 2.

Table 1 Payoff matrix for message forwarding game

		Node Y	
		F	D
Node X	F	$r - b, r - b$	$-r - b, r$
	D	$r, -r - b$	$-r, -r$

Table 2 Payoff matrix after transformation with $b = 1$

		Node Y	
		F	D
Node X	F	1, 1	$\frac{-1}{2r-1}, \frac{2r}{2r-1}$
	D	$\frac{2r}{2r-1}, \frac{-1}{2r-1}$	0, 0

The communication between the nodes in a session is continuous meaning: it is a reiterated transmission of an enormous number of messages. Thus, apparently it seems straightforward to model the interaction design as an iterative Prisoners Dilemma. However, this is not workable. The message forwarding game work in a strepitous environment. A node may misinterpret the message dropping due to an error in the communication channel due to the refusal by its neighbors to forward the messages. So, we denote by ψ the probability of error (noise) in the perception of strategies used by other nodes. Recall that at any stage of the game we have two nodes X and Y. Each node $i \in \{X, Y\}$ takes an action $a_i \in \{F, D\}$. A node cannot see the actions of other nodes directly, however, it can observe a private signal $\varphi = \{\varphi_X, \varphi_Y\} \in \{c, d\} \times \{c, d\}$ after both nodes have chosen their actions. Here, c and d denote good and bad signals, respectively. We call $\varphi = \{\varphi_X, \varphi_Y\}$ the signal profile from the two nodes. Note that each node's signal φ_i, $i \in \{X, Y\}$ is a stochastic variable; its distribution is affected not only by their actions but also by the noise in the environment. Each kind of signal φ_i occurs with a positive probability $\delta (\varphi_X, \varphi_Y | a)$, where $a = \{a_X, a_Y\}$ is the action profile from the pair of nodes $\{X, Y\}$. In each stage, if node X plays $a_X = F$ (or $a_X = D$) but Y observes $\varphi_Y = d$ (or $\varphi_Y = c$), it means an error has occurred. For example, if both nodes choose F, then $\delta(c, c|(F, F)) = 1 - \psi, \delta(c, d|(F, F)) = \psi$ for some $\psi \in [0, 1]$. Based on the action and the observed signal, the private outcome for player $i \in \{X, Y\}$ at each stage of the game can be defined as $(a_i, \varphi_i) \in \{F, D\} \times \{c, d\}$. Since stochastic changes to the environment, and the opponent's action, are jointly involved in the signals, the realized stage payoff for each node depends on the action it chooses and the message he receives, which is denoted as $u_i(a_i, \varphi_i)$ for $i \in \{X, Y\}$. Based on Table 2, the realized stage payoff of X and Y is ordered as $u_X(F, c) = u_Y(F, c) = 1, u_X(F, d) = u_Y(F, d) = \frac{-1}{2r-1}$, $u_X(D, c) = \frac{2r}{2r-1} = u_Y(D, c), u_Y(D, d) = u_Y(D, d) = 0$.

According to the general structure [6], the expected stage payoff for each node over noisy signal is defined as

$$f_i(a) = \sum_{\varphi} u_i(a_i, \varphi_i)\delta(\varphi|a) \tag{1}$$

etc. The expected payoff under different action profiles (F, F), (F, D), (D, F), and (D, D) are denoted as R, S, T, P, receptively. The expected stage payoff is then calculated using Eq. (1) and we have the values

$$
\begin{cases}
R = \frac{1}{2r-1}(-\psi + (2r - 1)(-\psi + 1)) \\
S = \psi - \frac{-\psi+1}{2r-1} \\
T = \frac{2r(-\psi+1)}{2r-1} \\
P = \frac{2r\psi}{2r-1}.
\end{cases}
\tag{2}
$$

The expected stage payoff vector for node X is denoted by $S_X(t) = (R, S, T, P)$ and node Y's payoff is denoted by the vector $S_Y(t) = (R, T, S, P)$. Denote by $s_X(t)$ the payoff received by a node X at time t. Assuming that the social dilemma is repeated and given the strategy of all other nodes in the network, the expected payoff of X is

$$
U(X) = \lim_{T \to \infty} \frac{1}{T} \sum_{t=1}^{T} s_X(t).
\tag{3}
$$

The free rider nodes act to maximize the average payoff at the cost of other truthful nodes which cooperate in relaying messages. So, we need to enforce fair outcomes or prevent the free riders from taking over the network. This implies that every node in the networks should have a uniform average payoff U.

3.1 Message Forwarding Algorithm

Communication among nodes in OMCN as mentioned in the previous section is the iterative noisy noncooperative game. To model the game, we follow the Markov Process [3]. To model indirect reciprocal using the Markov Process, we refine the game such that the interaction between the nodes is divided into slots of fixed time intervals t. Accordingly, we define the strategy of node X at any time t as a set of four parameters, $p = (p_1, p_2, p_3, p_4)$. This parameter p is the probability of X forwarding relayed messages such that: $p_1 = P(F|(F_X(t - 1), F_Y(t - 1)))$, $p_2 = P(F|(F_X(t - 1), D_Y(t - 1)))$, $p_3 = P(F|(D_X(t - 1), F_Y(t - 1)))$ and $p_4 = P(F|(D_X(t - 1), D_Y(t - 1)))$; where $F_X(t - 1)$, and $D_X(t - 1)$ are the forwarding and dropping strategies adopted by node X at time $t - 1$; $F_Y(t - 1)$ and $D_Y(t - 1)$ are the forwarding and dropping strategies of Y perceived by X till time $t - 1$. The function $P(.)$ is the conditional probability. Similarly, the strategy used by node Y will be defined as a set of four parameters, $q = (q_1, q_2, q_3, q_4)$ with similar descriptions. If nodes X and Y interact under the noisy environment, the interaction between X and Y at any discrete time t can be represented by the Markov Process M.

$$M =$$

	$(F_X(t), F_Y(t))$	$(F_X(t), D_Y(t))$	$(D_X(t), F_Y(t))$	$(D_X(t), D_Y(t))$
$(F_X(t-1), F_Y(t-1))$	$p_1\psi q_2 +p_1 q_1 (-3\psi+1) +p_2\psi q_1 +p_2\psi q_2$	$p_1\psi(-q_2+1)+p_1(-3\psi+1)(-q_1+1)+p_2\psi(-q_1+1)+p_2\psi(-q_2+1)$	$\psi q_1(-p_2+1)+\psi q_2(-p_1+1)+\psi q_2(-p_2+1)+q_1(-p_1+1)(-3\psi+1)$	$\psi(-p_1+1)(-q_2+1)+\psi(-p_2+1)(-q_1+1)+\psi(-p_2+1)(-q_2+1)+(-p_1+1)(-3\psi+1)(-q_1+1)$
$(F_X(t-1), D_Y(t-1))$	$p_1\psi q_3 +p_1\psi q_4 +p_2\psi q_4 +p_2 q_3 (-3\psi+1)$	$p_1\psi(-q_3+1)+p_1\psi(-q_4+1)+p_2\psi(-q_4+1)+p_2(-3\psi+1)(-q_3+1)$	$\psi q_3(-p_1+1)+\psi q_4(-p_1+1)+\psi q_4(-p_2+1)+q_3(-p_2+1)(-3\psi+1)$	$\psi(-p_1+1)(-q_3+1)+\psi(-p_1+1)(-q_4+1)+\psi(-p_2+1)(-q_4+1)+(-p_2+1)(-3\psi+1)(-q_3+1)$
$(D_X(t-1), F_Y(t-1))$	$p_3\psi q_1 +p_3 q_2 (-3\psi+1) +p_4\psi q_1 +p_4\psi q_2$	$p_3\psi(-q_1+1)+p_3(-3\psi+1)(-q_2+1)+p_4\psi(-q_1+1)+p_4\psi(-q_2+1)$	$\psi q_1(-p_3+1)+\psi q_1(-p_4+1)+\psi q_2(-p_4+1)+q_2(-p_3+1)(-3\psi+1)$	$\psi(-p_3+1)(-q_1+1)+\psi(-p_4+1)(-q_1+1)+\psi(-p_4+1)(-q_2+1)+(-p_3+1)(-3\psi+1)(-q_2+1)$
$(D_X(t-1), D_Y(t-1))$	$p_3\psi q_3 +p_3\psi q_4 +p_4\psi q_3 +p_4 q_4 (-3\psi+1)$	$p_3\psi(-q_3+1)+p_3\psi(-q_4+1)+p_4\psi(-q_3+1)+p_4(-3\psi+1)(-q_4+1)$	$\psi q_3(-p_3+1)+\psi q_3(-p_4+1)+\psi q_4(-p_3+1)+q_4(-p_4+1)(-3\psi+1)$	$\psi(-p_3+1)(-q_3+1)+\psi(-p_3+1)(-q_4+1)+\psi(-p_4+1)(-q_4+1)+(-p_4+1)(-3\psi+1)(-q_3+1)$

An entry in M is calculated considering the parameter $\psi \in [0, 1]$. For example, the entry in $(F_X(t-1), F_Y(t-1))$th row and the $(D_X(t), F_Y(t))$th column is an inclusive collection of chances of X and Y's to forward or drop messages, i.e., the chance of dropping for node X after correctly perceiving earlier Y's strategy and the chance of forwarding by Y after correctly perceiving X's previous move is $(1 - 3\psi)(1 - p_1)q_1$; the chance of dropping for node X after wrongly perceiving Y's strategy and the chance of forwarding after Y correctly perceives X's strategy is $\psi(1 - p_2)q_1$. The entry also includes the chance of forwarding by Y when it wrongly perceives X's strategy and the chance of dropping by X by correctly perceiving Y's strategy, i.e., $\psi(1 - p_1)q_2$; and finally, when both X and Y perceive each other's previous strategy wrongly, i.e., $\psi(1 - p_2)q_2$. Similarly, for other entries also we calculate the influence of ψ during the interaction of nodes X and Y. Let $\zeta = M - I_4$ whose determinant is 0.

Let $adj(\zeta)$ be the adjoint of ζ such that $adj(\zeta)\zeta = 0_{4,4}$ (by Cramer's rule). Let π be the limiting distribution of M and let w be the limiting row vector, without normalization such that w is a vector satisfying $wM = w$, i.e., $w\zeta = 0$ (since ζ is a singular matrix) and $\pi = \frac{w}{w\cdot 1}$. Changing the last column of of the singular matrix ζ into player X's stage payoff vector $S_X \in \{R, S, T, P\}$, we can get a new matrix ζ'. Adding the first column into the second and the third column of ζ' gives us a new form of the determinant.

$$det(\zeta') = \begin{bmatrix} \cdots & -2p_1\psi + p_1 + 2p_2\psi - 1 & -2\psi q_1 + 2q_2\psi + q_1 - 1 & R \\ \cdots & 2p_1\psi - 2p_2\psi + p_2 - 1 & -2\psi q_3 + q_3 + 2q_4\psi & S \\ \cdots & -2p_3\psi + p_3 + 2\psi p_4 & q_2 + 2\psi q_1 - 2\psi q_2 - 1 & T \\ \cdots & 2p_3\psi - 2p_4\psi + p_4 & 2\psi q_3 - 2\psi q_4 + q_4 & P \end{bmatrix}$$

The resulted determinant has a peculiar feature as follows. Its second column $\hat{p} = (-2p_1\psi + p_1 + 2p_2\psi - 1, 2p_1\psi - 2p_2\psi + p_2 - 1, -2p_3\psi + p_3 + 2p_4\psi, 2p_3\psi - 2p_4\psi + p_4)$ depends only on X's strategy p and the noise ψ; Similarly its third column $\hat{q} = (-2\psi q_1 + 2\psi q_2 + q_1 - 1, -2\psi q_3 + 2\psi q_4 + q_3, 2\psi q_1 - 2\psi q_2 + q_2 - 1, 2\psi q_3 - 2\psi q_4 + q_4)$ depends only on Y's strategy q and the noise ψ. Following the Zero-determinant strategy (ZD), X can get a pre-specified extortionate share of the total mutual defection or cooperation payoffs, in the long run, depends on the value of K as explained below. This is by setting a unilateral relationship between its payoff and opponent's payoff. We express such a linear relationship following [5] as

$$\hat{p} = \phi((S_X - K) - \chi(S_Y - K) \tag{4}$$

where $\chi \geq 1$ is called extortion factor and K is either P or R.

When K is P, the extortion strategies ensure that either X receives a higher payoff than Y or otherwise, both nodes receive the payoff for mutual dropping, i.e., $s_X = s_Y = P$. Whereas when K is R, strategies ensure that both nodes receive the payoff for mutual forwarding or otherwise X receives a lower payoff than Y. To enforce such relation, node X needs to choose the strategies that satisfied Eq. 4. The value of ϕ is small and chosen in such a way that the parameters p_1, p_2, p_3, and p_4 have valid values.

Using an extortionate strategy, node X can get an extortionate share, the value of which depends on Y's strategy. If Y's strategy is $q(1, 1, 1, 1)$, i.e., if it is all forwarding, then both X and Y get the maximum payoff. However, if Y's strategy is $q(0, 0, 0, 0)$, i.e., all dropping, then the maximum payoff received by X and Y will be P when $K = P$ or value ranges between R and S such that $s_X < s_Y$ when $K = R$. Because of the noisy environment, we want to relax the value of K to R. This is done to show the softness in handling opponent's node as compared to the harsh condition put up by $K = P$. Based on the above discussion, we propose Algorithm 1 in the following.

4   Simulation Analysis

We take a world of size 4500×3500 m^2 for the simulation. We formed three groups (two pedestrian groups and a group in a car) of the total nodes having 40 nodes each, which are distributed randomly. Each node in the cluster has a single short-range single interface system (maximum range of 10 m with the transmission speed of 2 Mbps). Along with these groups, there are also three routers which route messages from any clusters. They also have two interface systems. The two interfaces comprise

---

Algorithm 1: Message forwarding algorithms

   Result: F or D
1 OtherNode;
2 Decision $p(p_1, p_2, p_3, p_4)$;
3 **while** *True* **do**
4   Set OtherNodeCurrentStrategy = getStrategy(OtherNode);
5   Set CurrentNodeStrategy = getStrategy(CurrentNode);
6   **if** *OtherCurrentStrategy == 'F' AND CurrentNodeStrategy == 'F'* **then**
7 | return 'F' with Probability p_1;
8   **else**
9   **if** *OtherCurrentStrategy == 'D' AND CurrentNodeStrategy == 'F'* **then**
10 | return 'F' with Probability p_2;
11   **else**
12   **if** *OtherCurrentStrategy == 'F' AND CurrentNodeStrategy == 'D'* **then**
13 | return 'F' with Probability p_3;
14   **else**
15 | return 'F' with Probability p_4;
16   **end**
17   **end**
18   **end**
19 **end**
20 return D

Table 3 Different Parameters used for the simulation

Parameters	Value
Routing algorithm	Epidemic router
Buffer size of nodes	5 MB
Buffer size of router	50 MB
Wait time of the group	0–120 s
Message TTL	300 s
Walking speed	0.5–1.5 km/s
Car speed	2.5–13 km/s
Router movement speed	7.0–10.0 km/s
Speed of interface	2 and 10 Mbps
Range of interface	10 and 1000 m
Number of nodes per group	41
Buffer of routers	50 MB
χ, ϕ	2, 0.2

long-range, high-speed signal interface with a range of 1000 m (with the transmission speed of 10 Mbps) and a short-range signal interface with the signal interface of 10 m (with the transmission speed of 2 Mbps). Details of the parameters defined for the simulation are given in Table 3.

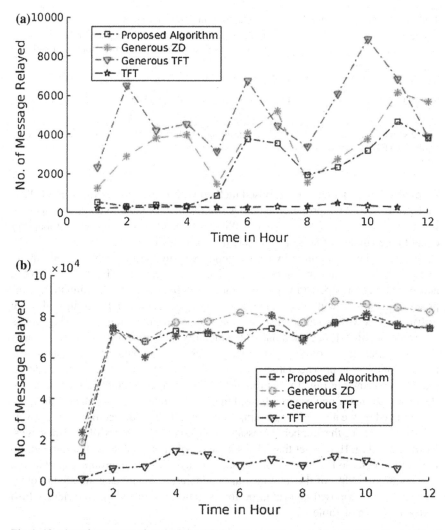

Fig. 1 Number of messages relayed originating from the selfish group when the network consists of the group following memory-one strategies (**a**) and from the group following memory-one strategies in the presence of selfish group (**b**) in the network environment with ψ=0.1

In the simulation, groups are assigned with different strategies, i.e., All Forward (allf), All Drop (alld), Generous Zero-determinant (GenZD), Generous Tit for Tat (GTFT), and Tit for Tat (TFT) strategies. We simulate the different combinations of these policies to study the behavior of nodes. We also assume that every node determines the action of the other nodes by itself. To do so, we change the ONE simulator [1] such that every node can learn the policy followed by its interacting nodes. The policy acquire is stored in a history table. Every time a node interacts with other nodes its history table gets updated, so it only saves the latest strategy

Table 4 Comparison of rate of messages relayed from different sources

Strategy	% of messages relayed originating from selfish nodes	% of messages relayed originating from other nodes
Proposed algorithm	19.34	31.07
GTFT	32.04	33.77
GenZD	45	31.17
TFT*	2.815	3.98

[a]Table shows that TFT performed well in controlling selfish riders. However, it exploits the nodes following TFT's strategy.

followed. We implement the proposed method with $r = 2$. Nodes following GTFT, TFT, GenZD, and the proposed methods use the history table to determine its plan for the next course of action. We also flipped the learned strategy with probability equal to the amount of noise to simulate the noise environment.

To check the exploitation by selfish nodes, we simulate OMCN network taking two groups having an equal number of nodes. To compare the TFT, GTFT, GenZD, and our algorithm's performance, we run four simulations. Each simulation comprises a group following one of the memory-one strategies and a group following selfish strategy. We also run four simulations with each simulation comprising groups following truthful strategy and one of the memory-one strategies. For every simulation, we collect traces of how many times a message originating from the groups is relayed for every hour. The statistics of the collected information are plotted against the time as shown in Fig. 1a, b. From Fig. 1a, we made an interesting observation. We observed that if one group followed the TFT's strategy, the number of the message relayed by the network originating from the selfish group declined to the lowest value i.e., reducing the number of messages relayed to only 2.815% of total messages. However, in Fig. 1b, we get that TFT's nodes exploit each other in the same group forcing to decrease the number of messages relayed from the group, i.e., reducing to relaying of 3.98% of the total messages generated. Again, as compared to other strategies, our proposed algorithm performs better in controlling free rider selfish nodes as shown in Table 4.

5 Conclusion

Every node is a potential router in OMCN. Lack of resources like power, buffer, etc., make nodes reluctant to relay other node's messages. However, they exploit the network resources to relay their messages. TFT and its variants like GTFT or GenZD control such exploitation. But uncertainty in a communication channel reduces TFT, GTFT, and GenZD's performance. To address uncertainty, the proposed algorithm considers network error (noise) in the decision-making. We showed through rigorous analysis and simulations that as compared to various popular strategies, our algorithm works well.

In this work, we have prefixed the value of expectation against different strategies. However, if all the nodes learned its expected payoff online in advance and evolved to meet the changing environment, the whole scenario might have changed. Analysis of such behavior will be the future course of this work.

Acknowledgements This work is supported by UKIERI Grant 184-15/2017(IC). The authors express their acknowledgment to the anonymous reviewers for their encouraging comments.

References

1. Keränen, A., Ott, J., Kärkkäinen, T.: The ONE simulator for DTN protocol evaluation. In: SIMUTools '09: Proceedings of the 2nd International Conference on Simulation Tools and Techniques, ICST, New York, NY, USA (2009)
2. Misra, S., Pal, S., Saha, B.K.: Distributed information-based cooperative strategy adaptationin opportunistic mobile networks. IEEE Trans. Parallel Distrib. Syst. **26**(3), 724–737 (2015). https://doi.org/10.1109/TPDS.2014.2314687
3. Nowak, M.A., Sigmund, K.: The alternating prisoner's dilemma. J. Theoret. Biol. **168**(2), 219–226 (1994). https://doi.org/10.1006/jtbi.1994.1101. http://www.sciencedirect.com/science/article/pii/S0022519384711015
4. Pelusi, L., Passarella, A., Conti, M.: Opportunistic networking: data forwarding in disconnected mobile ad hoc networks. IEEE Commun. Mag. **44**(11), 134–141 (2006). https://doi.org/10.1109/mcom.2006.248176
5. Press, W.H., Dyson, F.J.: Iterated prisoners dilemma contains strategies that dominate any evolutionary opponent. Proc. Natl. Acad. Sci. **109**(26), 10,409–10,413 (2012). https://doi.org/10.1073/pnas.1206569109
6. Sekiguchi, T.: Efficiency in repeated prisoner's dilemma with private monitoring. J. Econ. Theory **76**(2), 345–361 (1997). https://doi.org/10.1006/jeth.1997.2313. http://www.sciencedirect.com/science/article/pii/S0022053197923139
7. Shevade, U., Song, H.H., Qiu, L., Zhang, Y.: Incentive-aware routing in DTNs. In: 2008 IEEE International Conference on Network Protocols, pp. 238–247 (2008), https://doi.org/10.1109/ICNP.2008.4697042
8. Socievole, A., Rango, F.D., Caputo, A., Marano, S.: Simulating node selfishness in opportunistic networks. In: 2016 International Symposium on Performance Evaluation of Computer and Telecommunication Systems (SPECTS), pp. 1–6 (2016). https://doi.org/10.1109/SPECTS.2016.7570516
9. Stewart, A.J., Plotkin, J.B.: Extortion and cooperation in the prisoners dilemma. Proc. Natl. Acad. Sci. **109**(26), 10,134–10,135 (2012). https://doi.org/10.1073/pnas.1208087109
10. Tang, C., Li, X., Wang, Z., Han, J.: Cooperation and distributed optimization for the unreliable wireless game with indirect reciprocity. Sci. China Inf. Sci. **60**(11), 110,205 (2017). https://doi.org/10.1007/s11432-017-9165-7
11. Yin, L., Lu, H., Cao, Y., Gao, J.: Cooperation in delay tolerant networks. In: 2010 2nd International Conference on Signal Processing Systems, vol. 1, pp. V1-202–V1-205 (2010). https://doi.org/10.1109/ICSPS.2010.5555572
12. Zhang, X., Wang, X., Liu, A., Zhang, Q., Tang, C.: Reputation-based scheme for delay tolerant networks. In: Proceedings of 2011 International Conference on Computer Science and Network Technology (2011). https://doi.org/10.1109/iccsnt.2011.6182124
13. Zhu, H., Lin, X., Lu, R., Fan, Y., Shen, X.: Smart: a secure multilayer credit-based incentive scheme for delay-tolerant networks. IEEE Trans. Veh. Technol. **58**(8), 4628–4639 (2009). https://doi.org/10.1109/TVT.2009.2020105

A Study of DC–DC Converters with MPPT for Standalone Solar Water-Pumping System

Baishyanar Jana, Shriraj Dhandhukiya, Ramji Tiwari and N. Ramesh Babu

Abstract In this paper, a DC/DC converter strategy is employed for solar water-pumping application. Maximum power point tracking (MPPT) control strategy is incorporated in order to extract maximum available power from the solar irradiation. In rural areas, the fields will be normally located far from the village, so it will not be able to extend the connection from the grid to such places. Thus, here standalone PV systems will be appropriate for powering pumps and other loads on the fields. In this paper, the performance of various DC–DC converters like boost, CUK, and SEPIC with MPPT techniques like P and O, fuzzy and NN (RBFN) is studied to identify the more suitable converter for the pumping application.

Keywords Standalone photovoltaic · DC–DC converter · Maximum power point technique · Fuzzy logic controller · Neural network

1 Introduction

Energy demand is growing day by day and due to the use of conventional energy sources like using coal in thermal power plants results in a lot of pollution. Thus, due to the concern over the environment pollution, utilization of nonconventional energy sources for generating power is necessary [1]. Only by tapping these abundant supplies of energy and converting efficiently into a form which can be utilized, we can slowly decrease the use of conventional sources of energy [2]. The photovoltaic

B. Jana · S. Dhandhukiya · R. Tiwari
School of Electrical Engineering, Vellore Institute Technology (VIT) University, Vellore, Tamil Nadu, India
e-mail: baishyanar.jana@gmail.com

S. Dhandhukiya
e-mail: shrirajd003@gmail.com

N. R. Babu (✉)
M. Kumarasamy College of Engineering, Karur, Tamil Nadu, India
e-mail: nrameshme@gmail.com

© Springer Nature Singapore Pte Ltd. 2019
J. Kalita et al. (eds.), *Recent Developments in Machine Learning and Data Analytics*,
Advances in Intelligent Systems and Computing 740,
https://doi.org/10.1007/978-981-13-1280-9_35

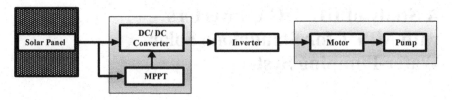

Fig. 1 Standalone PV water-pumping system

array converts solar energy into electricity. The photovoltaic systems can be used either as standalone systems or grid-connected systems.

Standalone PV water-pumping system works independently without being connected to the grid. A standalone PV water-pumping system is installed in places, where grid connection is not viable and where constant high irradiance is available [3].

The major parts of a standalone PV Water-pumping system are solar panel, DC–DC converter, motor-pump set. The major disadvantage of the PV system is that the efficiency of energy conversion is low. Thus, the maximum energy should be extracted from the PV module. For a solar panel, there exists only one maximum power point (MPP) and it varies with climatic conditions. The tracking of the MPP at varying irradiance and temperature is done by the MPPT algorithm, but its implementation in the circuit is done by DC–DC converters. Commonly used DC–DC converters are Buck, Boost, Buck–Boost, CUK, SEPIC, etc. [4]. In this paper, the MPPT techniques like P and O, Fuzzy and NN (RBFN) with DC–DC converters like Boost, CUK, and SEPIC are studied. The output voltage from a solar panel is normally a low voltage. For pumping application which uses a three-phase AC induction motor, a high voltage is required to drive the load. So, the DC–DC converters need to step-up the voltage. Thus, buck and buck–boost converters are not suitable for pumping application (Fig. 1).

2 System Overview

Standalone PV water-pumping system mainly has a solar panel, DC–DC converter, MPPT algorithm, and motor-pump set. The rating of the motor (load) was needed to fix the rating of the solar panel. In this paper, three 3-phase AC induction motors of 0.75 hp connected in parallel are used. The motor ratings are: 415 V, 50 Hz, 0.75 hp, 1.7 A, 2800 rpm. The AC induction motor is fed a voltage of 415 V. Since the output voltage from the PV panel is DC, the output from the DC–DC converter needs to be fed to an inverter which delivers an AC voltage of 415 V to the load. In this paper, a 3-phase inverter with sinusoidal PWM to find the value of the output voltage from the DC–DC converter was used.

$$V_i = M * \sqrt{3} * (V_{\text{conv}}/2) \tag{1}$$

where V_i = Rated motor voltage, M = Modulation Index and V_{conv} is the converter output voltage.

The modulation index of value 0.8 is considered in this study so as to get smoother three-phase sinusoidal AC waveform. Thus, the output of converter voltage is termed as

$$V_{\text{conv}} = (V_i * 2)/\left(\sqrt{3} * M\right) \tag{2}$$

Once the output voltage from the converter and the power rating of the solar panel is known, the design of various converter topologies can be done.

3 DC–DC Converters

3.1 Boost Converter

In a boost converter, output voltage is larger than the input voltage. Satisfying performance of the boost converter is obtained when the duty cycle is less than 0.8 [5].

$$D = V_i/(V_i - V_o) \tag{3}$$

where V_o = Output voltage, D = Duty cycle, $V_{\text{conv}} = V_o$ is the converter output voltage. Boost converter can boost only up to 2–3 times of the input voltage. So, to get an output voltage of 600 V, V_1 of around 240 V was selected which gives the duty ratio of about 0.6. This value of D is most suitable for the smooth operation of the converter. The circuit was designed for a ripple in current and voltage to be around 1%.

The boost converter is designed to process a power of 1894 W. The values of inductor (L) and capacitor (C) were designed according to the following formula [6]:

$$L = ((V_i - V_o) * (1 - D))/(\Delta I * f) \tag{4}$$

where ΔI represents the ripple in inductor current and f represents switching frequency.

$$C = D/R * (\Delta V_o/V_o * f) \tag{5}$$

where ΔV_o represents the ripple in output voltage and R represents the load resistance. The values of L and C considered in this study are 37 mH and 6.3 μF, respectively. Other converters like CUK and SEPIC have good voltage gain than boost converter.

3.2 CUK Converter

In CUK converter, the magnitude of the output voltage can be either larger or smaller than that of the input and there is a polarity reversal on the output. In the pumping application, CUK converter is designed such that it operates in boost mode. The voltage gain of CUK converter is better than that of the boost converter [7].

$$V_o = (-V_i * D)/(1 - D) \tag{6}$$

The CUK converter is designed to process a power of 2165 W. Having a better voltage gain, the input voltage to the converter can be made less than 240 V so as to get an output of 600 V from the converter efficiently. The ripple in output voltages and currents was limited to 1%. Energy transfer in a boost converter is associated with the inductor. But in the case of CUK converter, it depends on the capacitor C_1. The inductances and capacitance are designed according to the formulas [8] as given below

$$L_1 = (V_i * D)/(f * \Delta I L_1) \tag{7}$$

$$L_2 = (V_i * D)/(f * \Delta I L_2) \tag{8}$$

$$C_1 = (V_0 * D)/(Rf * \Delta V C_1) \tag{9}$$

$$C_2 = (1 - D)/\left((\Delta V_O/V_O) * 8 * L_2 * f^2\right) \tag{10}$$

The values of L_1, L_2, C_1, and C_2 considered in this study are 14.45 mH, 62.44 mH, 8 μF, and 0.015 μF, respectively.

3.3 SEPIC Converter

Single-ended primary inductance converter (SEPIC) is similar to CUK converter. The SEPIC converter can also produce an output voltage that is either greater or less than the input voltage but with no polarity reversal [9].

$$V_o = (V_i * D)/(1 - D) \tag{11}$$

The SEPIC converter is designed to process a power of 2165 W. The voltage gain of SEPIC converter is also better than the boost converter so that the input voltage to the converter can be much less than 240 V. The ripple in output voltages and currents were limited to 1%. The inductances and capacitance are designed according to the formulas [10] as given below

$$L_1 = (V_i * D)/(f * \Delta I L_1) \tag{12}$$

$$L_2 = (V_i * D)/(f * \Delta I L_2) \tag{13}$$

$$C_1 = (V_0 * D)/(Rf * \Delta V C_1) \tag{14}$$

$$C_2 = (V_0 * D)/(Rf * \Delta V_0) \tag{15}$$

The values of L_1, L_2, C_1, and C_2 considered in this study are 15 mH, 19.6 mH, 8 μF, and 10 μF, respectively.

4 Maximum Power Point Tracking Algorithms

MPPT algorithm is used to track maximum power from the PV irradiance. This paper mainly deals with P and O, fuzzy logic, and RBFN algorithm.

4.1 Perturb and Observe Method

The broadly used technique for MPPT is perturb and observer [11]. The current and voltage of solar panel are measured and perturb operation is performed, as in this the voltage and current of PV panel are changed by comparing with the previous panel output power. If there is an increase in panel output power then the next perturbation will remain the same, and if there is a decrease in panel output power then next perturbation will be changed [9]. With different converter circuits, maximum power point can be achieved by applying appropriate duty cycle, which is generated by this algorithm.

4.2 Fuzzy Logic Controller

Many methods and algorithms have some disadvantages like high ripple factor, inaccurate power for a load, etc. Fuzzy logic can handle this problem as it comes under the advanced controlling technique [1]. The membership function is the set of input and output, which is further needed for fuzzy operation. Fuzzy deals with the complex as well as the nonlinear system. Fuzzy can be divided into three procedures—Fuzzification, Inference, Defuzzification [9]. The main approach is to design the membership function with voltage and current as a linguistic variable which further produces desired duty cycle.

4.3 Radial Basis Function Network-Based Neural Network

The fuzzy technique is tough to implement with a microcontroller. Whereas neural networks (RBFN) is well adapted for the microcontroller [10]. The neural network which uses radial basis network for activation purpose is known as radial basis function network. It works on the basis of the difference between the input and the prototype vector. RBFN works in two stages. In the first stage, the unsupervised technique is used and parameters are monitored by radial basis function. In second stage supervised technique is used for training purpose [11]. In this paper, RBFN is used to control the duty cycle, which is applied to the converter circuit to maximize the output power.

5 Simulation Results

The main requirements for a water-pumping system to run efficiently are, the maximum power extracted from the solar panel should be delivered to the motor and this power delivered should be smooth. The motor runs more efficiently to drive the pump set, when the ripple in the power delivered is low. Another major requirement for the system to operate efficiently is that, as the irradiance or temperature changes, the maximum power point tracking algorithm must be fast and efficient enough to track the maximum power point. Thus, MPPT methods like P and O, Fuzzy and NN (RBFN) are implemented with various DC–DC converter topologies like Boost, CUK, and SEPIC. The study is carried out under varying irradiances of 600 W/m^2, 800 W/m^2, and 1000 W/m^2, respectively.

The output power versus time with various MPPT techniques for different DC–DC converter topologies is shown in Fig. 2. From Fig. 2a, it is evident that for a boost converter when the irradiance is about 600 W/m^2 Fuzzy MPPT technique is giving higher average output power compared to other techniques. But the amount of ripple observed in the case of fuzzy is very high compared to other MPPT techniques. When the irradiance changes suddenly, it is observed that the P and O and NN techniques give smooth transition than that given by fuzzy technique. Even at the irradiance level of 1000 W/m^2 fuzzy technique gave higher ripple than NN and P and O method.

Despite the average power being higher for fuzzy technique, the amount of ripple observed is higher than other techniques. Since the boost converter can only boost up the voltage about 2–3 times the input voltage, the input voltage needs to be high which is difficult to achieve in a PV system. Other DC–DC converters like CUK and SEPIC have higher voltage gains and will be more efficient in applications involving PV modules. From Fig. 2b, it is observed that the ripple in output power for CUK converter is much less with NN than other MPPT techniques. When the irradiance changes suddenly, the fuzzy method becomes more transient than other methods. Also, the ripples in the fuzzy method are higher than that in P and O and NN method. Another DC–DC converter which is similar to CUK converter is the SEPIC converter.

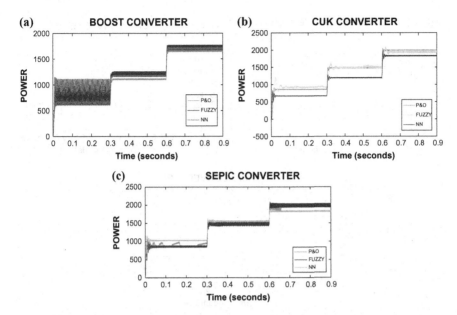

Fig. 2 Output power with MPPT techniques like P and O, FUZZY and NN (RBFN) for (a) Boost converter (b) CUK converter (c) SEPIC converter

Table 1 Output voltage and power parameters for various MPPT techniques with different DC–DC converters at 1000 W/m² irradiance

DC–DC converter/MPPT	Boost converter		CUK converter		SEPIC converter	
Output at rated values	Voltage (V)	Power (W)	Voltage (V)	Power (W)	Voltage (V)	Power (W)
P and O	566	1680	604	1990	577	1996
FLC	577	1723	590	1967	575	1986
RBFN	572	1705	575	1821	555	1846

From Fig. 2c, it can be observed that the P and O method's output varies continuously from one range to other and thus it is not a smooth output. For the fuzzy method, the initial transients are more and the ripples are also more. But in the case of the NN method, the transients are less and the output is smooth. Figure 3 shows the voltage with various MPPT techniques for different DC–DC converter topologies. For the fuzzy method, the ripples in output voltage are higher than other techniques. The ripples in output voltages are not desirable since for pumping application, the load is a motor which drives the pump. As they have mechanical parts, these transients in voltage profile are highly undesirable. Table 1 shows the voltage and power output of various converters and MPPT methods.

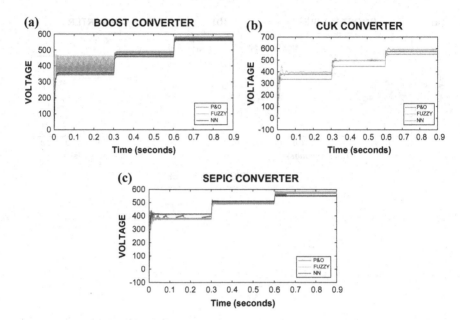

Fig. 3 Output voltage with MPPT techniques like P and O, FUZZY and NN (RBFN) for **a** Boost converter **b** CUK converter **c** SEPIC converter

6 Conclusion

For water-pumping using standalone PV system, the maximum power must be extracted from the PV for better efficiency and the motor driving the pump set should have a smooth operation. For that to happen, the ripple in the power delivered must be minimum and the system should be able to track the MPP whenever a change in irradiance and temperature occurs. From the above results, we can say that among P and O, fuzzy, and NN (RBFN) MPPT methods, the NN (RBFN) method has much fewer ripples than other techniques. When the irradiance level is changed suddenly then NN (RBFN) method gives a smooth transition and is less transient than other methods. In fuzzy MPPT technique, the ripple in output voltage is higher than that for other MPPT techniques. For converters, we can say that SEPIC and CUK are better than boost converter because the boost converter has less voltage gain than these converters. In PV applications, the voltage available from the panel will be low. So for pumping application where a high voltage is required, converters with high voltage gains give better results.

References

1. Tiwari, R., Babu, N.R.: Recent developments of control strategies for WECS. Renew. Sustain. Energy Rev. **66**, 268–285 (2016)
2. Tiwari, R., Babu, N.R.: Comparative analysis of pitch angle strategies for PMSG based WECS. Int. J. Intel. Syst. Appl. **9**, 62–73 (2017)
3. Tiwari, R., Babu, N.R., Sanjeevikumar, P.: Coordinated control strategies for a permanent magnet synchronous generator based wind energy conversion system. Energies **10**, 1493–1509 (2017)
4. Saravanan, S., Babu, N.R.: MPPT algorithms for photovoltaic system—a review. Renew. Sustain. Energy Rev. **57**, 192–204 (2016)
5. Tiwari, R., Babu, N.R.: Fuzzy logic based MPPT for PMSG in WECS. IFAC-PapersOnLine **49**, 462–467 (2016)
6. Kumar, K., Babu, N.R., Prabhu, K.R.: Design and analysis of an integrated Cuk-SEPIC converter with MPPT for standalone Wind/PV hybrid system. Int. J. Renew. Energy Res. **7**(1), 96–106 (2017)
7. Kumar, K., Babu, N.R.: Design and analysis of modified single P&O MPPT control algorithm for a standalone hybrid solar and wind energy conversion system. Gazi Univ. J. Sci. **30**(4), 296–312 (2017)
8. Tiwari, R., Babu, N.R.: RBFN based maximum power point strategy with SEPIC converter for standalone PMSG based WECS. In: Proceedings of Innovation Power and Advanced Computing Technology (i-PACT), Vellore, India, pp. 1–6 (2017)
9. Tiwari, R., Babu, N.R., Sanjeevikumar, P.: Fuzzy logic-based pitch angle controller for PMSG-based wind energy conversion system. Advances Smart Grid Renew. Energy, pp. 277–286. Springer, Singapore (2018)
10. Saravanan, S., Babu, N.R., Sanjeevikumar, P.: Comparative analysis of DC/DC converters with MPPT techniques based PV system. Advances Power System Energy Management, pp. 275–284. Springer, Singapore (2018)
11. Kumar, K., Prabhu, K.R., Babu, N.R., Sanjeevikumar, P.: A novel six-switch power converter for single-phase wind energy system applications. Advances in Smart Grid and Renewable Energy, pp. 267–275. Springer, Singapore (2018)

Edge Detection Technique Using ACO with PSO for Noisy Image

Aditya Gautam and Mantosh Biswas

Abstract In image processing, the edges of an image are those pixels whose intensity values changes drastically. Various techniques have been applied which rely on the ant colony optimization (ACO) for edge detection and the threshold value calculated for edge detection technique is either user defined or taken as the mean value of the obtained pheromone matrix. Other challenges to deal with edge detection are the presence of noisy environment, so to deal with it, in this paper, we define adaptive threshold value based on particle swarm optimization (PSO) for edge detection to overcome the limitation of existing ACO-based edge detection techniques. The experiment results have shown that the proposed technique has performed better under noisy environment over Sobel, Canny, and ACO-based technique both for objective criteria, i.e., restored edge images and subjective criteria, i.e., PSNR, precision, recall, and F-measure for test images in addition to reducing the execution time.

Keywords Edge detection · Noise · Ant colony optimization
Particle swarm optimization

1 Introduction

The image processing is described as the operation to improve, correct, analyze, manipulate, or render an image. There are a number of application areas where image processing is used; images from satellites, space probes, and aircraft are enhanced by using image processing. The appropriate analysis tools are provided by image processing even in those cases where data has nothing to do with image processing.

A. Gautam (✉) · M. Biswas
Computer Engineering Department, National Institute
of Technology Kurukshetra, Kurukshetra, Haryana, India
e-mail: aditya_31603109@nitkkr.ac.in

M. Biswas
e-mail: mantoshbiswas@nitkkr.ac.in

© Springer Nature Singapore Pte Ltd. 2019
J. Kalita et al. (eds.), *Recent Developments in Machine Learning and Data Analytics*,
Advances in Intelligent Systems and Computing 740,
https://doi.org/10.1007/978-981-13-1280-9_36

The use of image processing is increasing day by day due to the number of advantages it provides [1]. In the past, structure and properties of objects were estimated by identifying relevant features in images. Edges are also one of the features present in the image. The local changes that are indicated by edges are one of the important features in analyzing images. The edge is represented by the boundary present within two regions of an image. Due to increase in its requirement, edge detection has become a continuous active research area [2]. Various edge detection techniques have been proposed by researchers. In order to provide the perception of various ranges of the edges, there are separate individuals designed. The edge orientation, edge structure, and noise environment are some of the different aspects. The edge detection operator can be chosen through these aspects. Edge detection aims to extract important features like corners, lines, and curves which can be further used in high level computer vision algorithm for recognition.

Various researchers over the years have come up with techniques to solve the edge detection problem. The techniques like the conventional edge detectors which are the most initial ones to be used for detection of edges and are still used extensively in the industry, but are based upon complex mathematical operations. Some of traditional edge detectors are: Roberts Edge Detection: In 1965, this method was introduced by Lawrence Roberts and an image simple, quick computation, 2D spatial gradient measurement had been performed using it. The high spatial frequency regions are emphasized using this method that is corresponding to edges. The grayscale image is given as an input to operator and output will be used most commonly. The estimated complete magnitude is the output pixel values in every point. Sobel Edge Detection: In 1970, this method was introduced by Sobel and the Sobel approximation to the derivative was used to find image edges. The points or pixels on which there is highest gradient, those edges pixels are preceded [3]. The image 2D spatial gradient quantity is performed by this technique that highlights the high spatial frequency regions corresponding to different edges. The estimated absolute gradient magnitude is generally found using this technique at each point in input grayscale image. Prewitt Edge Detection: In 1970, this method was introduced by Prewitt and it was considered to be the appropriate method to gauge the orientation or magnitude of an edge. It is restricted to be used in only eight possible directions [4]. There is higher imperfection of the direction estimates as per the analysis. Kirsch Edge detection: In 1971, this method was introduced by Kirsch and a single mask was considered to define mask of this technique or rotating it to eight main compass directions. Robinson Edge detection: In 1977, this method was introduced by Robinson that was similar to Kirsch masks. As this method only relies on 0, 1 or 2 coefficients, it is very easy to be implemented within the applications [5]. The directional analysis or gradient direction in which there is axis along zeros, the masks are seen to be symmetrical. The results are to be computed on four masks only. The result from initial four masks is negated in order to compute the results of rest of the four masks in the image. Marr–Hildreth Edge Detection: In 1980, this method was introduced by Marr–Hildreth that detected the edges in continuous curves digital images where a fast variation was there in digital images [6]. The LoG function has been used as an operator to convolve images and it is easy to approximate it by DoGs. In order to find the edges, zero crossings are significantly discovered in the filtered results. Discrete wavelet technique (DWT), support vector machine (SVM), least square (LS), least square support vector machine (LS-SVM) [7], fuzzy based [8] and meta-heuristics which involves ant colony optimization (ACO), particle swarm optimization (PSO) and bee colony optimization (BCO) have also been applied to solve the edge detection problem but most of these techniques have been worked over the noiseless environment.

There have been various researchers who have used ACO technique for edge detection process, but to the best of our knowledge, we still believe it seems to be yet a fully explored area. ACO can be classified into two categories: one is for direct detection of edges and other is for postprocessing purpose, where after a conventional edge detector has detected the edges of an image, there may be some broken edges which need to be identified. In our work, ACO has been used for direct edge detection. We know noise plays a crucial role in image processing and moreover, edge detection becomes a lot tougher when noisy environment is taken into consideration. Zhuang [9] was the first one to use ACO for direct detection of edges and ant colony system (ACS) to come up with a perceptual graph which could extract the edges from an image, but only simple edges could be detected and in addition to it, no noisy environment was taken into consideration. Later Nezamabadi [10] used ant system (AS) for edge detection where the relationship between the image size and algorithm parameters was established, but noise was not considered. Tian [11, 12] applied ant colony optimization where a pheromone matrix was obtained and later a threshold value was computed which was the mean of the matrix and further also used an adaptive threshold to obtain the final binary image. Xiaochen Liu [13] came up with a new heuristic function, user-defined threshold and also took noise into consideration. Yara Khaluf [14] suggested that instead of random initialization, a proper way of ant distribution yields a better result. The performance of conventional edge detectors degrades when noise is taken into consideration in addition to complex mathematical operations involved. Marr et al. and Canny [15] have used a Gaussian filter to smoothen the image in presence of noise, but this step results in blurring which effects the precision of edge detection [16]. The performance of these edge detectors is also affected when noise is taken into consideration. In our work, we have considered the ant colony optimization technique with particle swarm optimization (PSO) to find out the threshold value to detect the edge in a noisy environment which overcomes the limitation of above existing techniques. We use dynamic optimization function based on PSO which ensures an effective threshold value for edge detection. Sections 1, 2, 3, and 4 include Introduction, Proposed Methodology, Experiment Results, and Conclusion, respectively.

2 Proposed Methodology

In our proposed work, we have considered edge detection as an optimization problem and have applied an optimization algorithm which seems to be an effective way to deal with this problem, with the help of ACO to achieve edge features by converting the intensity values of the pixels into pheromone values which are laid by the ants in every iteration and finally studying these pheromone value which lead to these edges. The input to the algorithm is a two-dimensional grayscale image with noise (Gaussian) on which we have a group of artificial ants which would be placed over randomly chosen nodes and for a specific number of iterations, the algorithm will run and in each iteration, an ant will be chosen and will move for specified number of steps and in its movement it will lay pheromone over the nodes in addition to update the process and finally a pheromone matrix is obtained on which a threshold is applied which will be calculated through PSO to get the required output. Therefore, our proposed edge detection method for noisy environment is based on ACO and PSO, which are described below.

2.1 Ant Colony Optimization (ACO)

In order to solve the optimization problems, a meta-heuristic algorithm was introduced by researchers called ACO. Researchers used the ant system as a base to generate this algorithm and actual behavior of ants was considered as a base in this algorithm. A local data related problem is considered here in which the parallel search is made over various constructive computational threads and quality of previously obtained results is presented on the basis of information gathered from dynamic memory structures [17]. A randomized heuristic construction is implemented by ants in ACO and function is to be provided within the artificial pheromone trails and the probabilistic decisions are generated within this algorithm. On the basis of heuristic information, the problems of input data are resolved here and the traditional construction heuristics are extended within this algorithm. In our ACO-based proposed technique, there are mainly four steps involved: (i) Initialization process, in which ants are placed over the image in a random manner and the pheromone values along with the heuristic information is computed, (ii) Construction process of node transition rule, in which ants are chosen and moved for a certain number of steps, probabilistically in addition to deciding the admissible range of the ants, (iii) Pheromone update process, in which comparison is done between the threshold value which will be computed through PSO and heuristic value, and (iv) finally, termination criterion, which specifies the number of iteration after which the algorithm will come up with the result.

(i)   Initialization process: A$M \times N$ sized Image I is taken, where M is height, N is width, and m indicates the numbers of ants that are randomly distributed over the image I, such that at most one ant can be present at any pixel. The pheromone value at all the pixels is considered to be 0.0001 which is initialized in T_{init}. The heuristic information will be computed offline for all the pixels.

(ii) Construction process of node transition rule: A stochastic approach is utilized to select kth ant from the m artificial ants in the nth construction step and on image I, there is a continuous mobility of this ant from node (r, s) towards its neighboring node (i, j). The node transition rule is followed and the equation is generated by below function.

$$P^n_{(r,s),(i,j)} = \begin{cases} \dfrac{\left(\tau^{(n-1)}_{(i,j)}\right)^{\alpha}\left(\eta_{(i,j)}\right)^{\beta}}{\sum_{(i,j)\in\Omega(r,s)}\left(\tau^{(n-1)}_{(i,j)}\right)^{\alpha}\left(\eta_{(i,j)}\right)^{\beta}} & \text{if } (i, j) \in \Omega(r, s) \\ 0 & \text{otherwise} \end{cases} \tag{1}$$

Here, $\tau^{(n-1)}_{(i,j)}$ and $\eta_{(i,j)}$ are the pheromone and heuristic information values for pixel (i, j), respectively. α and β are the influence coefficients of pheromone and heuristic information, respectively, where α indicates the intensity of pheromone while β is for visibility. Ω denotes the neighborhood of pixel and every pixel has to be chosen from its neighborhood. The already chosen pixels will not be chosen again.

The heuristic information for all the pixels will be calculated offline with the size of heuristic information region being 5×5 i.e. 24. The previous researchers have also tried with regions of 8, and 16, but the gradient response on the edge is higher for the region size of 5×5 [13]. The heuristic information is a parameter to be considered because it will act as an indicator for the ant that in its vicinity where it has to move and it is quite similar to finding out the gradient of an image. The admissible range of the ant's movement is taken as 8. The heuristic information is calculated according to [13].

(iii) Calculation of threshold value: The threshold value is calculated using particle swarm optimization (PSO) which produces an effective value dynamically and hence will make the edge detection process better. The update process specifies which nodes will have the deposition of pheromone trails i.e. where the pheromones will be updated if certain conditions are met and these conditions are based upon threshold value obtained from PSO.

(iv) Termination criterion: The algorithm can be stopped after predefined numbers of iterations are completed. The number of iterations in our work has been three but it is not hard-coded and can be changed according to needs. Normally, it has been seen that after a given number of iterations usually, results do not change.

2.2 Particle Swarm Optimization (PSO)

Eberhart et al. accredited PSO algorithm, which is a meta-heuristic algorithm based on population inspired by nature, with birds flocking and fishes schooling, and has been mimicked by this algorithm. The fitness function of this algorithm is to improve the solution after starting from a randomly distributed set of particles [18] and the particles are moved within the search space. This method also helps in providing interparticle communications. The set of points are generated within this algorithm.

The initial velocity vector is assigned here for each of the generated points, with the position of each particle changed in iterative manner with the help of velocity vectors. However, some random factors are utilized to adjust the velocity vectors. In our work, the motive behind using PSO is to get a single-valued threshold. In order to obtain threshold value using PSO algorithm, the pixel intensity and pixel number are pivotal in ACO and the pixel intensity of each particle in the swarm can be updated by below equation:

$$v_i(t + 1) = wv_i(t) + c_1r_1[\hat{x}_i(t) - x_i(t)] + c_2r_2[g(t) - x_i(t)] \qquad (2)$$

Here, $v_i(t)$ is the velocity of particle i at time t. $x_i(t)$, location of particle i at time t. There are some user-defined coefficients which are represented as w, c_1, and c_2. Velocity update, some random values are regenerated which are represented by r_1 and r_2. $\hat{x}_i(t)$, is the best candidate solution at particle i at time t. $g(t)$ is swarm's global best candidate solution at time t. The position of each particle is updated after the intensity for each particle is computed. The new threshold is applied to the previous position of particle in order to update the position which is provided by below function:

$$x_i(t + 1) = x_i(t) + v_i(t + 1) \qquad (3)$$

Until the stopping condition is satisfied, the process keeps repeating. The aforementioned condition or the predefined condition that can be placed includes the number of iterations that are preset within the PSO algorithm, the number of iterations left since the last update of the global best candidate solution or the target fitness value that is predefined. Then, $x_i(t + 1) = t$ where "t" is the threshold value for the pheromone update. The pheromone is updated when all the m artificial ants have moved within each of the construction process based on below function:

$$\tau_{(i,j)}(\text{new}) = (1 - \rho)\tau_{(i,j)}(\text{old}) + \sum_{k=1}^{m} \Delta\tau_{(i,j)}^k \qquad (4)$$

Here, pheromone evaporation rate is denoted by ρ. The boundless pheromone deposition is avoided with the help of ρ and can help in restraining the ants from selecting similar routes which can cause stagnation of the algorithm. For the kth ant, the pheromone is deposited at pixel (i, j) which is denoted by $\Delta\tau_{(i,j)}^k$ by below function:

$$\Delta\tau_{(i,j)}^k = \begin{cases} \eta_{(i,j)} & \text{if pixel } (i, j) \text{ is visited by the } k\text{th ant and } \eta_{(i,j)} \geq t \\ 0 & \text{otherwise} \end{cases} \qquad (5)$$

Here, t is the threshold value to update pheromone. During the mobility of ants in these edges, the pheromone is deposited with the help of comparison between heuristic information value and threshold value (t) (Fig. 1).

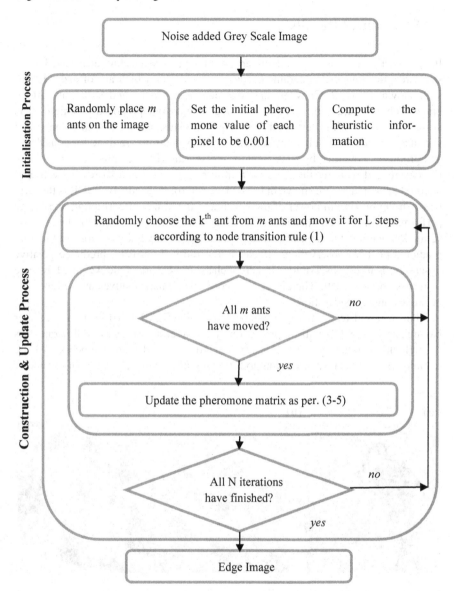

Fig. 1 Flowchart of ACO with PSO based edge detection technique

3   Experiment Results

In our experiment results, the effectiveness of the proposed technique is tested using standard gray images: cameraman, house, and pepper which are of sizes 128 × 128, shown in Fig. 2. Conventional edge detectors, Canny and Sobel, and ACO-based technique are considered and compared with our proposed technique. The noise considered while testing these images is Gaussian noise with mean 0 and variance of 0.01. Both subjective and objective evaluations are done for considered and proposed technique by peak signal-to-noise ratio (PSNR), precision, recall and F-measure and their results are shown in Tables (1, 2, 3 and 4) respectively and restored edge images are shown in Figs. (4, 5, and 6), respectively. PSNR is the ratio between the maximum power of the signal and the power of corrupting noise. The performance of the technique is considered as better if that technique has high PSNR value. Precision is the ratio of correctly predicted positive observations to the total predicted positive observations. Recall is the ratio of correctly predicted positive observations to all observations in actual class. F-measure is the weighted average of precision and recall. Therefore, this score takes false positives as well as false negatives into consideration.

The proposed technique is implemented in MATLAB R2013a on a computer having RAM of 8 GB with i5 processor by considering various initial parameters through initialization process, construction process, and update process. All the parameters have been set according to following procedure. The value of parameter

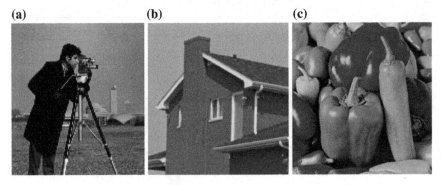

Fig. 2 Test images: **a** Cameraman, **b** House, and **c** Pepper

Table 1 PSNR results for Sobel, Canny, ACO, and the proposed technique on test images cameraman, house, and pepper with variance 0.01 of Gaussian noise

Images	Sobel	Canny	ACO [13]	Proposed
Cameraman	17.89	16.78	18.05	21.34
House	18.89	17.78	20.05	24.34
Pepper	19.23	21.18	22.03	25.45

Table 2 Precision results for Sobel, Canny, ACO, and the proposed technique on test images cameraman, house, and pepper with variance 0.01 of Gaussian noise

Images	Sobel	Canny	ACO [13]	Proposed
Cameraman	0.795	0.807	0.827	0.864
House	0.798	0.807	0.828	0.868
Pepper	0.788	0.796	0.817	0.878

Table 3 Recall results for Sobel, Canny, ACO, and the proposed technique on test images cameraman, house, and pepper with variance 0.01 of Gaussian noise

Images	Sobel	Canny	ACO [13]	Proposed
Cameraman	0.3789	0.3678	0.3201	0.2878
House	0.3889	0.3778	0.3301	0.2778
Pepper	0.4089	0.4178	0.3801	0.2978

Table 4 F-measure results for Sobel, Canny, ACO, and the proposed technique on test images cameraman, house, and pepper with variance 0.01 of Gaussian noise

Images	Sobel	Canny	ACO [13]	Proposed
Cameraman	0.7345	0.7435	0.7782	0.8035
House	0.7345	0.7435	0.7782	0.8068
Pepper	0.7445	0.7635	0.7882	0.8135

m is $\lfloor \sqrt{M \times N} \rfloor$ and τ_{init} is set to 0.0001 which is set in initialization process. In the construction process, four parameters α, l, L, and β are arranged. α is the pheromone intensity control factor, l is memory length of ant, L is steps taken by ant while moving, and β is the heuristic information of control factor. The proposed model has been affected by the parameters L and l, and they mainly depend on the size of the image as they manipulate the communication between the ants and moving distance of an ant. Edge detection is low if the sizes of L and l are small due to large size of an image but it consumes more processing time. It is necessary to create a transaction between the image size and its two parameters, i.e., L and l. During the experiments, L is $\lfloor 3 \times \sqrt{M \times N} \rfloor$ and l is equal to $\lfloor 2 \times (M + N) \rfloor$. Parameter α determines the relative weight of pheromone trail while β is used to determine the heuristic information. In order to get desired results, the values of two parameters are set to $\alpha = 2$ and $\beta = 2$. The pheromone update is controlled by two parameters that are ρ and t. The evaporation rate of pheromone is faster if the value of p is higher that leads to less detection of edges. Less noise has been detected if the value of t is larger. In order to get the desired result, the values of two parameters are set to $\rho = 0.02$ and t will be calculated using PSO algorithm.

(a)   **(b)**   **(c)**

(d)   **(e)**   **(f)**

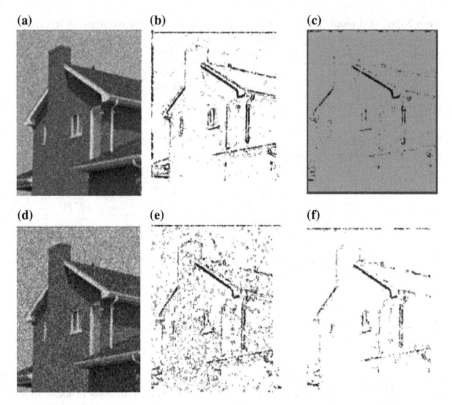

Fig. 3 House **a** Noisy image with noise variance 0.01, Edge image of proposed **b** $t = 0.1$ **c** $t = 0.12$ **d** Noisy image with noise variance 0.03, Edge image of proposed **e** with $t = 0.12$ **f** $t = 0.14$

Tables 1, 2, 3, and 4 shows the PSNR, precision, recall and F-measure results of Sobel, Canny edge detectors, ACO [13], and the proposed technique for edge detection with Gaussian noise added to test images having variance of 0.01, respectively. From Tables 1, 2, 3, and 4, we can infer that the proposed technique which uses PSO to find the threshold value is better compared to the Sobel, Canny, and ACO for all test images.

Figure 3(b–c) and (e–f) shows the result of house edge image with noise variance 0.01 and 0.03 of Gaussian noise, for Fig. 3a, d respectively with different values of t, which comes through PSO algorithm. Here, we use dynamic obtained threshold value through PSO which brings better edge detection results and enhances the binary decision which is taken after the pheromone matrix is obtained.

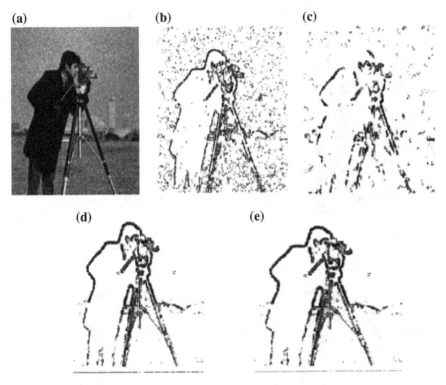

Fig. 4 Cameraman **a** Noisy image with noise variance 0.01, Edge image **b** Sobel **c** Canny **d** ACO, and **e** Proposed

Objective results of noisy test images (Figs. 4, 5 and 6a) are shown in Figs. (4, 5, and 6(b–e)) for Cameraman, House, and Pepper, respectively, and these edge images give a clear indication about the effectiveness of the proposed technique over the considered edge detection techniques. Smoothing procedure is not applied in both these detectors, Sobel and Canny, in order to remove the noise before applying the edge detection procedure. We can easily make out through the results that the visual quality of our proposed technique is better compared to other techniques and this betterment can be seen due to the PSO algorithm which provides us with a single dynamic threshold value in addition to the ACO-based technique.

Running time of proposed technique is also quite low thanks to ant system which only involves single update. The execution time of our proposed technique came out to be 21.22 s which is quite low when compared to the existing technique whose execution time is 23.02 s.

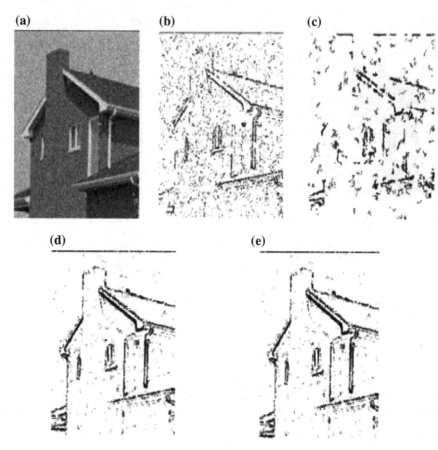

Fig. 5 House **a** Noisy image with noise variance 0.01, Edge image **b** Sobel **c** Canny **d** ACO, and **e** Proposed

(a) (b) (c)

(d) (e)

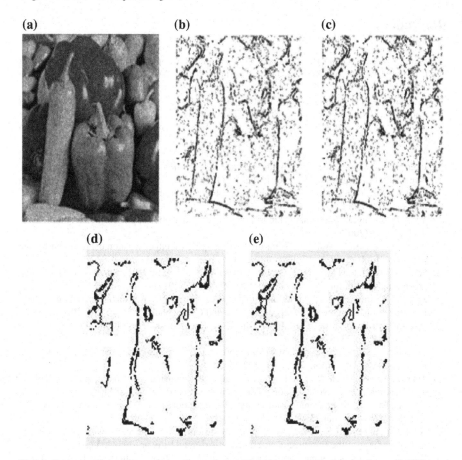

Fig. 6 Pepper **a** Noisy Image with noise variance 0.01, Edge image **b** Sobel **c** Canny **d** ACO, and **e** Proposed

4 Conclusion

The proposed edge detection technique using ACO with PSO based threshold outperforms over Sobel, Canny, and ACO in terms of PSNR, precision, recall and F-measure as well as restore edge images in a noisy environment. The inclusion of PSO to find out the threshold seems to be a better option compared to the existing threshold calculation techniques. Further research could be done on the use of this technique on other kinds of noises.

References

1. Hingrajiya, K.H., Gupta, R.K., Chandel, G.S.: An ant colony optimization algorithm for solving travelling salesman problem. Int. J. Sci. Res. Publ. **2**(8), 1–6 (2012)
2. Muthukrishnan, R., Radha, M.: Edge detection techniques for image segmentation. Int. J. Comput. Sci. Inf. Technol. **3**, 259–267 (2011)
3. Rao, K.N., Rao, P.S., Rao, A.A., Sridhar, G.R.: Sobel edge detection method to identify and quantify the risk factors for diabetic foot ulcers. Int. J. Comput. Sci. Inf. Technol. **5**, 39–46 (2013)
4. Zheng, Y.-Y., Rao, J.-L., Wu, L.: Edge detection methods in digital image processing. In: International Conference on Computer Science & Education, pp. 471–473 (2010)
5. NagaRaju, C., NagaMani, S., rakesh Prasad, G., Sunitha, S.: Morphological edge detection algorithm based on multi-structure elements of different directions. Int. J. Inf. Commun. Technol. Res. **1**, 37–43 (2011)
6. Yang, H., Zhang, J.: Mathematical morphology in edge detection application. J. Liaoning Univ. **32**(1), 50–53 (2005)
7. Zhang, L., Zhou, W., Wang, B.: Filtering SAR imagery for edge detection using support value transform. In: International Joint Conference on Neural Networks, pp. 1–8 (2015)
8. Hien, N.M., Binh, N.T., Viet, N.Q.: Edge detection based on fuzzy C means in medical image processing system. Int. Conf. Syst. Sci. Eng. **21**, 12–15 (2017)
9. Zhuang, X.: Image feature extraction with the perceptual graph based on the ant colony system. In: International Conference on Systems, Man and Cybernetics **7**, 5364–5359 (2004)
10. Nezamabadi-Pour, H., Saryazdi, S., Rashedi, E.: Edge detection using ant algorithms. Soft. Comput. **10**(7), 623–628 (2006)
11. Tian, J., Yu, W., Xie, S.: An ant colony optimization algorithm for image edge detection. Evolutionary Computation, 751–756 (2008)
12. Tian, J., Yu. W., Chen, L., Ma, L.: Image edge detection using variation-adaptive ant colony optimization. Lecture Notes in Computer Science, vol. 6910, 27–40 (2011)
13. Liu, X., Fang, S.: A convenient and robust edge detection method based on ant colony optimization. Opt. Commun. **353**, 147–157 (2015)
14. Khaluf, Y., Gullipalli, S.: An efficient ant colony system for edge detection in image processing. In: Proceedings of the European Conference on Artificial Life, pp. 398–405 (2015)
15. Canny, J.: A computational approach to edge detection. Trans. Pattern Anal. Mach. Intell. **PAMI-8**(6), 679–698 (1986)
16. Qiu, P.H.: Jump surface estimation, edge detection, image restoration. J. Am. Stat. Assoc. **102**, 745–756 (2007)
17. Hasan, R.A., Mohammed, M.A., Tapus, N., Hammood, O.A.: A comprehensive study: ant colony optimization (ACO) for facility layout problem. In: Networking in Education and Research (2017)
18. Tandon, A., Raja, R., Chouhan, Y.: Image segmentation based on particle swarm optimization technique. Int. J. Sci. Eng. Technol. Res. **3**(2), 257–260 (2014)

Improved Convolutional Neural Networks for Hyperspectral Image Classification

Shashanka Kalita and Mantosh Biswas

Abstract Classification is the process of setting class labels to pixels based on some obtained properties. Hyperspectral images (HSI) have very high dimensionality, which results in higher cost and complexity for analyzing and classifying them as superfast processors and large storage devices are required. Moreover, due to limited training samples and labeled data, classification remains an arduous task. Many methods have been presented till now for classification of HSI based on traditional methods that use handcrafted features beforehand, principal component analysis and its variations, decision trees, random forests, SVM-based methods, and neural networks, but most of these consider only the spectral information for classification resulting in low classification accuracy. Nowadays, increasing spatial resolution of HSI demands obtaining spatial data for further improving classification performance. We, therefore, present a classification method which obtains spectral as well as spatial features using convolutional neural network (CNN) model and then a logistic regression (LR) classifier that uses the activation function softmax for predicting classification results. Our proposed method is compared with considered techniques and tests on HSIs, namely, Indian pines and Pavia University, which have shown better performance regarding parameters such as overall accuracy (OA), average accuracy (AA), and kappa coefficient (K).

Keywords Hyperspectral image · Classification · Convolutional neural networks · Logistic regression · Folded-Principal component analysis

S. Kalita (✉) · M. Biswas
Computer Engineering Department, National Institute of Technology, Kurukshetra, Haryana, India
e-mail: shashanka_31603117@nitkkr.ac.in

M. Biswas
e-mail: mantoshbiswas@nitkkr.ac.in

© Springer Nature Singapore Pte Ltd. 2019   397
J. Kalita et al. (eds.), *Recent Developments in Machine Learning and Data Analytics*,
Advances in Intelligent Systems and Computing 740,
https://doi.org/10.1007/978-981-13-1280-9_37

1 Introduction

With the latest advancement in satellite and airborne sensor technology, the application of HSI in the fields of geoscience, biomedical imaging, agriculture, astronomy, and physics have vastly increased. Images captured by hyperspectral remote sensing systems contain a lot of pixels and a rich spectral response is contained by each pixel that represents the pixel's area of light absorption. It is possible to identify various objects or materials in a scene from their spectral information, as they absorb, reflect, and scatter electromagnetic waves differently. Identifying different objects in a scene based on their reflectance properties can be regarded as a classification task. However, a lot of challenges are faced while analyzing HSI which hampers classification accuracy. HSI has very high dimensionality and it lacks sufficient number of samples for training, thus leading to "The curse of dimensionality" [1].

Classification refers to extracting information from intensity values of pixels present in the image and categorizing them into a particular class. In the last decade, a lot of work has been done for information extraction and using those information for performing the classification of HSI. There are two types of information for a pixel that can be extracted, spectral and spatial. Traditional classifiers like k-nearest neighbor [2], which has a maximum likelihood criterion [3] and logistic regression [4] can obtain spectral information needed for classification. Some dimension reduction methods like principal component analysis (PCA) [5] and independent component analysis (ICA) [6] are applied to extract spectral features considering the fact that pixels belonging to the same class tend to show the same spectral signatures in most of the bands, so correlated bands can be removed thus reducing spectral dimension of HSI. Then, those extracted features are used by some classifiers for predicting classification results. Support vector machine (SVM)-based methods does not require feature reduction techniques in order to classify HSI data as they can be directly applied on the original data [7]. They gave better classification accuracies and thus remained the revolutionary method before the concept of deep learning (DL) was established. Nowadays, with the improvement of sensors, spatial resolution of HSI has increased, and thus including spatial features improve classification accuracy. All the above existing methods take into account only spectral features and fail to conceive the spatial variability observed in high-resolution data leading to low performance. Moreover, PCA and ICA methods, and classifiers such as linear SVMs and LR (single layered) are not deep making them unable to learn complex features about the data. To enhance classification accuracy, obtaining both spectral and spatial features are necessary, which can be procured separately or can be fused together to produce joint features. Different spatial filters such as morphological profiles [8], entropy [9], low-rank representation [10], and attribute profiles [11] can acquire spatial dependence beforehand and later combined with spectral information to carry out classification. A series of 3D Gabor filters [12] and 3D wavelet filters [13], which are generated at different frequencies and scales can be applied to HSI to procure combined spectral and spatial features. However, most of the conventional methods build on handmade features and shallow learning models and requires knowledge of the considered domain. The design of these features can be very time consuming and nonoptimal.

DL has brought a revolution in the world of classifying HSI data, as they can extract much deeper features unlike the rest of the methods. Different DL architectures, which can be supervised or unsupervised have been designed and implemented for feature extraction and classification. Unsupervised deep neural architectures such as stacked auto-encoders (SAE) [14], stacked denoising auto-encoders (SDAE) [15], and deep belief networks (DBN) [16] have shown much better results than the previous methods regarding classification accuracy. They can learn deep features automatically layer after layer, thus eliminating the need for manually designed features. However, the input patches taken from the HSI data has to be leveled to one dimension before being fed to these types of models resulting in considerable loss of spatial information. Moreover, unsupervised techniques do not make use of labeled information which is a disadvantage. Supervised methods make use of labeled information while learning features. Each training example in the supervised method is a pair consisting of the input and its true output value. Various supervised methods have been put forward till now including the latest CNNs. CNN, introduced in [17–19], improves over other DL approaches, as they can efficiently extract spatial features, and also due to weight sharing, it needs less parameters to learn. In methods [17–19], dimension reduction is performed at first using techniques like PCA, and then fed to the CNN model as input. But the drawback is that, due to applying PCA, though the spatial information remains intact, spectral information is lost and so the model may not acquire full benefit of the joint spectral–spatial details. Semi-supervised methods consider both unlabeled and labeled samples for training and aim to improve the classification performance even if less labeled data is present. In [20], a semi-supervised classification method is presented to acquire both spectral–spatial features from HSI. In [20], they analyzed the problem of efficiently obtaining both spectral and spatial features associated with the feature extraction phase using a SAE and a CNN model to obtain spectral and spatial features, respectively. However, due to plenty of layers and a large number of units in each of those layers, a lot of parameters are required to be trained in the SAE model which increases computation time and complexity.

In this paper, in order to alleviate the above issues, we present a classification method which uses CNN for spectral–spatial feature extraction and then uses a LR classifier responsible for estimating classification results. CNN is an improvement over other DL models due to its effectiveness in better assumption of locality of a pixel in an image and also it requires fewer parameters to train due to weight sharing. Our method can extract both spectral and spatial features and fuse them together for better classification accuracy. Also, considering the issue of scarcity of training samples a virtual sample generation method proposed in [21] is used. Experimental results on two HSIs show that our approach improves over considered DL methods [14 and 18]. Section 1 introduces the problems related to classification of HSI and techniques presented till now to solve some of those issues. A complete description of our presented work is given in Sect. 2 and the test results in Sect. 3.

2 Proposed Work

Our work involves classifying the pixels present in the HSI dataset into various classes based on some properties using convolutional neural networks for feature extraction and LR for classification. We have used a similar approach as [22], where the spectral features are obtained by a one-dimensional CNN (1-D CNN) and the spatial features by a two-dimensional CNN (2-D CNN). In our approach, the dimension of the input is reduced by folded-PCA which performs better than standard PCA [23] before being fed as input to the 2-D CNN. Moreover, we design the architectures of both the CNN models in such a way that it gives better classification results. Also, virtual samples are generated according to [21], which solves the problem of limited training samples to some extent. In the following sections, the details of our approach will be explained and Fig. 2 shows the flowchart of our approach.

2.1 Folded-PCA

HSI is a very high-dimensional dataset where, for each pixel, there are more than hundreds of reflectance-related information in the spectral bands. However, the contrast between pixels' spectral responses belonging to the same class is very little, i.e., they have the same reflectance properties in most of the bands. Therefore, a dimension reduction technique can be applied (removing the correlated bands but retaining the original information almost intact) to the HSI in order to lower the complexity of calculation and training period. Folded-PCA is a technique used for dimension reduction and feature extraction of high-dimensional datasets like HSI [23]. In this technique, each pixel vector of hyperspectral data is first folded into a matrix, which is then used for computation of partial covariance matrices. These partial covariance matrices are then combined to form the final covariance matrix for eigen decomposition and then projection of data. The N-dimensional pixel vector, where N denotes total bands present, is folded into a matrix containing X rows and Y columns. If there are odd number of bands, zeroes are added at the end of every row such that the matrix has segments of the same size as Y'. Let V_i be the matrix of dimension $X \times Y$ formed by folding the ith pixel vector. Its covariance matrix is given by

$$\mathrm{Cov}_i = V_i^\mathrm{T} V_i \tag{1}$$

where Cov_i is of dimension $Y \times Y$. The final covariance matrix, $\mathrm{Cov}_{\mathrm{final}}$, for the entire HSI is computed by adding these partial covariance matrices generated for every pixel vector. $\mathrm{Cov}_{\mathrm{final}}$ is then decomposed in the form of

$$\mathrm{Cov}_{\mathrm{final}} = AGA^\mathrm{T} \tag{2}$$

where the eigenvalues of Cov_{final} forms the diagonal matrix G and the corresponding eigenvectors comprise the orthonormal matrix denoted by A. The final covariance matrix's size is $Y \times Y$, which is very much less than that obtained using standard PCA which is $N \times N$. The orthonormal matrix A is reduced to size $q \times Y$, where the total retained eigenvalues (discarding the small ones) are denoted by q. Finally, the projected result is given by

$$Z_i = V_i \times A \qquad (3)$$

Unlike conventional PCA, folded-PCA takes into account both global and local structures of the data. It needs less than 1% of memory compared to conventional PCA and the overall computation cost is also reduced [24]. After applying folded-PCA, the reduced input will be provided as input to a 2-D CNN model for extraction of spatial features.

2.2 Convolutional Neural Networks

CNN is a supervised DL-based method that has been widely used in analysis and processing of visual imagery outperforming existing DL methods in detection and classification tasks. A typical CNN architecture comprises of an input layer, a layer comprising of the various output classes (output layer), and in between them there are multiple hidden layers. The hidden layers comprise of convolution layers, downsampling or pooling layers, normalization layers, and fully connected (FC) layers. The final FC layer is linked to the output layer. During the training process, a CNN can automatically learn features from input data layer after layer, where high-level features are learnt from lower level features. These features are detected by filters in the output of the previous layer. To capture different features, in our approach, we have used a 1-D CNN for extraction of spectral features and a 2-D CNN for spatial feature extraction and then combined the extracted features for final classification using a LR classifier. The architecture of a typical CNN is shown in Fig. 1 consisting of two convolution layers each followed by a downsampling or pooling layer and finally linked to output layer is the FC layer.

Spectral feature extraction using 1-D CNN. In our approach, we have extracted the spectral features using a 1-D CNN directly from the input data. In 1-D CNN, 1-D filters are utilized in place of 2-D filters and these filters can adequately detect important information along the spectral dimension. We extract a small patch size of $R \times R \times N$ for each pixel from the input HSI and the patch is then converted into individual 1-D vectors of size N to be fed to the CNN model. The input vectors go through a convolution operation with the 1-D filters to generate feature maps after being passed through an activation function. We have used leaky ReLU as the activation function for efficient computation [25], which permits a tiny gradient when the neuron is inactive and is expressed as

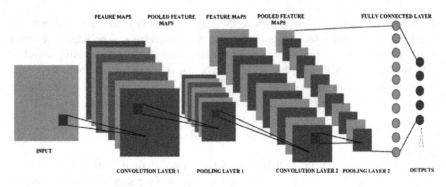

Fig. 1 Architecture of a typical CNN

$$f(n) = \max(0.01 \times n, n) \tag{4}$$

where n is the input to the activation function. A neuron's value n_{ij}^c positioned at c of the feature map j in layer i is

$$n_{ij}^c = f\left(b_{ij} + \sum_m \sum_{p=0}^{P_i-1} w_{ijm}^p n_{(i-1)m}^{c+p}\right) \tag{5}$$

where the bias of feature map j in layer i is b_{ij} and m indicates the feature map in layer $(i-1)$ linked to the present. The weight's value positioned at p linked to mth feature map is w_{ijm}^p and P_i is length of the filter along the spectral dimension. After the convolution layer, we use a max pooling layer which combines patches of the neuron clusters into a single neuron to be fed as input to the next convolution layer. This layer aims to decrease the feature space, thus reducing the complexity of computation. After several convolution and pooling layers, the feature maps generated were provided as input for the FC layer whose principle is identical to traditional multilayer perceptron (MLP) neural network, where every neuron in one layer is linked to every neuron in the other layer. Finally, loss is calculated by LR for a mini batch of input data. This loss helps to update filter values while training, such that the loss is minimized in successive iterations and is computed as

$$\text{loss} = -\frac{1}{m}\sum_{i=1}^m o_i \log(z_i) + (1 - o_i)\log(1 - z_i) + \frac{\lambda}{2m}\sum_{j=1}^W w_j^2 \tag{6}$$

Here, the mini batch size is m. o_i and z_i represent ith actual and predicted label in the batch, respectively. The last term, $\frac{\lambda}{2m}\sum_{j=1}^W w_j^2$ in the loss function is a L2 regularization model, which is used to solve the problem of overfitting. W denotes total number of weights and regularization parameter λ controls trade-off between two goals in order to keep the hypothesis simple to prevent overfitting. The first goal is to fit the training data well and other one to keep the parameters small. LR uses

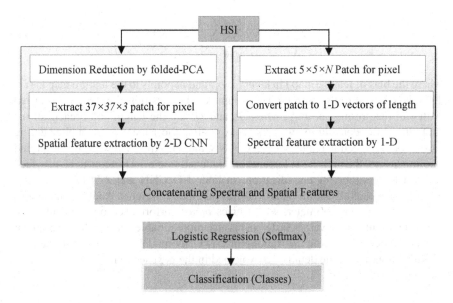

Fig. 2 Flowchart of the proposed approach

softmax [26] as the activation function which outputs a vector of class probabilities summing up to one, thereby, generalizing LR to C classes. After training the 1-D CNN model to extract spectral features, we train the 2-D CNN model for obtaining spatial features. Details of the 2-D CNN model will be explained in the next section.

Spatial feature extraction using 2-D CNN. 2-D CNN utilizes 2-D filters to acquire the spatial features of HSI. The architecture of our 2-D CNN model is similar to the 1-D model which contains the input layer and several convolution and pooling layers, the FC layer, and finally the output layer. Activation function used is the same as the one used during spectral feature extraction (Fig. 2).

As HSI has very high dimensionality incurring very high complexity while processing and due to high correlation among spectral bands we have first used folded-PCA which reduces spectral dimension of our high-dimensional dataset. The spatial information remains unchanged after feature reduction [18], and, therefore, we can apply 2-D CNN model to extract the spatial features on the reduced input. We have taken into consideration the first three principal components after reduction according to [22], i.e., the spectral dimension of the input is now reduced to three. After dimension reduction, we extract a $R \times R \times 3$ patch centered at the pixel to be classified and then feed it to the 2-D CNN model, which then generates feature maps after convolution operation with filters and going through the activation function, i.e., leaky ReLU. A neuron's value n_{ij}^{cd} positioned at (c, d) of feature map j in layer i is

$$n_{ij}^{cd} = f\left(b_{ij} + \sum_m \sum_{p=0}^{P_i-1} \sum_{q=0}^{Q_i-1} w_{ijm}^{pq} n_{(i-1)m}^{(c+p)(d+q)}\right) \tag{7}$$

where the bias of feature map j in layer i is b_{ij} and m points to the feature map in $(i-1)$th layer linked to the current. The weight's value positioned at (p, q) associated with mth feature map is w_{ijm}^{pq}, where P_i and Q_i are the length and breadth of the 2-D filter. After convolution we apply max pooling the same way we applied during extraction of spectral features by 1-D CNN model. After a few convolution and pooling layers, the high-level features are flattened into a 1-D vector to be provided to the LR classifier for classification. In both 1-D CNN and 2-D CNN, we use adaptive gradient algorithm (AdaGrad) [27] to update weights. AdaGrad maintains a learning rate per parameter, which gives better performance on problems with sparse gradients. It converges faster and is also more reliable than stochastic gradient descent. After completing the training of 2-D CNN model, we proceed to the final classification phase, the details are explained in the next section.

2.3 Classification Using the Extracted Spectral and Spatial Features

In this phase, to classify an input pixel, the acquired spectral and spatial features of the pixel by the already trained 1-D and 2-D models, respectively, go through a max pooling operation in order to lower the dimensions and learnable parameters. After that they are concatenated together to produce joint spectral–spatial features and then fed to a LR classifier, which uses softmax function for final classification of the pixel.

2.4 Generation of Virtual Samples

To solve the issue of less available training samples for HSI, we use a method for generation of virtual training samples proposed in [21]. The new sample t_n is generated by

$$t_n = \alpha_m x_m + \beta n \tag{8}$$

where α_m is the disturbance in light intensity varying under different atmospheric conditions, x_m is a cube taken from the HSI, whereas n is the random Gaussian noise, and β controls n's weight.

3 Experimental Results

We have simulated our work on two HSIs, Indian pines and Pavia University, according to the defined approach and compared our results with [14 and 18]. All the experiments have been conducted using Python language (version 3.5) in an Anaconda environment and using tensorflow framework on a 64-bit machine, 2.4 GHz processor, 8 GB of memory, and additional 8 GB swap space. A detailed explanation of evaluation parameters, datasets, preprocessing of data, architecture design of CNN models, and results are provided below.

3.1 Evaluation Parameters

The evaluation parameters help to test the proposed work's performance in comparison to the considered techniques and the parameters based upon which the comparisons are done are OA, AA, and K. OA denotes the ratio between total test samples accurately classified to the total available test samples. AA denotes the mean accuracy of classes, whereas K indicates the degree of agreement and is calculated by weighing the measured accuracies.

3.2 Datasets

Indian Pines: AVIRIS acquired this dataset in 1992 [28]. The size of the image is 145×145 and consists of 224 spectral bands having a range of wavelength 0.4–2.5 μm and 20 m spatial resolution. The water absorption bands are not considered lowering the number of bands to 200. The ground truth data provides 16 different land cover classes. There are 10,249 labeled pixels which constitute nearly 50% of the total number of pixels, each pixel belonging to one of the 16 classes.

University of Pavia: This dataset was obtained by ROSIS over the city of Pavia in Italy in 1988 [28]. This dataset covers a university area having a size of 610×340 pixels and 103 spectral bands. The dataset's wavelength range is 0.43–0.86 μm and spatial resolution is 1.3 m per pixel. The ground truth information provides nine different classes. The count of labeled pixels in this dataset is 42,776 out of 207,400 total pixels.

3.3 Data Preprocessing

For achieving better results from learning models, it is required to preprocess the data in order to convert it into a proper format. During this stage, we first load the dataset and normalize it in the range $(-0.5, 0.5)$. Then, we make training and testing splits randomly, where 75% of the labeled samples constitute the training set and rest 25% as the test set. After making splits, we set a threshold of 200, which indicates the maximum number of samples per class. If available samples of some classes are less than our selected threshold, we include the whole 75% of it for training, else for classes having more number of samples, we select only 200, ignoring the rest. Thus, for some classes, only 20–25% of labeled samples are used for training. For each generated virtual sample, α_m is set uniformly between [0.9, 1.1] and β is 1/25 according to [21]. Patches centering the pixel to be classified are extracted for the datasets. For 1-D CNN, patch size of $5 \times 5 \times N$ is extracted. For the 2-D CNN model, patch size of $37 \times 37 \times 3$ is extracted for each pixel, 3 is the number of retained principal components after applying folded-PCA.

3.4 CNN Architecture Design

The optimization of the CNN models was done by trial and error approach for setting the number of convolution layers, filters and size of each filter, batch size, learning rate, and epoch values.

1-D CNN: The architecture comprises of three convolution layers each followed by a max pooling layer for both datasets. The pooling window size used is 2×2 with stride 1. Filters' sizes used for first, second, and third layers are 5, 4, and 3, respectively, and the number of filters for all layers is set to be 36. Initial learning rate for Indian Pines is set to be 0.005 and for Pavia University is 0.01 and batch size used for both is 100. Number of training epochs is set as 300 and 200 for Indian Pines and Pavia University, respectively.

2-D CNN: The model uses three convolution layers for Indian Pines and two for Pavia University datasets, where after each layer max pooling is applied with same size and stride as in 1-D CNN. For, Indian pines, the count of filters utilized in first, second, and third layers are 36, 36, and 72, respectively. Each filter is of size 5×5, 6×6, and 4×4 in first, second, and third layers, respectively. For Pavia scene, 36 and 72 filters are used in first and second layers, respectively, with each filter of size 5×5 and 6×6 in first and second layers, respectively. Initial learning rate set is 0.01 and training epochs are set to be 100 for both.

3.5 Experimental Results and Validation

To test the efficacy of our approach, we compare our results against two other DL techniques proposed in [14], which uses SAE and [18] which is based on CNN. Tables 1 and 2 display the results OA, AA, and K of [14, 18], and the proposed approach, respectively, on datasets taken into account and it shows that our approach improves over the considered techniques in terms of the evaluation parameters.

Table 1 Classification results of OA, AA, and K in % of [14, 18], and the proposed for Indian Pines

Parameters	Chen et al. [14]	Makantasis et al. [18]	Proposed
OA	97.78	98.63	98.87
AA	97.11	97.90	98.23
K	97.54	98.32	98.61

Table 2 Classification results of OA, AA, and K in % of [14, 18], and the proposed for Pavia University

Parameters	Chen et al. [14]	Makantasis et al. [18]	Proposed
OA	98.49	99.51	99.60
AA	97.75	98.99	99.12
K	98.05	99.20	99.42

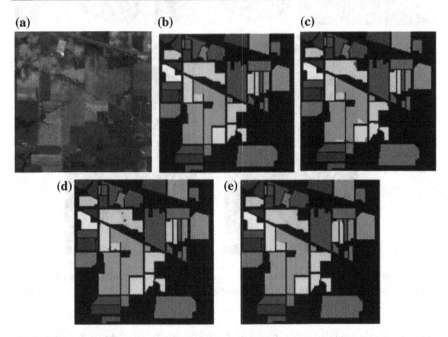

Fig. 3 Indian Pines **a** image (false color), **b** ground truth; classification maps of **c** Chen et al. [14], **d** Makantasis et al. [18], and **e** proposed approach

Classification Maps. Figure 3a shows a false color image of Indian Pines, (b) the ground truth, (c–d) classification maps of [14 and 18], respectively, and (e) classification map of our approach. From the classification maps (c–e) and comparing them with the ground truth image, it is seen that our approach performs better in classifying the pixels into classes compared to considered techniques.

(a) (b) (c)

(d) (e)

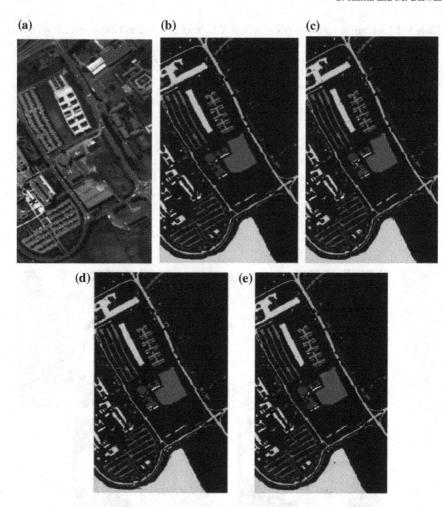

Fig. 4 Pavia University **a** image (false color), **b** ground truth; classification maps of **c** Chen et al. [14] **d** Makantasis et al. [18], and **e** proposed approach

Figure 4a shows false color image of Pavia University, (b) the ground truth, (c–d) classification maps of [14 and 18], respectively, and (e) classification map of the proposed approach. The classification maps show that, our approach succeeds in correctly classifying the pixels into various classes compared to the considered techniques.

4 Conclusion

We have presented, in this paper, a method for HSI classification considering both spectral and spatial features. Both 1-D and 2-D CNN models are individually trained at first using the training set. Before training the 2-D CNN model, we remove correlation among bands by folded-PCA. During the classification phase, the spectral and spatial features of pixel to be classified are extracted by the already trained 1-D and 2-D CNN models, respectively, and then concatenated together. Then, with the help of LR, final classification of the pixel is performed. In our approach, architectures of both the CNN models are designed such that it gives very good classification results. We have also generated virtual samples to solve the issue of limited training samples and tests on two HSIs taken under consideration, which have shown that our approach performs better than the considered methods in terms of OA, AA, and K. Incorporating our work with different CNN architectures is left as future work.

References

1. Donoho, D.L.: High-dimensional data analysis: the curses and blessings of dimensionality. AMS Math Challenges Lect **13**, 178–183 (2000)
2. Samaniego, L., Bardossy, A., Schulz, K.: Supervised classification of remotely sensed imagery using a modified, k-NN technique. IEEE Trans. Geo. Rem. Sens. **46**, 2112–2125 (2008)
3. Ediriwickrema, J., Khorram, S.: Hierarchical maximum-likelihood classification for improved accuracies. IEEE Trans. Geo. Rem. Sens. **35**, 810–816 (1997)
4. Li, J., Dias, J.M.B., Plaza, A.: Semisupervised hyperspectral image segmentation using multinomial logistic regression with active learning. IEEE Trans. Geo. Rem. Sens. **48**(10), 4085–4098 (2010)
5. Rodarmel, C., Shan, J.: Principal component analysis for hyperspectral image classification. Surveying Land Inf. Sys. **62**(2), 115 (2002)
6. Robila, S.A., Varshney, P.K.: Feature extraction from hyperspectral data using ICA. In: Advanced Image Processing Techniques for Remotely Sensed Hyperspectral Data, pp. 199–216 (2004)
7. Melganiand, F., Bruzzone, L.: Classification of hyperspectral remote sensing images with support vector machines. IEEE Trans. Geo. Rem. Sens. **42**(8), 1778–1790 (2004)
8. Ghamisi, P., Mura, M.D., Benediktsson, J.A.: A survey on spectral–spatial classification techniques based on attribute profiles. IEEE Trans. Geo. Rem. Sens. **53**, 2335–2353 (2015)
9. Tuia, D., Volpi, M., Mura, M.D., Rakotomamonjy, A., Flamary, R.: Automatic feature learning for spatio-spectral image classification with sparse SVM. IEEE Trans. Geo. Rem. Sens. **52**(10), 6062–6074 (2014)
10. Jia, S., Zhang, X., Li, Q.: Spectral–spatial hyperspectral image classification using regularized low-rank representation and sparse representation-based graph cuts. IEEE J. Sel. Topics App. Earth Obs. Rem. Sens. **8**, 2473–2484 (2015)
11. Mura, M.D., Villa, A., Benediktsson, J.A., Chanussot, J., Bruzzone, L.: Classification of hyperspectral images by using extended morphological attribute profiles and independent component analysis. IEEE Trans. Geo. Rem. Sens. Letters **8**, 542–546 (2011)
12. Shen, L., Jia, S.: Three-dimensional Gabor wavelets for pixel-based hyperspectral imagery classification. IEEE Trans. Geo. Rem. Sens. **49**(12), 5039–5046 (2011)
13. Qian, Y., Ye, M., Zhou, J.: Hyperspectral image classification based on structured sparse logistic regression and three-dimensional wavelet texture features. IEEE Trans. Geo. Rem. Sens. **51**, 2276–2291 (2013)

14. Chen, Y., Lin, Z., Zhao, X., Wang, G., Gu, Y.: Deep learning-based classification of hyperspectral data. IEEE J. Sel. Topics App. Earth Obs. Rem. Sens. **7**(6), 2094–2107 (2014)
15. Vincent, P., Larochelle, H., Lajoie, I., Bengio, Y., Manzagol, P.A.: Stacked denoising autoencoders: learning useful representations in a deep network with a local denoising criterion. J. Mach. Learn. Res. **11**, 3371–3408 (2010)
16. Chen, Y., Zhao, X., Jia, X.: Spectral–spatial classification of hyperspectral data based on deep belief network. IEEE J. Sel. Topics App. Earth Obs. Rem. Sens. **8**(6), 1–12 (2015)
17. Yue, J., Zhao, W., Mao, S., Liu, H.: Spectral–spatial classification of hyperspectral images using deep convolutional neural networks. Rem. Sens. Lett. **6**(6), 468–477 (2015)
18. Makantasis, K., Karantzalos, K., Doulamis, A., Doulamis, N.: Deep supervised learning for hyperspectral data classification using convolutional neural networks. In: IEEE International Geoscience and Remote Sensing Symposium, pp. 4959–4962 (2015)
19. Liang, H., Li, Q.: Hyperspectral imagery classification using sparse representations of convolutional neural network features. Rem. Sens. **8**(2), 99 (2016)
20. Yue, J., Mao, S., Li, M.: A deep learning framework for hyperspectral image classification using spatial pyramid pooling. Rem. Sens. Lett. **7**(9), 875–884 (2016)
21. Chen, Y., Jiang, H., Li, C., Jia, X., Ghamisi, P.: Deep feature extraction and classification of hyperspectral images based on convolutional neural networks. IEEE Trans. Geo. Rem. Sens. **54**(10), 6232–6251 (2016)
22. Zhang, H., Li, Y., Zhang, Y., Shen, Q.: Spectral-spatial classification of hyperspectral imagery using a dual-channel convolutional neural network. Rem. Sens. Lett. **8**(5), 438–447 (2017)
23. Zabalza, J., Ren, J., Yang, M., Zhang, Y., Wang, J., Marshall, S., Han, J.: Novel folded-PCA for improved feature extraction and data reduction with hyperspectral imaging and SAR in remote sensing. ISPRS J. Photogrammetry Rem. Sens. **93**, 112–122 (2014)
24. Setiyoko, A., Dharma I.G.W.S., Haryanto, T.: Recent development of feature extraction and classification hyperspectral images: a systematic literature review. J. Phys.: Conf. Ser. **801**(1) (2017)
25. Maas, A.L., Hannun, A.Y., Ng, A.Y.: Rectifier nonlinearities improve neural network acoustic models. In: 30th International Conference on Machine Learning, p. 28 (2013)
26. Memisevic, R., Zach, C., Hinton, G., Pollefeys, M.: Gated softmax classification. In: Advances in Neural Information Processing Systems, p. 23 (2010)
27. Duchi, J., Hazan, E., Singer, Y.: Adaptive subgradient methods for online learning and stochastic optimization. J. Mach. Learn. Res. **12**, 2121–2159 (2011)
28. http://www.ehu.eus/ccwintco/index.php/Hyperspectral_Remote_Sensing_Scenes

Automated Vision Inspection System for Cylindrical Metallic Components

Krithika Govindaraj, Bhargavi Vaidya, Akash Sharma
and T. Shreekanth

Abstract The gap in manufacturing and assembly, in the automobile industry, is bridged by quality inspectors, who ensure that components manufactured are in proper condition to build a reliable system. However, if inspection is carried out by humans, it could lead to erroneous classification of defected components as good or vice versa, resulting in compromising of quality and wastage of material, respectively. Automation of this process is possible through the use of a vision-based inspection system and in this work such a system is proposed for cylindrical metallic components. The system uses computer vision and image processing methodologies such as morphology, edge and contour detection techniques to evaluate the errors effectively. Most systems evaluate errors stage by stage, which increases the consumption of space and cost of installation. The present inspection systems do not incorporate inspection of components that require rotation and some errors in small components cannot be detected. To not only address these issues but to also ensure reliability and efficiency, the proposed system, was tested for 250 samples, and was found to have an accuracy of 97%, which is comparable to the present systems.

Keywords Canny edge detection · Morphological erosion · Morphological dilation · Image segmentation · Hough transform · Adaptive thresholding OpenCV

K. Govindaraj (✉) · B. Vaidya · A. Sharma · T. Shreekanth
Department of Electronics and Communication,
Sri Jayachamarajendra College of Engineering, Mysore, India
e-mail: krithikagovindaraj@gmail.com

B. Vaidya
e-mail: bhargavivaidya17@gmail.com

A. Sharma
e-mail: akashshrm02@gmail.com

T. Shreekanth
e-mail: shreekanth_t@sjce.ac.in

© Springer Nature Singapore Pte Ltd. 2019   411
J. Kalita et al. (eds.), *Recent Developments in Machine Learning and Data Analytics*,
Advances in Intelligent Systems and Computing 740,
https://doi.org/10.1007/978-981-13-1280-9_38

1   Introduction

Components in the manufacturing industry undergo various operations before it is sent for assembly. The error is generated when the operations such as grinding, drilling, chamfering, etc., are performed improperly leading to the damage of the component, and these are the errors that need to be detected to ensure reliability of inspection. Vision systems not only play a role in customer satisfaction through quality assurance [1], but also ensure human resources are put to better use by taking on repetitive tasks, and performing with greater efficiency.

Existing vision systems have the capability of error detection for cylindrical components with a minimum diameter of 25 mm and are not capable of detecting the presence of outer-diameter burrs. Most systems incorporate heavy machinery which consumes floor space such as adjustable arms [2], belt drives, pulleys, and rotating wheels. Template matching is extensively used in the existing systems to identify errors, which is effective only if errors are uniform and localized.

The proposed system addresses the above shortcomings of the existing systems as it can evaluate errors in cylindrical components of minimum size 11 mm, by using minimal heavy machinery. The errors in the components are non-uniform and non-localized, due to which template matching becomes ineffective, and hence the algorithm in the proposed system is an upgrade over the algorithms used in existing systems.

The following sections explain the system with reference to the 11 mm dowel as the errors in this are generic, yet complex due to the size of the component, and can be extrapolated to many other components. The rest of the paper is divided into five sections. The second section elaborates on the typical errors of the dowel. The third section elaborates on the methodology involved in the processing of errors. The fourth section evaluates and interprets the results. The fifth section summarizes the findings and provides insight into the future scope of the system and the sixth section provides a description about the design.

2   Types of Errors

Errors can occur on the surface and within the cavities of the cylindrical component. Some errors are guaranteed to be present on one side of the component or at a particular location, whereas others can occur randomly and have different sizes and shapes.

2.1   Surface Errors

(a) Chamfer: Chamfer is an incline, usually of 45 degrees produced on the edges of the components to ensure that they can be held in place during assembly. There are various thicknesses of chamfers that need to be produced depending on the components, and a component to pass inspection, must have chamfers that lie within a tolerance.

(b) Outer burr: Excess metal on the edges of the component, which could lead to faster rusting.

(c) Grinding error: Grinding is used to produce a smooth and shiny surface through the removal of impurities, however, excessive heat during grinding can produce damaged surfaces.

(d) Outer-diameter damage: Damage produced by grinding machine leads to a damage on the chamfered part of the component.

2.2 Errors Within the Cavity

(a) Inner burr: In order to create a cavity, metal needs to be drilled out during which some metal is still left inside the cavity, this is an error because no other component can fit into this unless it is fully hollow.

(b) Components with variations in inner- and outer-diameter are also errors.

3 Methodology

In order to evaluate surface and cavity errors, the same precision and speed in capturing images are important. The techniques used to evaluate these errors, however, are different. Image acquisition ensures that the RGB image is suitable for preprocessing. The preprocessing stage helps in making the image suitable for extraction of the main parts of the dowel, upon which errors are evaluated in the processing stage. The segregation stage helps to separate defected components from the good components. Figure 1 shows the various stages involved in the process.

3.1 Image Aquisition

Real-time systems need to perform effectively without being constrained by the change in environment. This section elaborates on the methods used to ensure that images are obtained reliably for processing.

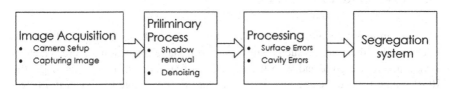

Fig. 1 Stages involved in the process

3.1.1 Camera Setup

The system uses a versatile microprocessor, i.e., Raspberry Pi using OpenCV and in order to incorporate parallel processing we use two of these microprocessors, each with its own CMOS camera with USB-type connector. The cameras have a resolution of five megapixels, and allow for capturing HD images. We first allow for autofocusing, so that the object is clearly detected, after which we manipulate the hardware into stopping autofocus, which is possible through the use of the V4L2 driver that comes preinstalled with the microprocessor. V4l2 controls any USB camera module attached to a Raspberry Pi. Video4Linux2, abbreviated as V4l2, allows for not only controlling autofocus but also brightness, saturation, white light, etc. Hence, through permutations of these properties, we can obtain an optimal image for various computer vision systems [3]. The proposed system uses images of size 960×720. Based on v4l2 setup of the camera, the sharpness measured on the scale of 1–10, must be set to 8, and brightness is set to 50% so that the strips of light reflected by the metallic surface is properly highlighted.

3.1.2 Capturing Image

Each component manufactured has a different size, and the camera as well as the lighting system needs to work with reference to a particular component. Subtle variation in the orientation, illumination, positions can lead to faulty detection of errors [4]. In order to obtain images of the dowel with highest quality, we adjust the light and camera in parallel with one another, and the image is taken in two angles, perpendicular and parallel to the component. Doing so will allow for simultaneously finding the errors on the surface and inside the cavity. Another challenge is the cylindrical shape of the dowel, which requires rotation as errors can be on any side. The camera must take pictures after every 0.05 s interval, which is the time taken to process the errors on the surface of one side and simultaneously process errors within the cavity which needs only one iteration. Synchronization of not only the two cameras but also the gating system, which segregates good from bad components, with the microprocessor is important. This is possible through the introduction of properly calculated delays.

3.2 Preliminary Processing

The system is not devoid of shadows and light cannot always be adjusted in order to ensure that there are no such disturbances, this stage ensures that there is a clean image before we go ahead with the detection of surface errors. In this stage, the RGB image from the camera is first converted to gray image as shown in Fig. 2a, and using global Otsu's thresholding, we convert the image to binary, as shown in Fig. 2b. Otsu's method iteratively reduces inter-class variance, as weighted sum of

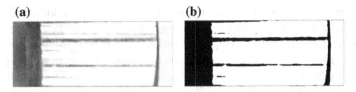

Fig. 2 **a** Gray image of dowel with shadow **b** Binary image of dowel with shadow

variances. The two classes mentioned here are the foreground and background. The foreground is usually the lighter part of the image and background is the darker part. Through this, the metallic part illuminated by the light is highlighted, considering the chamfer and shadow as background or darker parts. The equation is as given below [5]:

$$\sigma_w^2(t) = w_0(t)\sigma_0^2(t) + w_1(t)\sigma_1^2(t) \tag{1}$$

weights w0 and w1 are probabilities of two classes separated by a threshold t, and σ_0^2 and σ_1^2 are variances of two classes.

The image, as expected, has an edge that lies on the side of the shadow, i.e., the left side, which cannot be detected. To detect these type of edges, the same gray image from the camera, is subjected to adaptive thresholding using Otsu's method. This method generates the same output as the Otsu, but the entire picture is evaluated blockwise as shown in Fig. 3a. A mask is generated from the adaptive thresholded image using morphological erosion and is overlapped with the binary image generated by using global thresholding producing an image with properly detected boundaries of the dowel as shown in Fig. 3b.

3.3 Surface Error Detection

The preprocessed image is used as the main image, and we extract various boundaries to determine the dimensions of the various aspects such as chamfer and length of the component and errors such as outer burr and grinding error.

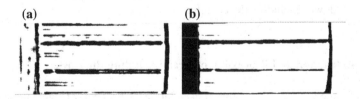

Fig. 3 **a** Adaptive thresholded image **b** Preprocessed image

The area incorporating the chamfer is usually dull, and the light does not illuminate this area, and using this to our advantage we compute the boundary of the lights to separate the dull surface from the shining surface. This is done using morphological closing in order to remove internal noise, and detecting the edges of white light using canny. Canny is one of the most reliable edge detection methods as it allows for proper detection of even the smallest edges [6, 7], which is very crucial in our application. The following steps are used for detecting edges using canny edge detector [8]:

- The image f (m, n) is convoluted with the Gaussian function G(m, n) given by Eq. 2, to get a smooth image [9]. The resultant is given by g(m, n) as shown in Eq. 3

$$G = \frac{1}{(2\pi\sigma^2)^{1/2}} e^{(-\frac{(m^2+n^2)}{2\sigma^2})} \tag{2}$$

$$g(m, n) = G(m, n) * f(m, n). \tag{3}$$

- First difference gradient operator is applied to obtain edge magnitude and direction. Let magnitude of the gradient be M(m, n) and angle be $\theta(m, n)$
- Non-maximal suppression is applied to the gradient magnitude.
- The non-maximal suppressed image is thresholded, and threshold M is given by

$$M_T(m, n) = \begin{cases} M(m, n) & \text{if } M(m,n) > T \\ 0 & otherwise \end{cases} \tag{4}$$

where T is chosen such that edges are retained and the noise is removed.
- Using non-maximal suppression, evaluate if the pixel has greater magnitude than its neighbors, if so retain the magnitude else set to zero.
- Finally, the previous result is thresholded using two different thresholds t1 and t2 (where t1 < t2) to obtain binary images B1 and B2 and continuous edges are obtained by linking edge segments in B2.

Once the boundary is obtained, the preprocessed image, and the image depicting the boundary of white lights is overlapped, called the processed image as shown in Fig. 4, to mark three main areas, which are as follows:

1. Area excluding the dowel.
2. Area between edge of dowel and inner boundary of chamfer.
3. Area of dowel excluding the chamfer.

3.3.1 Extraction and Processing of Area Excluding the Dowel

This is of importance as it takes into consideration the presence of outer burr as shown in Fig. 5. Outer burrs vary in sizes, and look like projections from the corners of the dowel. The edges of the dowel are extracted from the processed image by detecting the change in pixel level from white to black on either side. Extracting the ends of

Fig. 4 Processed image: image depicting the boundary of white lights is overlapped

Fig. 5 RGB image of dowel showing outer burr on the left side

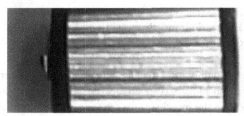

the image until the boundary, we obtain area excluding the dowel. This part must be devoid of any black pixels as it denotes the presence of outer burr. We compute the size of the outer burr, and classify the presence based on the size.

3.3.2 Extraction and Processing of Area Between Edge of Dowel and Inner Boundary of Chamfer

This part determines the width of the chamfer which is extracted by first detecting the edges of the dowel on either side. The boundaries of the lights marked by white lines in the processed image is detected and the image is cropped. Morphological operations are used in order to remove the curved surface and to approximate the image to the gray image as much as possible. Once the cropped surfaces showing the upper and lower chamfer, as shown in Fig. 6, on each side are obtained, we measure the length based on calibration.

Fig. 6 a Left side or upper chamfer of dowel shown in Fig. 1 **b** Right side or lower chamfer of dowel shown in Fig. 1

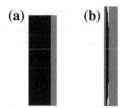

(a)   (b)

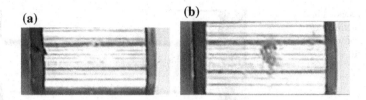

Fig. 7 **a** RGB image of dowel with outer-diameter damage; **b** Gray image of dowel with grinding error

3.3.3 Extraction and Processing of Area of Dowel Excluding the Chamfer

This part is mainly used to detect the grinding and outer-diameter damages as shown in Fig. 7a and Fig. 7b, respectively. Grinding errors, shown in Fig. 7, produce a discoloration on the metal surface that can vary from a light yellow to a dull black. In either case, due to dullness on the metallic surface produced by these errors, the light is absorbed and this area looks black in the binary image. Outer-diameter damages are usually found near the inner boundaries of the chamfer, and by increasing the size of these damages using morphology it can be detected. Grinding and outer-diameter damages can be differentiated based on size and location.

3.4 Errors Within Cavity

Inner burrs can be ring-like structures within the cavity or even small lumps of metal. A proper component when viewed through the cavity has no extra material and essentially looks like a circle as shown in Fig. 8. Hence, we use circular Hough transform [10] in order to detect the correctness of the component using the following steps:

1. Edges of the image are extracted using Canny edge detection technique [11].
2. Create an accumulator array of zeros, A[x, y, r] to hold values of center locations and radius according to image dimensions.
3. Compute for each edge point, (x_i, y_i):

   - For each candidate radius r :
   - For each candidate x for center horizontal location:
   - Find each possible y verifying the equation:

$$(xi - x)^2 + (yi - y)^2 = r^2 \tag{5}$$

   - Increment the cell: A[x, y, r]=A[x, y, r] + 1

4. The local maxima in A[x, y, r], denotes the center locations and the radii of the circles in the image.

(a)   (b)

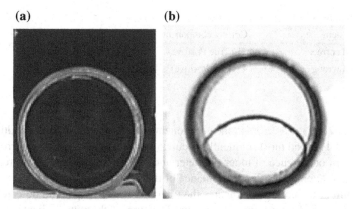

Fig. 8 **a** Small inner burr; **b** Ring within the cavity

Dynamically setting the size of the Hough circle to extract the inner surface of the dowel, we examine for the presence of red, blue, or green pixels. The presence of these pixels denotes an error.

4 Design

The experimental setup consists of cameras, a rotating mechanism, incoming path, and segregation system, which should be placed within an enclosed area so that there is no interference from external light. The minimum camera distance is 10 cm and the maximum is 30 cm, depending on the types of errors to be detected. The surface underneath the component must be bright, nonreflective, and opaque. The overall rate of evaluation is 300 dowels/min as the total time of evaluation is 0.2 s for one dowel. Each stage produces four results, obtained from evaluating the four sides at 0.05 s. Error on any side should lead to discarding of the component, so we use AND logic to evaluate the overall result. The segregation system consists of a servo, and is controlled using an AND gate, whose inputs are those obtained from the two microprocessors' output pins. The servo behaves as a latch, and moves up to allow the component to fall into the correct bin, if the output of the AND gate is high. The servo remains closed if the output is low, allowing for the plunging system to divert the component into the alternate path, and eventually into the wrong bin.

5 Results and Discussion

Dowels with errors that vary in intensity and sizes as well as dowels containing combinations of these errors were included in testing. Out of the 250 samples tested, 175 images were of improper components and 75 were images of proper components,

Table 1 Summary of system performance

Human/System	Correct components	Defected components
Interpreted correct	71 (True Positive)	3 (True Negative)
Interpreted wrong	4 (False Positive)	172 (False Negative)

it was found that most errors in the component were with respect to the width of the chamfer. The second most frequently occurring errors were the grinding errors. The frequency of occurrence of inner-diameter and outer-diameter burrs were relatively low. Table 1 summarizes the performance of the system.

The sum of the columns indicates the total number of proper and defected components as judged by a human inspector. The sum of the rows indicates the total number of components interpreted as correct and incorrect by the system. The system is said to perform correctly when the human and the system produce the same output. Hence, the accuracy is measured using the following formula:

$$Accuracy = \frac{(true - positive + false - negative)}{(total\ number\ of\ samples)} \qquad (6)$$

which is found to be 97.2%. False positives and true negatives can occur due to displacement of the camera-lighting setup caused by vibration of the mechanical setup, which can be solved by using stable mounts, or if the error is very subtle. Associativity of one error with another helps detect subtle errors; one such example is that of the outer burr. If this error is too small to be detected as an outer-diameter burr, it means the metal is bent inwards and it can be detected as an inner-diameter burr.

6 Conclusion

The main aim of the system is to broadly classify the components, and ensure that defected components are not classified as correct components. Classification based on errors plays an important role only if the errors in the components can be removed, since these errors cause permanent damage to the components, the main objective of the system is to separate bad components from the good. Associativity of one error to another holds importance as it allows for detection of errors at one stage or another, allowing for increased accuracy of the entire system. The price is cut down because of the low cost of the processor, camera, and microcontroller. The speed is increased as there is no requirement for storage of a database and use of computationally intensive techniques. The system can be implemented using machine learning, and be made interactive by introducing an interface to provide better user experience.

References

1. Neogi, N., Mohanta, D.K., Dutta, P.K.: Review of vision-based steel surface inspection systems. EURASIP J. Video Image Process. (2014)
2. Hashim, H.S., Abdulla, S.N.H.S., Prabuwono, A.S.: Automated visual inspection for metal parts based on morphology and fuzzy rules. In: International Conference on Computer Applications and Industrial Electronics (ICCAIE), Kuala Lumpur, Malaysia, 2010
3. Yinli, L., Hongli, Y., Pengpeng, Z.: The implementation of embedded image acquisition based on V4L2. In: International Conference on Electronics, Communications and Control (ICECC), Ningbo, China, 2011
4. Joo, Y.-B., Huh, K.M., Soek, H.C.: Robust and consistent defect size measuring method in automated vision inspection system. In: IEEE International Conference on Control and Automation, Christchurch, New Zealand, 2009. ICCA 2009
5. Al-Bayati, M.: Automatic thresholding techniques for optical images. Int. J. (SIPIJ) **4**(3) (2013)
6. Aggarwal, H., Maini, R.: A study and comparison of various image edge detection techniques. IJCSI Int. J. Image Process. (IJIP) **3**(1) (1993)
7. Katiyar, S.K., Arun, P.V.: Comparative analysis of common edge detection techniques in context of object extraction. IEEE TGRS **50**(11b) (2014)
8. Muthukrishnan, R., Radha, M.: Edge detection techniques for image segmentation. Int. J. Comput. Sci. Inf. Technol.(IJCSIT) **3**(6) (2011)
9. Basu, M.: Gaussian based edge detection methods-a survey. IEEE Trans. Syst. Man Cybern. Part C Appl. Rev. **3** (2002)
10. Duda, R.O., Hart, P.E.: Use of the Hough transformation to detect lines and curves in pictures. Commun. ACM **15**, 11–15 (1972)
11. Shivakshan, G.T, Chandrashekar, C.: A comparison of various edge detection techniques used in image processing. IJCSI Int. J. Comput. Sci. Issues **9**(5)(1) (2012)

A Hybrid Model for Optimum Gene Selection of Microarray Datasets

Shemim Begum, Ashraf Ali Ansari, Sadaf Sultan and Rakhee Dam

Abstract Selection of genes is one of the most onerous tasks for the study of microarray data, which is accounted because of the higher number of features, rising up to tens of thousands. Feature selection is a crucial step for proper analysis and classification of microarray data. Filter methods are pre-processing algorithms that are independent of the type of classifiers used. Wrapper methods predict the advantages of adding or removing a feature from the dataset by introduction of the induction algorithm and cross validation. In our proposed technique we have tried for significant reduction of the dimensionality of the feature set namely, Leukaemia, Prostate Cancer and DLBCL datasets by passing it to various filters namely, T-test, Bhattacharyya and ReliefF. The further reduction in dimension is done in the second layer with the Mutual Information Maximisation (MIM) filter, which is further optimised by the Adaptive Genetic Algorithm (AGA).

Keywords Microarray · Dimensionality reduction · Feature selection · Mutual information maximisation · Adaptive genetic algorithm

1 Introduction

The emergence of microarray data analysis has enabled the researchers to evaluate the expression level of large number of genes concurrently only in one microarray exper-

S. Begum (✉) · A. A. Ansari · S. Sultan · R. Dam
Department of Computer Science and Engineering, Government College of Engineering
and Textile Technology, Berhampore, Murshidabad, West Bengal, India
e-mail: shemim_begum@yahoo.com

A. A. Ansari
e-mail: ali.ansariashraf0341@gmail.com

S. Sultan
e-mail: sadafsultan3@gmail.com

R. Dam
e-mail: dam.rakhee@gmail.com

© Springer Nature Singapore Pte Ltd. 2019   423
J. Kalita et al. (eds.), *Recent Developments in Machine Learning and Data Analytics*,
Advances in Intelligent Systems and Computing 740,
https://doi.org/10.1007/978-981-13-1280-9_39

iment [1, 2]. The experimental results thus obtained can be variedly used in medical diagnosis and prognosis. The main application is the classification of tissue-samples into normal and cancerous. Although in most of the instances, a large number of genes are generally insignificant to the clinical applications [3, 4]. Feature selection is one of the most widely used methods used by the machine learning community for the investigation of the smallest subset of the features that provide minimum generalisation error. The objectives of feature selection processes are firstly, to improve the model with the feature subset with respect to the whole dataset with lower generalisation error. Secondly, to provide an exposure and good understanding of the methods for data generation [5, 6]. A good dimensionality reduction technique, T-test removes the irrelevant, redundant and noisy data; pertaining all the relevant genes. While, the Bhattacharyya filter methods providing a restriction on the upper bound of the classification error to obtain the informative genes from the large dimension of microarray datasets. The ReliefF filter methods are the most robust on attribute evaluation. In an evolutionary process [7] the choice of a good classifier [8] is crucial for high-quality results, especially when evaluating the fitness of the individuals. The major problem, for instance, is that some classifiers could mislead the searching process by overestimating or prioritising some of the poor genes by getting stuck at some local minima. Second, reduction in the dimensionality may lead to the loss of important features. The ensemble method for filtration may tackle the issue efficiently by using the union of the results of the different types of ranking filters. But, the filter techniques fail to provide better accuracy with minimum number of features. So, the optimisation techniques like AGA are opted for obtaining better accuracy with lesser number of features.

2   Filtration Methods

2.1   T-Test Method

The T-test method is an inferential test modelled to show the acceptance or rejection of the hypothesis [9]. The T-test method is used independently for comparison between the mean of two samples to check whether there is a difference between the samples [10]. The T-test generally provides the perceived differences between the sample mean due to the sampling error and it is represented in the form of the p-values, smaller the value of the p, higher is the doubt on the validity on the null hypothesis.

2.2   Bhattacharya Method

This method of feature selection mainly aims on finding a matrix θ of dimension less than that of the original dimension $(m \times n)$ and is linearly transformed, i.e. it maps the original space of dimension n into a feature space of dimension q with lower upper bound of error probability. The details of this method are explained [11].

2.3 ReliefF

The Relief class of algorithms was mostly avoided for they succumbed to very noisy and incomplete data. The ReliefF algorithm is more robust and handles noisy data with ease. The algorithm starts by randomly choosing a feature and then tries to find its k nearest neighbours belonging to the same class named as the nearest hits and it also finds the closest k neighbours from the other class called the nearest miss, with the results of the hits and misses it updates the weight of the all the attributes [12].

3 Feature Selection and Optimization Methods

3.1 MIM

Mutual information is one of the information theoretic methods that helps in the measurement of the relevance between two features or between the features and the class. The main application of maximisation of mutual information is to separate out the specific features with higher prediction rates for the class and with lower correlation with the other features. Mutual information measures the amount of decrease in the entropy of a variable with respect to the information available due to the other variables and can be stated as the relative entropy between the product distribution and the joint distribution [13]. Maximisation of mutual information mainly focuses on the proper selection of the features with higher relevance to the class and lower with the features.

3.2 AGA

Genetic algorithms [14] can be viewed as a process that tries to mimic the evolutionary method for selecting the most robust list of genes for the future generation based on their fitness values with the main motto "survival of the fittest". The algorithm achieves the goal of selection of the best by implementing the methods of mutation and crossover. Generally, mutation and crossover occur with a certain probability given by P_m and P_c respectively. An improper set of values of the P_m and P_c may lead to issues such as non-convergence or 'premature convergence' in the search. In case of AGA, the chances of mutation and crossover are computed with the obtained values of the fitness [15].

4 Proposed Methodology

The proposed work mainly focuses on dimensionality reduction of any dataset, successfully to produce a highly diminished set of features containing only the relevant features, which are given by various types of the heuristic filters and are further optimised using AGA, later the optimised feature subset are applied through three classifiers to obtain the efficiency of our proposed method. Three classifiers have been implemented namely the SVM classifier with linear kernel, gamma as 0.001, KNN with number of neighbours as five and Regression Tree.

Proposed method can be described in the following steps as illustrated in Fig. 1:

- First pass the full dataset to the filters *namely*, T-test, Bhattacharya and ReliefF and obtain the union of the result sets given by the named filters methods respectively.
- Then pass the resultant feature set to MIM for further reduction, giving a list of important and highly relevant features.
- The above found features are then optimised by passing into AGA.
- The reduced set of optimal features is passed to various kinds of classifiers for further study.

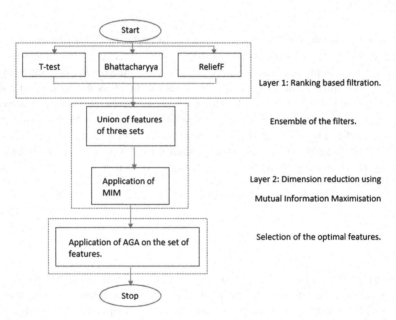

Fig. 1 Flowchart of the proposed approach

5 Results and Discussion

The obtained results from the application of the proposed approach on Leukaemia, Prostate Cancer, DLBCL that are available in [16], provides us satisfactory results. Further assurance is done by a comparative study with the closely related single layer approaches, i.e. ensemble with AGA for all the datasets. Here, the 10-fold cross validation results for all the classifiers are depicted in Table 1. This comparison brings out an evident fact that the proposed approach performs quite better than other single layer approaches; especially there is a perceptible increase in performance while using the SVM classifier.

After carrying out the proposed approach with the selected genes the performance measure like *accuracy, sensitivity, specificity, precision, recall, f_measure, gmean* is done for all the datasets in Fig. 2 for the dataset. The obtained results cement the fact that the given features are independent of the implementation of any type of classifier.

Table 1 (a), (b) and (c): The performance of the proposed two-layered method with single layer approach methods is carried out with the SVM, KNN and R-Tree classifiers respectively for all the three datasets

Dataset	Criteria	Ensemble + AGA	Ensemble + MIM + AGA
(a) SVM as classifier			
Leukaemia	No. of features	11	11
	Accuracy	95.83	100
Prostate cancer	No. of features	7	7
	Accuracy	92.16	96.08
DLBCL	No. of features	5	5
	Accuracy	88.31	100
(b) KNN as classifier			
Leukaemia	No. of features	11	11
	Accuracy	97.22	100
Prostate cancer	No. of features	7	7
	Accuracy	93.14	87.25
DLBCL	No. of features	5	5
	Accuracy	93.51	97.40
(c) R-Tree as classifier			
Leukaemia	No. of features	11	11
	Accuracy	90.28	95.83
Prostate cancer	No. of features	7	7
	Accuracy	83.65	90.93
DLBCL	No. of features	5	5
	Accuracy	89.71	92.36

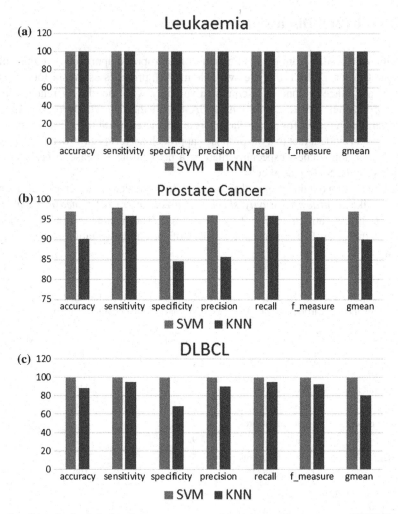

Fig. 2 a, b, and **c** Statistical performance measurement on leukaemia, prostate and DLBCL with different classifiers

The given approach has been compared with a few of the related works on the same datasets. The comparison Table 2 with the SVM, KNN and Regression Tree as classifiers respectively, shed some more light and provides us with an insight for a newer feature selection approach. The given approach successfully produces better results even when compared with hybrid methods.

Leukaemia: The maximum accuracy achieved in the related works is 98.61%, while the given approach gives accuracy as high as 100%, *DLBCL*: A maximum accuracy of 100% is observed which is equivalent to the hybrid model, while the given approach outperformed the other models. *Prostate Cancer*: The proposed methodology suggests an accuracy of 96.08% which is closer to the previous work.

Table 2 (a), (b) and (c): Comparison of the proposed method with other existing methods

Methods	Number of genes	Accuracy (%)
(a)		
IG-SVM-hybrid method [17]	3	98.61
ERGS-NBC [18]	10	98.61
ERGS-SVM [18]	10	93.06
Proposed method	11	100
(b)		
IG-SVM-hybrid method [17]	3	100
ERGS-NBC [18]	10	94.79
ERGS-SVM [18]	10	92.71
Fisher-RG-SVMRFE [19]	15	96.80
Proposed method	5	100
(c)		
IG-SVM-hybrid method [17]	3	96.08
ERGS-NBC [18]	10	94.12
ERGS-SVM [18]	10	89.22
Fisher-RG-SVMRFE [19]	14	95.90
Proposed method	7	96.08

6 Conclusion

In this paper, the given method has been developed after the several reviews of the existing methodologies [20] for data abstraction techniques and microarray data reduction. The given approach explores into the heuristic search methods and to sieve out the redundancies whatsoever with the union of the three kinds of filters. The MIM acts as a second layer of filtration for the data with co-relation as the measurement yardstick. The optimisation is done with the AGA to find a set of most strong features. The approach has been verified with three available datasets showing that the most robust selected features always indicated lower error rates irrespective of the classifiers applied. Future works shall be considered with the various kinds of optimisation algorithms such as the Ant Colony Optimisation (ACO), Particle Swarm Optimisation (PSO) for more better and dynamic accumulation of the features and further reduction of the computational cost.

References

1. Heller, M.J.: DNA microarray technology: devices, systems, and applications. Annual Rev. Biomed. Eng. **4**, 129–153 (2002)
2. Li, S., Li, D.: DNA microarray technology. In: DNA Microarray Technology and Data Analysis in Cancer Research, pp. 1–9 (2008)
3. Kumar, A., Kumar, S., Venkatesh, D., Prabhakaran, C., Ravi Prakash, D., Chakraborty, S.: Identification of genes associated with tumorigenesis of meibomian cell carcinoma by microarray analysis. Genomics **90**, 559–566 (2007)
4. Wang, A., An, N., Chen, G., Li, L., Alterovitz, G.: Improving PLS-RFE based gene selection for microarray data classification. Comput. Biol. Med. **62**, 14–24 (2015)
5. Kohavi, R., John, G.H.: Wrappers for feature subset selection. Artif. Intell. **97**, 273–324 (1997)
6. Brahim, A.B., Limam, M.: Robust ensemble feature selection for high dimensional data sets. In: International Conference on High Performance Computing and Simulation (HPCS), pp. 151–157 (2013)
7. Guyon, I., Elisseeff, A.: An introduction to variable and feature selection. J. Mach. Learn. Res. **3**, 1157–1182 (2003)
8. Liu, H., Li, J., Wong, L.: A comparative study on feature selection and classification methods using gene expression profiles and proteomic patterns. Genome Inform. **13**, 51–60 (2002)
9. Yu, L., Liu H.: Feature selection for high-dimensional data: a fast correlation-based filter solution. In: Twentieth International Conference on Machine Learning, pp: 856–863 (2003)
10. Duda, R.O., Hart, P.E., Stork, D.G.: Pattern classification, 2nd edn. Wiley, New York (2001)
11. Guorong, X., Peiqi, C., Minhui, W.: Bhattacharyya distance feature selection. In: 13th International Conference on Pattern Recognition, pp. 195–199 (1996)
12. Robnik-Sikonja, M., Kononenko. I.: An adaptation of relief for attribute estimation in regression. In: ICML'97 Proceedings of the Fourteenth International Conference on Machine Learning, pp: 296–304 (1997)
13. Cover, T.M., Thomas, J.A.: Elements of information theory, Chapter 2. Wiley, New York (1991)
14. Davis, L.: Handbook of Genetic Algorithms. Van Nostrand Reinhold (1991)
15. Srinivas, M., Patnaik, L.M.: Genetic algorithm: a survey. IEEE Trans. Comput. **27**, 17–26 (1994)
16. [Online] Available: http://www.biolab.si/sup/bi-cancer/projections/
17. Gao, L., Ye, M., Lu, X., Huang, D.: Hybrid method based on information gain and support vector machine for gene selection in cancer classification. Genom. Proteomics Bioinform. **15**, 389–395 (2017)
18. Mohammadi, A., Saraee, M.H., Salehi, M.: Identification of disease-causing genes using micro array data mining and Gene Ontology. BMC Med. Genomics **4**, 4–12 (2011)
19. Chandra, B., Gupta, M.: An efficient statistical feature selection approach for classification of gene expression data. J. Biomed. Inform. **44**, 529–535 (2011)
20. Vege, S.H.: Ensemble of feature selection techniques for high dimensional data. Master's Thesis and Specialist Projects (2012)

Lagrangian Twin-Bounded Support Vector Machine Based on L2-Norm

Umesh Gupta and Deepak Gupta

Abstract In this paper, a new convergent, flexible and easy technique is proposed for twin-bounded support vector machine (TBSVM) with the dual formulation. Here, a strongly convex objective function is constructed for proposed Lagrangian twin bounded support vector machine (LTBSVM) in consideration with L2-norm of the vector of slack variables in place of L1-norm vector of the slack variable. Twin support vector machine (TSVM) and TBSVM are used to solve quadratic programming problem (QPP) but in this proposed scheme, we consider an iterative method to solve a pair of linearly convergent equations. Further, a comparative analysis of generalized performance for TSVM, TBSVM, and proposed LTBSVM are tabulated for real world and as well as for synthetic datasets in consideration both the cases linear and nonlinear. This comparative analysis signifies the genuine improvement for the proposed method in terms of generalization performance and learning speed.

Keywords Support vector machine · Quadratic programming problem
Lagrangian function · Twin support vector machine · Twin-bounded support vector machine

1 Introduction

In the era of the computational field, traditional methods have many drawbacks like local minima problem, low convergence rate, weight adjustment etc. In order to overcome these drawbacks, Vapnik has introduced a novel approach termed as support vector machine (SVM) [1–3]. In SVM, Vapnik [1] uses the concept of structural risk minimization (SRM) principle instead of using empirical risk minimization principle

U. Gupta · D. Gupta (✉)
Computer Science and Engineering, National Institute of Technology, Yupia,
Arunachal Pradesh, India
e-mail: deepakjnu85@gmail.com

U. Gupta
e-mail: er.umeshgupta@gmail.com

© Springer Nature Singapore Pte Ltd. 2019   431
J. Kalita et al. (eds.), *Recent Developments in Machine Learning and Data Analytics*,
Advances in Intelligent Systems and Computing 740,
https://doi.org/10.1007/978-981-13-1280-9_40

in the traditional method like the artificial neural network. SVM is a one of the most popular classifier which is widely recognized by many researchers in different field image processing [4], video processing [4], medical imagery [5], automobile sector [6], human–computer interaction [7], biotechnology [8], chemical engineering [9], activity detection [7] and many more. SVM is a powerful classifier which is effective for binary classification and multi-class classification. Although SVM has very high generalization capability, it has been suffered higher training cost which is one of the major disadvantages. In fact, SVM faces computational overhead for large dataset due to a computation of large size QPP.

In order to further improvement of generalization performance and learning speed, number of improved variants of SVM have been proposed by researchers in the field of machine learning, e.g., v-support vector machine (v-SVM) [10], Sequential minimal optimization (SMO) [11], LIBSVM [12], Fung and Mangasarian [13] has proposed proximal support vector machine (PSVM) which generates parallel individual hyperplane for each class as nearer as possible with equal distance from one another. To handle the heteroscedastic noise structure, Hao [14] has suggested parametric margin based on v-support vector machine (Par-v-SVM). Generalized eigenvalues proximal SVM (GEPSVM) in 2006 by Mangasarian & Wild [15] which finds two non-parallel hyperplanes having lower computational complexity. Further, Jayadeva et al. [16] are introduced twin support vector machine (TSVM) that solves two smaller sized QPPs that make training cost lesser than the classical SVM. Kumar and Gopal [17] has proposed smooth twin support vector machine (STSVM) improves the computing performance by solving smooth unconstrained problems. Shao et al. [18] have proposed twin-bounded support vector machine which is improved version of twin support vector machine where regularization factor is added in the objective functions to implement the SRM principle. Peng [19] has been proposed twin parametric margin support vector machine (TPMSVM) by inspired the work Par-v-SVM and twin support vector machine (TSVM) that attains better generalization for different noise cases. Wang et al. [20] has proposed Genetic Algorithm based model on smooth twin parametric margin support vector machine (GASTPMSVM) gives the path to select the different features as well as parameters in an efficient way.

Kumar and Gopal [21] has proposed a least squares version of twin support vector machine (LSTSVM). In this method, it solves two linear equations for large size without any external optimization which take less computational time. Ye et al. [22] have proposed weighted twin support vector machine with local information (WTSVM) where WTSVM searches the similarity between two pairs of samples and gives effective classification with less computational cost. Mangasarian and Musicant [23] has proposed implicit Lagrangian support vector machine (LSVM) which is computationally faster than basic SVM. Related to this work, the reader may see [20–25].

In this paper, an improved version of TBSVM is developed termed as Lagrangian twin bounded support vector machine (LTBSVM) based on L2-norm which is solved by simple linearly convergent iterative approach instead of solving QPP. Our proposed method reduces the CPU training cost by using a linearly convergent iterative scheme instead of solving the QPPs as in TSVM and TBSVM. Further, the numerical experiment is demonstrated to check the effectiveness of proposed LTBSVM with TBSVM and TSVM in terms of accuracy and CPU training cost.

2 Related Work

Given U be the $m \times n$ dimension matrix as train data, where n is the number of features and m is the number of samples. Let us consider, for positive class (class +1) the matrix X_1 represents the l_1 data samples of dimension $m_1 \times n$ and for negative class (class −1) the matrix X_2 with l_2 data samples of dimension $m_2 \times n$ such that $U = [X_1; X_2]$.

2.1 Twin Support Vector Machine (TSVM)

An improved SVM-based classification technique, i.e., twin support vector machine (TSVM) [16] finds two hyperplanes. One hyperplane is nearer to one class (either +1 or −1) of data points and is the farthest from the other class of data points correspondingly. In nonlinear TSVM, a data point x maps into a higher dimensional feature space $\varphi(x)$ to find a pair of non-parallel hyperplanes $\left[K(x^t, U^t) \; e_1 \right] \begin{bmatrix} w_1 \\ b_1 \end{bmatrix} = 0$ and

$\left[K(x^t, U^t) \; e_2 \right] \begin{bmatrix} w_2 \\ b_2 \end{bmatrix} = 0$ by solving the following pair of quadratic programming problem (QPP) as:

$$\min \frac{1}{2} \left\| \left[K(X_1, U^t) \; e_1 \right] \begin{bmatrix} w_1 \\ b_1 \end{bmatrix} \right\|^2 + C_1 e_2^t \rho$$

subject to,

$$-\left[K(X_2, U^t) \; e_2 \right] \begin{bmatrix} w_1 \\ b_1 \end{bmatrix} + \rho \geq e_2, \; \rho \geq 0 \tag{1}$$

and

$$\min \frac{1}{2} \left\| \left[K(X_2, U^t) \; e_2 \right] \begin{bmatrix} w_2 \\ b_2 \end{bmatrix} \right\|^2 + C_2 e_1^t \delta$$

subject to,

$$\left[K(X_1, U^t)\, e_1 \right] \begin{bmatrix} w_2 \\ b_2 \end{bmatrix} + \delta \geq e_1, \ \delta \geq 0, \tag{2}$$

where ρ and δ are the slack variables; the values of penalty parameters C_1 and C_2 are strictly positive; $e_1, e_2 \in R^{l_1, l_2}$ are ones of the vectors; $K(x^t, U^t) = (k(x, x_1), \ldots, k(x, x_m))$ is a row vector in R^m, where $k(x, x_i) = \varphi(x)^t \varphi(x_i) \in R$ for $i = 1, \ldots, m$.

Now consider the Lagrangian multiplier $\alpha_1, \alpha_2, \beta_1, \beta_2 \in R^{l_1, l_2, l_1, l_2}$ are applied on Eqs. (1) and (2); we get the Wolfe dual of the above problems (1) and (2) by using the Karush–Kuhn–Tucker (KKT) [26] conditions as

$$\max e_2^t \alpha_1 - \frac{1}{2} \alpha_1^t P (Q^t Q)^{-1} P^t \alpha_1$$

subject to

$$0 \leq \alpha_1 \leq C_1 \tag{3}$$

and

$$\max e_1^t \alpha_2 - \frac{1}{2} \alpha_2^t Q (P^t P)^{-1} Q^t \alpha_2$$

subject to

$$0 \leq \alpha_2 \leq C_2 \tag{4}$$

where, $Q = \left[K(X_1, U^t)\, e_1 \right], P = \left[K(X_2, U^t)\, e_2 \right].$

To solve the following pair of QPP as Eqs. (3) and (4) for α_1 and α_2, one can obtain the values as:

$$\begin{bmatrix} w_1 \\ b_1 \end{bmatrix} = -(Q^t Q + \varepsilon I) P^t \alpha_1 \tag{5}$$

$$\begin{bmatrix} w_2 \\ b_2 \end{bmatrix} = (P^t P + \varepsilon I) Q^t \alpha_2, \tag{6}$$

where, $I \in R^{(n+1) \times (n+1)}$ is an identity matrix and $\varepsilon > 0$ is the input parameter.

According to hyperplane closer to any class ($+1$ or -1), the class of the new test sample $x \in R^n$ can be obtained by the following formula:

$$\text{class } i = \min \left| \left[K(x^t, U^t)\, 1 \right] \begin{bmatrix} w_i \\ b_i \end{bmatrix} \right|, \quad i = 1, 2. \tag{7}$$

2.2 Twin-Bounded Support Vector Machine (TBSVM)

Shao et al. (2011) [18] has been proposed an improved method termed as twin bounded support vector machine (TBSVM) where the SRM principle is incorporated with the addition of regularization terms. The two non-parallel kernel generated surfaces $\left[K(x^t, U^t)\, e_1 \right] \begin{bmatrix} w_1 \\ b_1 \end{bmatrix} = 0$ and $\left[K(x^t, U^t)\, e_2 \right] \begin{bmatrix} w_2 \\ b_2 \end{bmatrix} = 0$ are obtained by solving the primal problems as

$$\min \frac{1}{2} \left\| \left[K(X_1, U^t)\, e_1 \right] \begin{bmatrix} w_1 \\ b_1 \end{bmatrix} \right\|^2 + C_1 e_2^t \rho + \frac{1}{2} C_3 \left[w_1\ b_1 \right] \begin{bmatrix} w_1 \\ b_1 \end{bmatrix}$$

subject to,

$$-\left[K(X_2, U^t)\, e_2 \right] \begin{bmatrix} w_1 \\ b_1 \end{bmatrix} + \rho \geq e_2,\ \rho \geq 0 \tag{8}$$

and

$$\min \frac{1}{2} \left\| \left[K(X_2, U^t)\, e_2 \right] \begin{bmatrix} w_2 \\ b_2 \end{bmatrix} \right\|^2 + C_2 e_1^t \delta + \frac{1}{2} C_4 \left[w_2\ b_2 \right] \begin{bmatrix} w_2 \\ b_2 \end{bmatrix}$$

subject to,

$$\left[K(X_1, U^t)\, e_1 \right] \begin{bmatrix} w_2 \\ b_2 \end{bmatrix} + \delta \geq e_1,\ \delta \geq 0, \tag{9}$$

where $C_1, C_2, C_3, C_4 \geq 0$ are input parameters. By introducing the Lagrangian multipliers $\alpha_1, \alpha_2, \beta_1, \beta_2 \in R^{l_1, l_2, l_1, l_2}$ and using optimality KKT conditions, we get the following dual problems as:

$$\max e_2^t \alpha_1 - \frac{1}{2} \alpha_1^t P(Q^t Q + C_3 I)^{-1} P^t \alpha_1$$

subject to

$$0 \leq \alpha_1 \leq C_1 \tag{10}$$

$$\max e_1^t \alpha_2 - \frac{1}{2} \alpha_2^t Q(P^t P + C_4 I)^{-1} Q^t \alpha_2$$

subject to

$$0 \leq \alpha_2 \leq C_2 \tag{11}$$

After computing the values of α_1 and α_2, we get

$$\begin{bmatrix} w_1 \\ b_1 \end{bmatrix} = -(Q'Q + C_3 I)P'\alpha_1 \tag{12}$$

$$\begin{bmatrix} w_2 \\ b_2 \end{bmatrix} = (P'P + C_4 I)Q'\alpha_2 \tag{13}$$

The class of the new test sample $x \in R^n$ can be obtained by following formula:

$$\text{class } i = \arg \min \frac{|K(x^t, U^t)w_i + b_i|}{\sqrt{w_i^t K(U, U^t)w_i}}, \quad i = 1, 2 \tag{14}$$

3 Proposed Lagrangian Twin-Bounded Support Vector Machine (LTBSVM)

Following the idea of TBSVM [18] and LSVM [23], we have suggested a new approach termed as Lagrangian twin-bounded support vector machine (LTBSVM) based on L2-norm to incorporate the advantages of both methods. In this approach, we have considered L2-norm of the vector of the slack variables in place of L1-norm vector of slack variable in the primal problems. In proposed LTB-SVM, two non-parallel kernel generated surfaces $\begin{bmatrix} K(x^t, U^t) \, e_1 \end{bmatrix} \begin{bmatrix} w_1 \\ b_1 \end{bmatrix} = 0$ and

$\begin{bmatrix} K(x^t, U^t) \, e_2 \end{bmatrix} \begin{bmatrix} w_2 \\ b_2 \end{bmatrix} = 0$ are determined by the following pair of primal problems:

$$\min \frac{1}{2} \left\| \begin{bmatrix} K(X_1, U^t) \, e_1 \end{bmatrix} \begin{bmatrix} w_1 \\ b_1 \end{bmatrix} \right\|^2 + \frac{1}{2} C_1 \rho^t \rho + \frac{1}{2} C_3 \begin{bmatrix} w_1 \, b_1 \end{bmatrix} \begin{bmatrix} w_1 \\ b_1 \end{bmatrix}$$

subject to,

$$- \begin{bmatrix} K(X_2, U^t) \, e_2 \end{bmatrix} \begin{bmatrix} w_1 \\ b_1 \end{bmatrix} + \rho \geq e_2, \tag{15}$$

and

$$\min \frac{1}{2}\left\|\left[\, K(X_2, U^t)\ e_2\,\right]\begin{bmatrix} w_2 \\ b_2 \end{bmatrix}\right\|^2 + \frac{1}{2}C_2\delta^t\delta + \frac{1}{2}C_4\left[\, w_2\ b_2\,\right]\begin{bmatrix} w_2 \\ b_2 \end{bmatrix}$$

subject to,

$$\left[\, K(X_1, U^t)\ e_1\,\right]\begin{bmatrix} w_2 \\ b_2 \end{bmatrix} + \delta \geq e_1, \tag{16}$$

In the consideration of the Lagrangian multipliers $\alpha_1 \in R^{l_1}$ and $\alpha_2 \in R^{l_2}$, the Lagrangian functions of Eqs. (15) and (16) can be formulated as

$$L = \frac{1}{2}\left\|\left[\, K(X_1, U^t)\ e_1\,\right]\begin{bmatrix} w_1 \\ b_1 \end{bmatrix}\right\|^2 + \frac{1}{2}C_1\rho^t\rho + \frac{1}{2}C_3\left[\, w_1\ b_1\,\right]\begin{bmatrix} w_1 \\ b_1 \end{bmatrix}$$
$$- \alpha_1^t\left(-\left[\, K(X_2, U^t)\ e_2\,\right]\begin{bmatrix} w_1 \\ b_1 \end{bmatrix} + \rho - e_2\right) \tag{17}$$

and

$$L = \frac{1}{2}\left\|\left[\, K(X_2, U^t)\ e_2\,\right]\begin{bmatrix} w_2 \\ b_2 \end{bmatrix}\right\|^2 + \frac{1}{2}C_2\delta^t\delta + \frac{1}{2}C_4\left[\, w_2\ b_2\,\right]\begin{bmatrix} w_2 \\ b_2 \end{bmatrix}$$
$$- \alpha_2^t\left(\left[\, K(X_1, U^t)\ e_1\,\right]\begin{bmatrix} w_2 \\ b_2 \end{bmatrix} + \delta - e_1\right) \tag{18}$$

The dual of primal problems (15) and (16) can be written as

$$\max_{\alpha_1 \geq 0} e_2^t\alpha_1 - \frac{1}{2}\alpha_1^t\left(\frac{I}{C_1} + P(Q^tQ + C_3I)^{-1}P^t\right)\alpha_1 \tag{19}$$

and

$$\max_{\alpha_2 \geq 0} e_1^t\alpha_2 - \frac{1}{2}\alpha_2^t\left(\frac{I}{C_2} + Q(P^tP + C_4I)^{-1}Q^t\right)\alpha_2, \tag{20}$$

where $Q = \left[\, K(X_1, U^t)\ e_1\,\right]$, $P = \left[\, K(X_2, U^t)\ e_2\,\right]$. The augmented vectors are calculated as

$$\begin{bmatrix} w_1 \\ b_1 \end{bmatrix} = -(Q^t Q + C_3 I)^{-1} P^t \alpha_1 \tag{21}$$

$$\begin{bmatrix} w_2 \\ b_2 \end{bmatrix} = (P^t P + C_4 I)^{-1} Q^t \alpha_2 \tag{22}$$

The dual QPPs (19) and (20) are of the form,

$$\min_{\alpha_k \geq 0} \frac{1}{2} \alpha_k^t Z_k \alpha_k - r_k^t \alpha_k \tag{23}$$

for $k = 1, 2$ respectively, where $Z_1 = \frac{I}{C_1} + P(Q^t Q + C_3 I)^{-1} P^t$, $Z_2 = \frac{I}{C_2} + Q(P^t P + C_4 I)^{-1} Q^t$ and $r_k = e_k$. Hence, the dual Eqs. (19) and (20) can be written in the following pair of classical complementary problems, [3] for $k = 1, 2$

$$0 \leq \alpha_k \perp (Z_k \alpha_k - r_k) \geq 0 \tag{24}$$

One can write the Eq. (24) in the following equivalent form for $\alpha_k > 0$,

$$(Z_k \alpha_k - r_k) = (Z_k \alpha_k - \beta_k \alpha_k - r_k)_+ \tag{25}$$

Now, we can solve above problem using iterative approach [23] as

$$\alpha_k^{i+1} = Z_k^{-1}(r_k + (Z_k \alpha_k^i - \beta_k \alpha_k^i - r_k)_+), \quad i = 1, 2 \tag{26}$$

where $0 < \beta_k < \frac{2}{C_k}$, $k = 1, 2$.

$$\alpha_1^{i+1} = \left(\frac{I}{C_1} + P(Q^t Q + C_3 I)^{-1} P^t \right)^{-1}$$
$$\left(r_1 + \left(\left(\frac{I}{C_1} + P(Q^t Q + C_3 I)^{-1} P^t \right) \alpha_1^i - r_1 - \beta_1 \alpha_1^i \right)_+ \right) \tag{27}$$

and

$$\alpha_2^{i+1} = \left(\frac{I}{C_2} + Q(P^t P + C_4 I)^{-1} Q^t \right)^{-1}$$
$$\left(r_2 + \left(\left(\frac{I}{C_2} + Q(P^t P + C_4 I)^{-1} Q^t \right) \alpha_2^i - r_2 - \beta_2 \alpha_2^i \right)_+ \right) \tag{28}$$

Once we compute the values of α_1 and α_2 by using above iterative schemes (27) and (28), using Eqs. (21) and (22), the class of the new test sample can be obtained by using Eq. (14).

4 Numerical Results

In the experimental part, we use real world and synthetic datasets such as Ripley's datasets [27] to validate the effectiveness by a comparative analysis of our method LTBSVM with TSVM, TBSVM. The experiments are performed with 64 bit, 3.40 GHz Intel© Core™ i7-3770 processor, 8 gigabyte RAM, Windows 10 operating system, MATLAB R2017a software. One prerequisite has been taken before the experiment i.e. normalization in the range of (0, 1) for all the datasets. Here, we are considering both linear and nonlinear case for LTBSVM, TSVM, and TBSVM where Gaussian kernel function is given by

$$K(x_i, x_j) = \exp\left(-\frac{1}{2\sigma^2}\|x_i - x_j\|^2\right)$$

We have obtained the kernel parameter σ from the set $\{2^{-5}, \ldots, 2^5\}$; $C_i(1, 2, 3, 4)$ are chosen from the set $\{10^{-5}, \ldots, 10^5\}$ using 10-fold cross validations for TSVM, TBSVM and LTBSVM. To check the performance of our proposed LTBSVM with TSVM and TBSVM, we have measured two parameters, i.e., classification accuracy for optimum parameters values and training time for linear and nonlinear.

Here, artificially generated Ripley's datasets [27] in R^2 is considered which has 1250 samples. We choose 250 samples to train the model whereas 1000 samples are taken to test the model for TSVM, TBSVM, and proposed LTBSVM. In Fig. 1a–c, the positive class data point is marked as "×" and the negative data point is marked as "+" symbol. The resultant classifiers are shown in Fig. 1 and the results in terms of accuracy and CPU training time with their optimum values are depicted in Table 1 for TSVM, TBSVM and proposed LTBSVM. One can conclude from Table 1 that our proposed LTBSVM is more reasonable than TSVM and TBSVM.

In order to check the behavior of our proposed LTBSVM with other reported methods, we have performed numerical experiment on 12 real-world datasets, namely Breast Cancer-Wisconsin, BUPA Liver, Cleve, Ionosphere, Votes, WPBC, CMC, German, Sonar, Heart-c, Splice and Vowel which are downloaded from UCI [28] for linear and as well as nonlinear cases. In Table 2, we have shown the classification accuracy and CPU training time with their optimal values for our proposed linear LTBSVM with linear TBSVM and linear TSVM where bold figures show best

Table 1 The comparative analysis of TSVM, TBSVM and LTBSVM on Ripley's dataset using Gaussian kernel

Dataset (train size, test size)	TSVM ($C_1 = C_2$, μ) Time	TBSVM ($C_1 = C_2$, $C_3 = C_4$, μ) Time	LTBSVM ($C_1 = C_2$, $C_3 = C_4$, μ) Time
Ripley (250×2, 1000×2)	91.2 $(10^0, 2^1)$ 0.124	91.6 $(10^{-4}, 10^5, 2^{-2})$ 0.1332	**92** $(10^{-3}, 10^3, 2^{-2})$ 0.1052

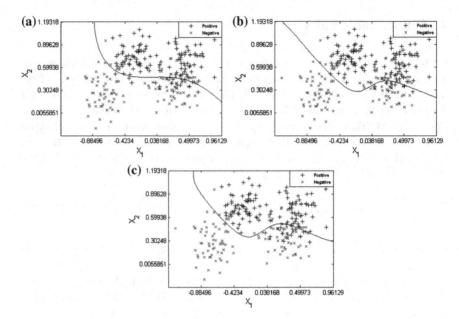

Fig. 1 Classification result of TSVM, TBSVM and LTBSVM on Ripley's dataset using Gaussian kernel

results in terms of testing accuracies. It is noticed that our suggested LTBSVM is comparatively better than TBSVM and TSVM in the respect of testing accuracy and CPU training time.

Further, we have performed numerical experiment for nonlinear case where Gaussian kernel is used and the results are depicted in Table 3. We have listed the testing accuracy and CPU training time with the values of their optimal parameters $C_1 = C_2$, $C_3 = C_4$ and μ in Table 3. It is observed that our proposed LTBSVM confirming the same conclusion similar to the linear case from Table 3.

5 Conclusion

In this paper, L2-norm of the vector of slack variables instead of L1-norm of the vector of slack variables is used in the primal problems of TBSVM to suggest a new approach LTBSVM where the solution of the problem easily is solved by using simple linearly convergent schemes instead of solving QPPs as in case of other reported methods. A comparative analysis of proposed LTBSVM with TSVM and TBSVM is indicated that our proposed LTBSVM significantly better in terms of accuracy and lower training CPU cost for most of the datasets which conclude the effectiveness and applicability of our proposed LTBSVM. In future, parameter selection may be one of the practical problems as proposed LTBSVM have four parameters which should be addressed.

Table 2 Comparative analysis of TSVM, TBSVM and LTBSVM using linear kernel on real-world datasets

Dataset (train size, test size)	TSVM $(C_1 = C_2)$ Time	TBSVM $(C_1 = C_2, C_3 = C_4)$ Time	LTBSVM $(C_1 = C_2, C_3 = C_4)$ Time
Breast Cancer-Wisconsin $(149 \times 9, 534 \times 9)$	97.191 (10^1) 0.0281	**97.5655** $(10^{-1}, 10^1)$ 0.02273	94.382 $(10^4, 10^3)$ 0.01687
BUPA Liver $(200 \times 6, 145 \times 6)$	**71.7241** (10^0) 0.03515	57.931 $(10^{-5}, 10^3)$ 0.03222	60.6897 $(10^{-1}, 10^3)$ 0.00427
Cleve $(150 \times 13, 147 \times 13)$	**82.9932** (10^{-1}) 0.02517	**82.9932** $(10^0, 10^2)$ 0.03204	80.2721 $(10^1, 10^{-5})$ 0.01594
Ionosphere $(150 \times 33, 201 \times 33)$	87.0647 (10^{-1}) 0.03117	**89.0547** $(10^1, 10^1)$ 0.01836	88.5572 $(10^1, 10^1)$ 0.01585
Votes $(200 \times 16, 235 \times 16)$	**94.8936** (10^{-1}) 0.03732	**94.8936** $(10^0, 10^0)$ 0.03195	94.4681 $(10^{-2}, 10^{-5})$ 0.00632
WPBC $(80 \times 32, 114 \times 32)$	**78.9474** (10^{-3}) 0.02631	**78.9474** $(10^0, 10^1)$ 0.01269	78.0702 $(10^1, 10^3)$ 0.00719
CMC $(200 \times 9, 1273 \times 9)$	70.6991 (10^0) 0.02925	74.1555 $(10^0, 10^3)$ 0.03674	**74.784** $(10^1, 10^5)$ 0.01994
German $(150 \times 24, 850 \times 24)$	69.2941 (10^0) 0.0283	**71.1765** $(10^1, 10^{-1})$ 0.02256	70.9412 $(10^0, 10^2)$ 0.00277
Sonar $(200 \times 60, 8 \times 60)$	**75** (10^{-1}) 0.03698	62.5 $(10^2, 10^0)$ 0.03279	**75** $(10^{-2}, 10^{-3})$ 0.00351
Heart-c $(200 \times 13, 97 \times 13)$	61.8557 (10^0) 0.05344	59.7938 $(10^0, 10^4)$ 0.03239	**69.0722** $(10^0, 10^0)$ 0.01549
Splice $(200 \times 60, 2975 \times 60)$	77.5798 (10^0) 0.03962	80.3361 $(10^0, 10^2)$ 0.02424	**81.479** $(10^{-5}, 10^2)$ 0.00364
Vowel $(200 \times 10, 788 \times 10)$	82.7411 (10^{-1}) 0.08082	90.1015 $(10^0, 10^1)$ 0.05695	**95.8122** $(10^4, 10^4)$ 0.02711

Table 3 Comparative analysis of TSVM, TBSVM and LTBSVM using Gaussian kernel on real-world datasets

Dataset (train size, test size)	TSVM ($C_1 = C_2$, μ) Time	TBSVM ($C_1 = C_2$, $C_3 = C_4$, μ) Time	LTBSVM ($C_1 = C_2$, $C_3 = C_4$, μ) Time
Breast Cancer-wisconsin (149×9, 534×9)	97.191 (10^0, 2^5) 0.0779	**97.5655** (10^{-1}, 10^{-2}, 2^5) 0.06681	97.3783 (10^1, 10^0, 2^4) 0.06664
Bupa Liver (200×6, 145×6)	64.8276 (10^0, 2^5) 0.13026	**68.2759** (10^{-5}, 10^1, 2^4) 0.15562	**68.2759** (10^{-1}, 10^1, 2^4) 0.10663
Cleve (150×13, 147×13)	79.5918 (10^{-4}, 2^1) 0.08394	79.5918 (10^{-5}, 10^{-5}, 2^1) 0.09327	**81.6327** (10^1, 10^0, 2^3) 0.07147
Ionosphere (150×33, 201×33)	**96.0199** (10^{-4}, 2^1) 0.12958	95.0249 (10^{-2}, 10^{-3}, 2^2) 0.10018	94.5274 (10^{-1}, 10^{-4}, 2^3) 0.09662
Votes (200×16, 235×16)	**95.3191** (10^{-2}, 2^5) 0.09166	**95.3191** (10^{-1}, 10^{-5}, 2^4) 0.15868	**95.3191** (10^{-2}, 10^{-3}, 2^4) 0.0919
WPBC (80×32, 114×32)	**78.9474** (10^{-5}, 2^{-5}) 0.02907	**78.9474** (10^{-5}, 10^{-5}, 2^{-5}) 0.04759	**78.9474** (10^{-5}, 10^{-5}, 2^{-5}) 0.02617
CMC (200×9, 1273×9)	74.5483 (10^{-5}, 2^0) 0.14949	74.7054 (10^0, 10^0, 2^2) 0.13999	**74.784** (10^1, 10^2, 2^3) 0.13324
German (150×24, 850×24)	**69.2941** (10^0, 2^4) 0.08267	**69.2941** (10^0, 10^{-4}, 2^4) 0.08163	**69.2941** (10^0, 10^{-1}, 2^3) 0.05904
Sonar (200×60, 8×60)	**87.5** (10^0, 2^1) 0.09471	**87.5** (10^0, 10^{-4}, 2^5) 0.16805	**87.5** (10^{-1}, 10^{-1}, 2^0) 0.08814
Heart-c (200×13, 97×13)	**62.8866** (10^{-5}, 2^3) 0.08508	**62.8866** (10^{-4}, 10^{-3}, 2^3) 0.16312	**62.8866** (10^{-1}, 10^{-4}, 2^5) 0.08299
Splice (200×60, 2975×60)	**82.3866** (10^{-3}, 2^3) 0.13746	80.2017 (10^0, 10^{-2}, 2^4) 0.1247	82.2857 (10^1, 10^0, 2^3) 0.1116
Vowel (200×10, 788×10)	91.4975 (10^{-5}, 2^0) 0.15455	92.3858 (10^{-5}, 10^{-5}, 2^0) 0.14503	**93.2741** (10^{-5}, 10^{-3}, 2^0) 0.13889

References

1. Cortes, C., Vapnik, V.: Support—vector networks. Spr Mach Learn **20**(2), 273–297 (1995)
2. Ding, S., Qi, B., Tan, H.: An overview on theory and algorithm of support vector machines. UESTC J. Univ. Electron. Sci. Technol. China **40**(1), 2–10 (2011)
3. Deng, N., Tian, Y., Zhang, C.: Support Vector Machines: Optimization based theory, Algorithms, and Extensions. Chapman and Hall/CRC Press, Boca Raton, Florida (2012)
4. Duan, L., Tsang, I.W., Xu, D., Maybank, S.J.: Domain transfer SVM for video concept detection. In: IEEE Conference on Computer Vision and Pattern Recognition, 2009. CVPR 2009, pp. 1375–1381. IEEE (2009, June)
5. Fung, G., Stoeckel, J.: SVM feature selection for classification of SPECT images of Alzheimer's disease using spatial information. Knowl. Inf. Syst. **11**(2), 243–258 (2007)
6. Martínez-de-Pisón, F.J., Barreto, C., Pernia, A., Alba, F.: Modelling of an elastomer profile extrusion process using support vector machines (SVM). J. Mater. Process. Technol. **197**(1–3), 161–169 (2008)
7. Sung, J., Ponce, C., Selman, B., Saxena, A.: Human activity detection from RGBD images. plan, activity, and intent recognition, from the 2011 AAAI workshop (WS-11-16) (2011)
8. Guo, J., Chen, H., Sun, Z., Lin, Y.: A novel method for protein secondary structure prediction using dual-layer SVM and profiles. Proteins Struct. Funct. Bioinf. **54**(4), 738–743 (2004)
9. Lee, C.J., Song, S.O., Yoon, E.S.: The monitoring of chemical process using the support vector machine. Korean Chem. Eng. Res. **42**(5), 538–544 (2004)
10. Schölkopf, B., Smola, A.J., Williamson, R.C., Bartlett, P.L.: New support vector algorithms. Neural Comput. **12**(5), 1207–1245 (2000)
11. Platt, J.: Probabilistic outputs for support vector machines and comparisons to regularized likelihood methods. Adv. Large Margin Classifiers **10**(3), 61–74 (1999)
12. Chang, C.C., Lin, C.J.: LIBSVM: a library for support vector machines. ACM Trans. Intell. Syst. Technol. (TIST) **2**(3), 27 (2011)
13. Fung, G.M., Mangasarian, O.L.: Multicategory proximal support vector machine classifiers. Mach. Learn. **59**(1–2), 77–97 (2005)
14. Hao, P.Y.: New support vector algorithms with parametric insensitive/margin model. Neural Networks **23**(1), 60–73 (2010)
15. Mangasarian, O.L., Wild, E.W.: Multisurface proximal support vector machine classification via generalized eigenvalues. IEEE Trans. Pattern Anal. Mach. Intell. **28**(1), 69–74 (2006)
16. Jayadeva, Khemchandani, R., Chandra, S.: Twin support vector machines for pattern classification. IEEE Trans. Pattern Anal. Mach. Intell. **29**(5), 905–910 (2007)
17. Kumar, M.A., Gopal, M.: Application of smoothing technique on twin support vector machines. Pattern Recogn. Lett. **29**(13), 1842–1848 (2008)
18. Shao, Y.H., Zhang, C.H., Wang, X.B., Deng, N.Y.: Improvements on twin support vector machines. IEEE Trans. Neural Netw. **22**(6), 962–968 (2011)
19. Peng, X., Wang, Y., Xu, D.: Structural twin parametric-margin support vector machine for binary classification. Knowl.-Based Syst. **49**, 63–72 (2013)
20. Wang, Z., Shao, Y.H., Wu, T.R.: A GA-based model selection for smooth twin parametric-margin support vector machine. Pattern Recogn. **46**(8), 2267–2277 (2013)
21. Kumar, M.A., Gopal, M.: Least squares twin support vector machines for pattern classification. Expert Syst. Appl. **36**(4), 7535–7543 (2009)
22. Ye, Q., Zhao, C., Gao, S., Zheng, H.: Weighted twin support vector machines with local information and its application. Neural Netw. **35**, 31–39 (2012)
23. Mangasarian, O.L., Musicant, D.R.: Lagrangian support vector machines. J. Mach. Learn. Res. **1**(Mar), 161–177 (2001)
24. Balasundaram, S., Kapil, N.: Application of Lagrangian twin support vector machines for classification. In: 2010 Second International Conference on Machine Learning and Computing (ICMLC), pp. 193–197. IEEE (2010)
25. Balasundaram, S., Gupta, D.: Training Lagrangian twin support vector regression via unconstrained convex minimization. Knowl.-Based Syst. **59**, 85–96 (2014)

26. Mangasarian, O.L.: Nonlinear programming. Soc. Ind. Appl. Math. J. Optim. **4**(4), 815–832 (1994)
27. Ripley, B.D.: Pattern recognition and neural networks. Cambridge University Press, Cambridge, UK (2007)
28. Murphy P.M., Aha, D.W.: UCI repository of machine learning databases. University of California, Irvine (1992). http://www.ics.uci.edu/~mlearn

An Improved K-NN Algorithm Through Class Discernibility and Cohesiveness

Rajesh Prasad Sarkar and Ananjan Maiti

Abstract The K-Nearest Neighbor (K-NN) is a primarily chosen method when it comes to the object classification, disease interpretation, and various other fields. In numerous cases, K-NN classifier uses the only parameter as K value, which is the number of nearest neighbors to decide the class of the instance and this appears to be insufficient. Within this study, we have looked at the initial K-Nearest Neighbor algorithm and also proposed modified K-NN algorithm to identify various ailments. Enhancing precision of the initial K-Nearest Neighbor algorithm, this specific suggested method consists of instance weights as an added parameter to determine the class of the example. This study presented a novel technique to assign weights, which utilizes the information from the structure of the data set and assigns weights to every instance relying on the priority of the instance in class discernibility. In this approach, we have included an additional metric "average density" together with "discernibility" to calculate an index which is used as a measure also with the value of K. The practice results obtained from UCI repository reveals that this classifier carries out much better than the traditional K-NN and preserve steady accuracy.

Keywords K-NN algorithm · Accuracy improvement · Weighted K-NN algorithm · Data mining · Classification · Discernibility

1 Introduction

Categorizing objects or object classification is a very important task in the field of medical disease diagnosis, robotic arm control tasks, autonomous vehicles, time

R. P. Sarkar (✉) · A. Maiti
University of Engineering & Management (UEM), Kolkata, India
e-mail: rajesh.mca.cu@gmail.com

A. Maiti
e-mail: ananjan.maiti@gmail.com

A. Maiti
Department of IT, Techno India College of Technology, Kolkata, India

© Springer Nature Singapore Pte Ltd. 2019
J. Kalita et al. (eds.), *Recent Developments in Machine Learning and Data Analytics*,
Advances in Intelligent Systems and Computing 740,
https://doi.org/10.1007/978-981-13-1280-9_41

445

series prediction, etc. The main objective of classification is to devise a model which could be applied to predict a class of an object. The mapper acts like a function to take instance as an input and maps it into a class from a finite group of classes. The task of classification of diseases [1, 2] is quite easy when there is a small number of attribute or features associated with the class. It is also seen that volume of test instances is very small but when it is required to classify a very high dimensional class. Human has limited capabilities to handle a huge volume of test cases. In such cases, the automation of classification [3, 4] is very much requited.

Many different classification algorithms are present, like Rule base classifier, K-Nearest Neighbor, Naive Bayesian, Neural Network, Support Vector machines, etc. This paper concerns K-Nearest Neighbor classifier, which is simple to understand, uses very few parameters to tune, and accurate. The mentioned attributes made K-NN very popular as a classifier though it is computationally expensive. Though K-NN has some advantages still this concept has a few inherent problems, which are why many researchers have proposed different version and extension of the idea and tried to make the process more accurate and efficient. In this paper, we have proposed an extended version of the K-NN algorithm [5, 6], which finds the relevancy of all instances and assigns a weight to each one according to their degree of relevance. When the training data set comprises with a high degree of irrelevant samples or instances, this process improves the accuracy of the algorithm by giving importance to the relevant cases as compared to the irrelevant ones. So in this method, the value of K and the associated weight both are the decision-making parameter. This instance weight is calculated by combining two measures "Discernibility" and "Average density", which are being derived from the structure of the data set. This proposed method is an extension of an existing concept, which is described briefly in this paper. The potential is also evaluated numerically through experiments using data sets from UCI repository [7].

The additional aspect of the study is shared as observed. The upcoming segment talks about a few of the very present significant literary works concerning K-NN algorithm using various machine learning strategies determining out their most motivating ideas as well as their drawbacks. Segment 3 also provides various stages of the method when it concerns evaluating the classifiers data set for several data sets. The outcomes of observations performed on these classifiers, as efficiently as on K-NN, are displayed in segment 4. Finally, in segment 5, we have indeed set up our investigation having different judgments.

2 Literature Review

It indicates that one of the essential classification algorithms is K-NN, as well as this, is also quick and easy to perform classification method. Many scientists come with one of the first selections for a classification study when there is little expertise about the frequency of the data. K-NN classification developed from the need to perform discriminant outlining when parametric measures of the probability distribution are difficult to calculate. Many studies have been performed by the researchers to enhance the utility of the algorithm. Gerhana et al. [8] recommended Case-Based Reasoning (CBR) model and it has been utilized to identify the problem in numerous situations. Their study was to discuss the execution of K-Nearest Neighbor. The research exposed that K-NN algorithm is suitable to be used in CBR model. The verdicts of this study are to determine the accuracy degree of automated response identification formation and explore the correlation answer in algorithm case. The screening with 10 algorithm questions using training data of 90 records with 3 classes: sequencing, option, and repetition, each of which has 30 data in every area, on the other hand, the examining result revealed that K-NN made an almost precise score. From the test, outcome presented that K-NN accuracy score acquired is 0.9 when the value of k is 5. This final result also presented the beneficial effect. Selvaluxmiy et al. [9] came out with a GPU based on the k-NN algorithm and included extra actions concerning searching, arranging as well as the parallel execution of the activity. The specified k-NN algorithm was executed on a GPU making use of CUDA and utilizing the parallel tasks. This k-NN classification algorithm possessed a pair of kernels; one was the estimation of distances of the kernel, and other was the nearest-neighbor search kernel. The program provides a guaranteeing outcome for the majority of the standard machine learning data set. It also carries some constraint when a process surpasses 12 GB whenever loaded into device memory to deal with huge data. Therefore, this might not function on high-dimensional data with the more considerable variety of training and screening records since it attempts to do all the operations at once keeping the entire training and testing data set within the memory. The outcomes were certainly appealing with 43,500 training records and 14,500 testing records and with 9 features. Jaiswal et al. [10] developed an automated system assortment handled through Cuckoo search along with K-NN learning algorithm. Relying on the value of k as well as the distance metric they assessed the efficiency of the algorithm. In addition, an objective function is designed to measure the accuracy of classification; and then by optimizing it, the parameters are determined. An efficient optimization method like Cuckoo search utilized to free up the value of k and choose a distance metric among Euclidean with other correlation metrics for K-NN classifier. The proposed method explored on various standard data sets gathered from UCI data sets such as Cleveland Heart, Ionosphere, Breast Cancer W, Diabetes, Lung, and Crab. As a result, it verified that the technique works out as well as reliable. Taneja et al. [11] at the same time researched K-NN along with its own various downsides as well as presented the idea from fuzzy computing in order to eliminate the disadvantage of original K-NN algorithm. The fuzzy-based method dealt with a trouble due to fuzzy membership, thus it incorporated considerable complication when it has to measure at every period of class. They changed fuzzy-based K-NN algorithm and included measurement of fuzzy membership in each preprocessing stage. Here, training data is taken into consideration and approached the centroid of the clusters. Consequently, MFZ-K-NN required 840 ms for $k = 3$ and 889 ms for $k = 10$. FZ-K-NN had certainly finished exact same activities with 858 ($K = 3$) and 904 ($k = 10$) milliseconds. The suggested algorithm decreases the complexity of time incredibly. They tested on UCI data set-Wine data set and carried out the algorithm with MATLAB and also NetBeans.

3 Methodology

The K-NN algorithm is a type of passive learning algorithms, which need not require any off-line training. K-NN algorithm reserves the entire training data set. It delays the efforts towards inductive generalization till classification time [5]. K-NN algorithm learns all the training set at the time of training and during the classification stage. For a particular testing instance, it directly searches through all the training set and calculates the distances between the testing example and all the training set. These ranges are then used to identify the nearest neighbors to classify the test instance. There are various methods to measure the distance, but among them, Euclidean [6] and Minkowskian [12] are the most popular. Finally to identify the class K-NN uses, the K-nearest neighbors need to vote among them to decide the class of the instance.

Let an instance(x) of a class be described by a vector of feature or attribute $(a_i) < a_1(x), a_2(x), a_3(x)...a_i(x)...a_m(m) >$ f m dimension where $a_i(x)$ denotes the value of the ith attribute. So the Euclidean distance between a training instance x and a test instance y will be as follows:

$$d(x, y) = \sqrt{\sum_{i=1}^{m} (a_i(x) - a_i(y))^2}$$

Depending upon the quality of data and dimensionality, one of the distance measures is determined. Once the distances with all training set are computed, K-NN classifier finds the K closest training instances as per their distances, and then the most common class among the K-nearest or closest training instances is identified as the class of the testing instance. The proposed method assigns a weight to each instance of the data set, i.e., if we have n number of cases present in the training data set, this technique computes n weights. These weights are then used to measure the importance or relevence of those instances in class representation. The more relevant ones get more weights than the trivial ones and eventually play the significantly more significant role than the less weighted cases. Last, these weights used as an extra parameter along with K neighbor for classification. In this way, our proposed method can significantly reduce the importance of noisy samples and leads to better classification results. The main idea of weight allocation is an extension of the technique developed [13–15].

The concept of discernibility index is extended, and a modified index of discernibility is determined (IDI; improved discernibility index). These include the average density of the hypersphere as additional information, which reduces the discernibility of tiny group of instances. These instances are significantly remote from the main cluster, and do not have a sufficient number of class instances to form a new

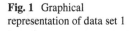

Fig. 1 Graphical
representation of data set 1

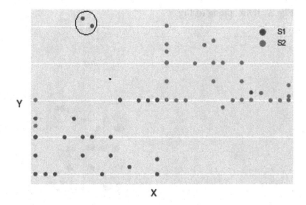

Fig. 2 Graphical
representation of data set 2

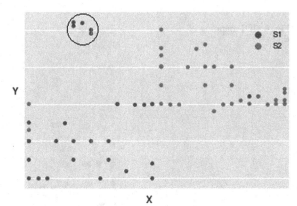

cluster of the same class. These phenomena could be regarded as noise in the data set. Reducing the discernibility of those cases minimizes the role of these instances in classification, and we achieve less noisy data set for classification. This approach leads to more accurate and generalized classification. For the ease of visualization, this has been demonstrated by the two sets of the two-dimensional data set. This data set consists of two classes ($S1$, $S2$) and two attributes (X, Y). We have calculated the ID and IDI for each of the data set to perceive discernibility map.

The above Figs. 1 and 2 represent the used two sets of data. In data set 2, we have included some more $S1$ instances inside the encircled region of data set 1 and created a new data set 2.

Figures 3, 4, 5 and 6 describes the estimated discernibility map including a color scale right side of each map; the darker color represents the greater value of the Index. Figures 3 and 4 shows the IDI and ID map for data set 1 and Figs. 5 and 6 shows the IDI and ID map for data set 2.

From the patterns, it is clear that the ID and IDI are very similar, but the contrast is visible at the encircled section. In ID calculation, we can observe that the encircled

Fig. 3 ID mapping of Fig. 1

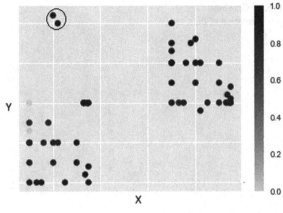

Fig. 4 IDI mapping of
Fig. 1

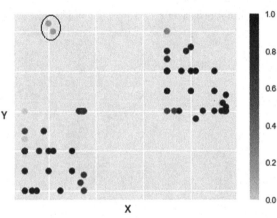

Fig. 5 ID mapping of Fig. 2

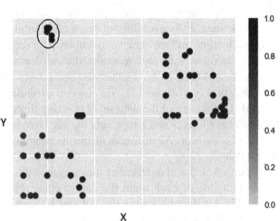

Fig. 6 DI mapping of Fig. 2

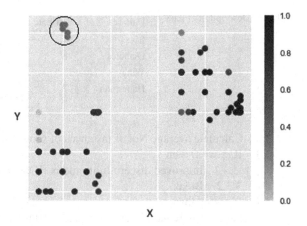

two $S1$ instances (Fig. 3) become very important (darker color) as only two $S1$ cases came inside the hypersphere.

But in IDI calculation (Fig. 4), we see a significant drop of discernibility index value as the proposed method considers not only the diversity of neighborhood but also the average density of hypersphere for the corresponding class. So, we see our approach is more practical as these two instances are far away from the central cluster of the class. We can give less importance to those. In the second data set, we increased the number of $S1$ instances of that near those two instances. As a consequence, we see an increase in discernibly index value (IDI) in Fig. 6 where Fig. 5 does not respond to the change in density, and its ID value is same as it was in Fig. 3. So our method is not only responsive to the diversity of the neighborhood but it also accounts the class distribution and leads to more generalization in classification.

3.1 Proposed Method

The proposed method consists of two main processes: (1) discernibility calculation of training set and (2) weighted K-NN for classification where IDI is used as weight. The algorithm for IDI calculation is given below.

Improved discernibility index algorithm (training data set):

1. Load training set in memory
2. For each class(C)

    2.1. Calculate the radius of the hypersphere (radius = standard deviation of the class instances)

3. For each instance (i)

    3.1. Count all the instances of same class C_i inside hypersphere excluding i, total no. of instances (n_i) excluding i

Table 1 Description of data sets

Data set	Attribute	Size	No. of class
Iris	4	150	3
Diabetes	8	768	2
Wine	13	178	3
Haberman	3	306	2

    3.2. Calculate average No. of instances of same class per hypersphere (a_i)

    3.3. If $c_i == 0$ then

        3.3.1 Improved discernibility index (IDI) = 0

        3.3.2. Break

    3.4. If $c_i < a_i$ then

        3.4.1. Improved discernibility index (IDI) = $(c_i/n_i) \times (c_i/a_i)$

    3.5 Else

        3.5.1 Improved discernibility index (IDI) = (c_i/n_i)

4. Save data set in file

Once the IDI is defined by the training set, the same data set is used to train the K-NN classifier. At the time of match, the test instance and closest K neighbors are selected by the Euclidean distance. Then, all the K neighbors have grouped accordingly to their class and calculated the sum of all IDIs for each group. The winner in IDI scores is chosen as the instance class. Below points are taken into account while implementing the proposed algorithm.

(a) As per the standard approach, 30% of the data set is used as the testing sample and 70% as the training sample.

(b) In K-NN algorithm, best "K" value results are only included in this below result section.

(c) As all numerical data sets are used for evaluation purpose, Euclidean distance has been used for distance measurement [16].

4 Result

The above actions were carried on a Pentium Atom workstation with 2 GB main memory operating on Windows 10. All algorithms were implemented in Python 3.6. The experiment is carried out on the below four standard data sets (Table 1) obtained from UCI repository. For all four data sets, 70% instances are used randomly as training set and other 30% is used as test set. We had executed the algorithm 10 times over each data set and taken the average of those as the final accuracy. The output is then compared with the traditional K-NN algorithm and shown in Table 2. The experiment shows the significant improvement of accuracy very clearly in the result set.

Table 2 Comparative study of the accuracy

Data set	K-NN	Proposed method
Iris	89.74	97.83
Diabetes	73.12	77.4
Wine	64.8	85.5
Haberman	72.42	79.6

5 Conclusion

We have presented the method that assigns weight to every instance of the training set, which is primarily a measure of distinguishability and coherency of the class. The method captures distinguishability of the class and also coherency of the class as a measuring weight so that it can capture both the neighborhood diversity as well as the class density of instance surroundings. This extended information helps proposed a method to identify the outliers by assigning low weights and yields considerable high accuracy. The evaluation of the method over four standard data sets confirms the claim. As a future work, fine tuning of the method as well as an extension of this concept to attribute weighting can be included in the further research.

References

1. Ahmed, S.S., Dey, N., Ashour, A.S., Sifaki-Pistolla, D., Bălas-Timar, D., Balas, V.E., Tavares, J.M.R.: Effect of fuzzy partitioning in Crohn's disease classification: a neuro-fuzzy-based approach. Med. Biol. Eng. Comput. 55(1), 101–115 (2017)
2. Kausar, N., Palaniappan, S., Samir, B.B., Abdullah, A., Dey. N.: Systematic analysis of applied data mining based optimization algorithms in clinical attribute extraction and classification for diagnosis of cardiac patients. In: Applications of intelligent optimization in biology and medicine, pp 217–231. Springer (2016)
3. Dey, N.: Classification and clustering in biomedical signal processing. IGI Global (2016)
4. Nath, S.S., Mishra, G., Kar, J., Chakraborty, S., Dey, N.: A survey of image classification methods and techniques. In: Control, Instrumentation, Communication and Computational Technologies (ICCICCT), 2014 International Conference on, 2014. IEEE, pp 554–557 (2014)
5. Wettschereck, D., Aha, D.W., Mohri, T.: A review and empirical evaluation of feature weighting methods for a class of lazy learning algorithms. Artif. Intell. Rev. 11, 273–314 (1997)
6. Zhang, M.-L., Zhou, Z.-H.: ML-KNN: A lazy learning approach to multi-label learning. Pattern Recogn. 40, 2038–2048 (2007)
7. Asuncion, A., Newman, D.: UCI machine learning repository (2007)
8. Gerhana, Y.A., Atmadja, A.R., Zulfikar, W.B., Ashanti, N: The implementation of K-nearest neighbor algorithm in case-based reasoning model for forming automatic answer identity and searching answer similarity of algorithm case. In: Cyber and IT Service Management (CITSM), 2017 5th International Conference on, 2017. IEEE, pp 1–5 (2017)
9. Selvaluxmiy, S., Kumara, T., Keerthanan, P., Velmakivan, R., Ragel, R., Deegalla, S.: Accelerating k-NN classification algorithm using graphics processing units. In: Information and Automation for Sustainability (ICIAfS), 2016 IEEE International Conference on, 2016. IEEE, pp 1–6 (2016)

10. Jaiswal, S., Bhadouria, S., Sahoo, A.: KNN model selection using modified Cuckoo search algorithm. In: Cognitive Computing and Information Processing (CCIP), 2015 International Conference on, 2015. IEEE, pp 1–5 (2015)

11. Taneja, S., Gupta, C., Aggarwal, S., Jindal, V.: MFZ-KNN—a modified fuzzy based K nearest neighbor algorithm. In: Cognitive Computing and Information Processing (CCIP), 2015 International Conference on, 2015. IEEE, pp 1–5 (2015)

12. Hechenbichler, K., Schliep, K.: Weighted K-Nearest-Neighbor Techniques and Ordinal Classification (2004)

13. Voulgaris, Z.: Discernibility Concept in Classification Problems. Citeseer (2009)

14. Voulgaris, Z., Magoulas, G.D.: Extensions of the k nearest neighbour methods for classification problems. In: Proceedings of the 26th IASTED International Conference on Artificial Intelligence and Applications (AIA), Innsbruck, Austria, February 11, 2008. pp 23–28 (2008)

15. Getis, A., Ord, J.K.: The analysis of spatial association by use of distance statistics. Geogr. Anal. 24, 189–206 (1992)

16. Hu, L.-Y., Huang, M.-W., Ke, S.-W., Tsai, C.-F.: The distance function effect on k-nearest neighbor classification for medical datasets. SpringerPlus 5, 1304 (2016)

Modified Energy-Efficient Stable Clustering Algorithm for Mobile Ad Hoc Networks (MANET)

S. R. Drishya and Vaidehi Vijayakumar

Abstract Cluster-Based Routing Protocol (CBRP) is popular and proven for energy efficiency in Mobile Ad hoc Networks (MANET). CBRP protocol divides the complete network into a number of clusters. Each cluster contains Cluster Head (CH) which maintains the cluster formation. Existence of CH improves routing performance in terms of reduction in routing overhead and power consumption. However, due to the mobility of the network, movement of the CH and cluster members, re-clustering is required and this increases overhead in the formation of clusters. The stability of the CH is an important factor for the stability of the cluster. Hence CH selection should be done efficiently such that the CH survives for a longer time. Existing CH selection algorithms use weight based approach which uses parameters like battery power, mobility, residual energy, and node degree to calculate the weight. Of all these parameters, mobility is an important factor in MANET and it has to be given more importance. Hence this paper proposes a Modified Energy-Efficient Stable Clustering (MEESC) algorithm in which node mobility is given more importance in weight calculation for the selection of CH. The proposed algorithm is simulated in NS3 and found to give better results in CH selection in terms of number of clusters formed and lifetime of the cluster head.

Keywords Clustering · Mobile ad hoc networks · Trust · Mobility · Weight based clustering

1 Introduction

MANET is a self-organized wireless and infrastructure-less network. Due to the movement of nodes, topology of the network is dynamic. MANET is used in military,

S. R. Drishya (✉) · V. Vijayakumar
VIT University, Chennai, India
e-mail: sr.drishya@gmail.com

V. Vijayakumar
e-mail: vaidehi.vijayakumar@vit.ac.in

© Springer Nature Singapore Pte Ltd. 2019
J. Kalita et al. (eds.), *Recent Developments in Machine Learning and Data Analytics*,
Advances in Intelligent Systems and Computing 740,
https://doi.org/10.1007/978-981-13-1280-9_42

Fig. 1 MANET
infrastructure

rescue operation and other places where mobile infrastructure is not reliable [1, 2]. Every node in MANET acts as a router. Maintaining routing and forwarding packets is a challenging task in the MANET. Due to dynamic topology of MANET, nodes go out of transmission range from each other which leads to frequent route failure [3].

Route failure leads to packet loss. To avoid this nodes have to rediscover the best path. This re-routing causes additional overhead in routing protocol for MANET [4]. Figure 1 shows a MANET infrastructure in which laptops and mobiles are nodes and dotted lines represent the connectivity of nodes with each other.

Other factors influencing the performance of MANET are: limited bandwidth, battery and heterogeneous communication links. These factors make routing a challenging task. Many researchers have suggested many routing protocols for MANET [5, 6]. Routing protocols are categorized as proactive, reactive, and hybrid. Proactive routing has a routing table and follows it for routing. So it is called table-driven protocol. Reactive protocols find the routing information when there is need of transmission. So it is called on-demand routing. Hybrid routing is the combination of both proactive and reactive routing protocols [7, 8].

Cluster-Based Routing Protocol is one of the hybrid routing protocols. In this protocol, nodes are organized as clusters. A cluster is a collection of nodes with similar parameters. One of the nodes in the cluster is elected as cluster head, which is responsible for packet transmission in the cluster [9]. Every node in the cluster is a cluster member and transfer packets to the cluster head. Cluster head on behalf of all members transmits the packets to the destination. The advantage of clustering algorithm is hierarchical organization of nodes. It is proved that clustering algorithm provides efficient bandwidth utilization in MANET [10, 11].

Nodes are formed into groups called clusters and have elected a coordinator or head. Routes are stored between clusters instead of nodes. A member in the cluster communicates to the cluster head [12]. One cluster head communicates with another cluster head in the network. If one cluster head is not in the communication range of other, then nodes between the clusters act as gateway nodes. The gateway node forwards the packets on behalf of its cluster head. Only cluster head and gateway nodes participate in the propagation of route formation messages [13]. This type

Fig. 2 Clustering in
MANET

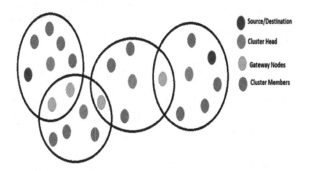

of transmission reduced the routing overhead in the dense networks and reduces the
scalability problems. In Fig. 2 nodes in the network are divided into 6 clusters that are
represented with circles. Each cluster has a cluster head. Nodes in red color represent
cluster head. Cluster members are represented with blue color. Gateway nodes are
represented in brown color.

A wide range of research is going on for stabilizing MANET. The lifetime of
the network depends on the stability of the clusters in the network [14]. One of the
methodologies is weight-based clustering which involves the weight calculation of
nodes using various parameters of nodes like residual energy in the battery, transmission range, degree of the node and mobility. Among all, mobility is the primary
reason for re-clustering in MANET [15]. Hence more weightage needs to be given
for mobility. This paper proposes a novel energy efficient weight based CH election
in which mobility is given more weight age. Moreover, trust factor is also considered
in weight calculation which helps to elect best node as cluster head.

Rest of the paper is organized as follows. Section 2 briefly surveys available weight
based clustering algorithms in MANET. Section 3 explains the metrics used for
proposed algorithm. Section 4 gives more detailed description of proposed algorithm.
Section 5 evaluates the performance and provides the simulation results. Section 6
concludes the paper and directs for future work.

2 Related Works

Clustering algorithms are dependent on certain metrics of the nodes like energy
level, battery, node degree, node mobility, position and direction and more. The
most crucial part in the clustering algorithm is to elect cluster head that will stay for
long time in the network [16]. The number of clusters in the network influences the
communication overhead and also latency. Clustering can be done in many ways,
weight is one of them. Weight based clustering can be done based on mobility, energy
and topology. Mobility based clustering is the primary concern. Figure 3 explains
different weight-based clustering algorithms in MANET that are already proposed
by many researchers.

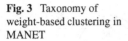

Fig. 3 Taxonomy of
weight-based clustering in
MANET

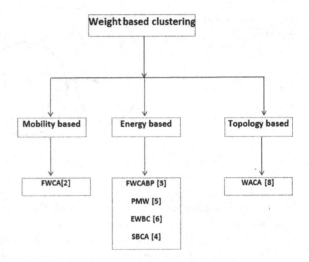

A Flexible Weight-Based Clustering Algorithm (FWCA) [6] uses different metrics to form clusters. It calculates node degree, remaining battery, transmission range, and mobility of nodes. It uses fuzzy logic to calculate the metrics and termed the result as weight. The node having the highest weight becomes the cluster head. If two nodes have same weight then it considers the lifetime of the communication link. A cluster head can handle certain number of members. A threshold is fixed for number of members in the cluster.

A Flexible weighted Clustering Algorithm based on Battery Power (FWCABP) [9] is the extension to FWCA. It is proposed to prevent nodes having less battery power becoming cluster head. Each node possesses its neighbor count, battery and node degree in its status. While forming the cluster, each node broadcasts their status information to its neighbors and then builds the neighbor table. The metrics considered to calculate the weight of the nodes are: node degree, mobility, distance between neighbor nodes and remaining battery power. The node having lesser weight value becomes cluster head. It invokes re-clustering when the battery reduces to some threshold value.

Adabi et al. proposed a weight based clustering algorithm called Score-Based Clustering Algorithm (SBCA) [7] to reduce the cluster count and increase the lifetime of the network. The metrics considered in the algorithm are remaining battery, degree of the node, number of neighbors. Through this metrics score of a node is calculated. Each node calculates the score and broadcast the value to the neighbors. The node having the highest score will become cluster head.

Another clustering algorithm is proposed by Xing et al. called PMW [12]. PMW stands for power, mobility, and workload. Weight of the node is calculated by considering the metrics power, mobility, and workload. These metrics can be collected locally and calculate the weight more easily. Node having least weight will become the cluster head. PMW extends the lifetime of the cluster head.

Efficient Weight Based Clustering Algorithm [17] is proposed to increase the usage of rare resources of the node such as bandwidth and energy. Their work is the first clustering algorithm that uses the energy to calculate the weight of the node. Its aim is to reduce the routing overhead. Weight of the node is calculated with residual energy in the battery, neighbor count, and distance with neighbor nodes. It allocates status of the nodes to NUL, CH, member node and gateway node. Initially status of all nodes set to NUL. Nodes calculate their weight and broadcast the value to neighbors. The node having higher weight is made as CH, i.e., node is elected as cluster head. Re-clustering is done when node leaves the cluster or cluster head consumes its battery below threshold value.

Matthias R. Brust et al., proposed WACA [18], a Hierarchical Weighted Clustering Algorithm. This algorithm focuses on reducing the re-clustering mechanism in the network by considering the stability factors of the node. It does not consider any metrics like node mobility, speed, or battery. It considers only topological characteristics and device parameters. This makes the algorithm easy to implement in the mobile devices. It is proved that their algorithm makes stable clustering scheme. However, in MANET, because of mobility frequent changes in topological parameters, the communication overhead for weight calculation increases.

Shayesteh and Karimi in [19], proposed another clustering algorithm which is based on the nodes weight. Selection of cluster head is done in two phases. A new weight function using fuzzy logic is proposed to calculate the weight of the node. In first phase using weight function, weight is calculated by the nodes. In second phase relative mobility is calculated and also the future mobility is predicted. Based on these two factors the final weight is calculated. Node with least weight becomes cluster head.

Though there are many clustering methods addressed in literature, there is a need to have novel clustering method to increase the stability of the cluster.

3 Metrics for Cluster Head Election

Different metrics are used for cluster head selection is given in this section.

(1) The trust level T_i of the node n_i: Initially every node is given equal trust value 1. This level is decreased by anomaly detection algorithm if a node is misbehavior. An abnormal node may be suspected node or malicious node. The trust level of the nodes is defined as [20]:

$$\left. \begin{array}{l} \text{Normal\_node} : 0.8 \leq T_i < 1 \\ \text{Suspected\_node} : 0.3 \leq T_i < 0.8 \\ \text{Malicious\_node} : 1 \leq T_i < 0.3 \end{array} \right\} \qquad (1)$$

Only normal nodes can participate in cluster head election. If the node is malicious node, it is excluded from every cluster and suspected node can be a cluster member but cannot participate in cluster head election as it can become malicious at any instant of time [20].

(2) Distance D_i of the node n_i from the neighbor node is given by: [21].

$$D_i = \sum_{j \epsilon N(i)} \{\text{dist}(i, j)\} \tag{2}$$

(3) Degree of the node n_i at time t, denoted as C_i is given by

$$C_i = |N(i)|$$
$$N(i) = \{n_j / \text{dist}(i, j) < tx_{\text{range}}\} i \neq j, \tag{3}$$

where dist(i ,j) is distance between two nodes n_i and n_j
tx_{range} is the transmission radius

(4) The residual energy Er_i of the node n_i after transferring k bits to a node n_j within a distance d is given by [22].

$$\text{Er}_i = E - (E_{Tx}(k, d) + E_{Rx\_\text{elec}}(k)), \tag{4}$$

where E is the current energy of the node
E_{Tx} is energy for transmit a message

$$E_{Tx}(k, d) = k E_{\text{elec}} + K E_{\text{amp}} d^2 \tag{5}$$

where E_{elec} is energy of electrons and E_{amp} is required amplified energy
E_{Rx_elec} is energy consumed for receive a message

$$E_{Rx\_\text{elec}}(K) = k E_{\text{elec}} \tag{6}$$

For each node that participating in cluster head election, calculate the weight of the node w_i as follows:

$$w_i = w_1 * T_i + w_2 * D_i + w_3 * C_i + w_4 * \text{Er}_i, \tag{7}$$

where w_1, w_2, w_3 and w_4 are coefficients to system criteria with

$$w_1 + w_2 + w_3 + w_4 = 1 \tag{8}$$

(5) The aim of the paper is to have a stable clusters in mobile ad hoc networks, so mobility of the node should be considered and the node having low mobility should be elected as cluster head. The mobility of the node M_i is calculated as: [21].

$$M_i = \frac{1}{T} \sum_{t=1}^{T} \sqrt{(x_t - x_{t-1})^2 + (y_t - y_{t-1})^2},$$ (9)

where (x_t, y_t) and (x_{t-1}, y_{t-1}) are coordinates at time t and $t-1$ respectively.

4 Modified Energy-Efficient Stable Clustering Algorithm (MEWCA)

4.1 Basis for Proposed Algorithm

1. Network is represented by an undirected graph $G(V, E)$ where each node is a vertex and links as edges.
2. All nodes are scattered in 2D plane.
3. No two nodes shares same coordinates in 2D plane.
4. Each cluster head can handle only certain number of members.
5. A cluster head can communicate efficiently with neighbor nodes without considering signal attenuation.
6. Energy consumed is less when communicating to nearby nodes and more for the far away nodes. Cluster head consumes more battery than a member node because it has to handle extra responsibility for cluster members.

4.2 Algorithm

1: Assign the values for coefficients w_1, w_2, w_3 and w_4.

2: **For** any node in the network $n_i \epsilon G$, find the list of neighbors $N(i)$

3: Calculate the mobility M_i of the node

4: Calculate its weight w_i : $w_i = w_1 * T_i + w_2 * D_i + w_3 * C_i + w_4 * Er_i$

5: Initialize status vector for all nodes with parameters

Status_vector (Id, CH, Mobility, Weight, Neighbor_list, Size, Nature)

6: Initialize CH=0, Size=0 and Nature="Null"

7: **Repeat**

8: **If** $N(i) \neq \emptyset$ **then**

9: Choose any node in the neighbor list $v \epsilon N(i)$

10: mobility (v) = min (mobility(u) /$u \epsilon N(i)$) **and** (weight(v) = max(weight(u) /$u \epsilon N(i)$) **or** weight(v) < Threshold_value)

11: **Else** n_i is cluster head for itself

12: Update the Status_vector of elected cluster head as CH=Id, Size=1, Nature=CH

13: Send CH_msg to neighbor nodes $N(CH)$

14: x=counts($N(CH)$)

15: **For** y= 1 to x **do**

    If($n_y \epsilon N(CH)$ receives the message and $n_y \rightarrow CH = 0$) **then**

    n_y sends Join_msg to the CH

    If (CH $\rightarrow$ size < Threshold_size) **then**

    CH sends Accept_msg to n_y

    CH $\rightarrow$ size = CH $\rightarrow$ size+1

    $n_y \rightarrow CH = CH \rightarrow ID$

    Else go to 7

    Endif

    Endif

Endfor

16: **Until** CH $\rightarrow$ size <Threshold_size

17. **End**

5 Simulation Analysis and Results

To evaluate the proposed algorithm, simulation is carried on NS-3 for evaluating the performance of proposed protocol. NS-3 is the discrete-event network simulator. It provides open simulation environment for network research.

Assume that mobile nodes used in the simulation are homogeneous and randomly distributed. In simulation, number of nodes varied from 5 to 100. At every point of time, nodes will moved randomly according to random way point model in 500 × 500 m^2 place. Number of simulations is 20. Figure 4a, b show cluster formation for 30 nodes at different time instances t_1 and t_2.

Table 1 represents the parameters considered in simulating MANET in NS3.

Clustering performance

To evaluate the cluster stability, the life time of the cluster is used as metrics. If the cluster lifetime is more, then the cluster is stable.

Evaluation Parameters

To evaluate the efficiency of the proposed clustering algorithm, the number of clusters formed for specified number of nodes and their life time are obtained from NS3.

The nodes of mobile ad hoc networks are randomly distributed all over the network. The transmission range of each node is considered as 300 meters. Number of nodes in the network varies from 5 to 100. Figure 5 shows the lifetime of the cluster with respect to number of nodes. The cluster elected by proposed algorithm is done wisely based on the weight calculated. Blue line represents the average cluster lifetime of the existed EWCA clustering method. Brown line represents the average life time of clusters for the proposed method. By observing the graph it is proved that life time of cluster is increased in the proposed method than the existing method.

Figure 6 shows the efficiency of the clustering algorithm for varying number of nodes. Efficiency is measured by the number of clusters formed for number of nodes in the network. Cluster reformation is lessened because of the increased lifetime of the availability of cluster head. It is seen that the number of clusters formed is more in existing clustering algorithm and it reduced in proposed clustering algorithm.

6 Conclusion

This paper proposes a stable and efficient clustering algorithm for mobile ad hoc networks. As the nodes are moving randomly in the mobile ad hoc network, a stable clustering scheme which can elect efficient cluster head is provided. The proposed

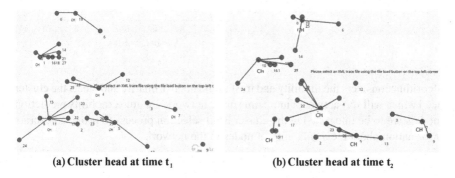

| (a) Cluster head at time t_1 | (b) Cluster head at time t_2 |

Fig. 4 Simulation scenario for cluster formation for 100 nodes

Table 1 Parameters for simulation

Parameter	Value
Number of nodes	5–30
Size of the network	500 × 500
The distribution of nodes	Random
Transmission power	300
Number of simulations	20

Fig. 5 Stability of clusters
(Number of nodes vs.
Average cluster lifetime)

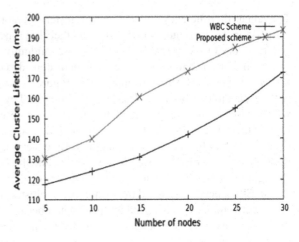

Fig. 6 Effectiveness of
cluster (Number of nodes vs.
Number of clusters formed)

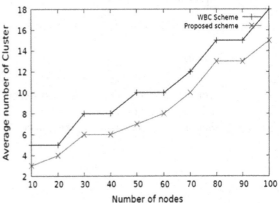

algorithm considers the mobility and trust of the node primarily and elects the cluster head which will stay active for more time in the network. In future, mobility prediction models are to be integrated in the cluster head selection process to provide accurate information about the availability of nodes in the network.

References

1. Bentaleb, A., Boubetra, A., Harous, S.: Survey of clustering schemes in mobile ad hoc networks. Commun. Netw. **5**(02), 8–14 (2013)
2. Hurley-Smith, D., Wetherall J., Andrew: SUPERMAN: Security using pre-existing routing for mobile ad hoc networks. IEEE Trans. Mobile Comput. **16**(10) (2017)
3. Geetha, V., Kallapur, P.V., Tellajeera, S.: Clustering in wireless sensor networks: performance comparison of LEACH & LEACH-C protocols Using NS2. Procedia Technolgy (C3IT-2012). Elsevier **4**, 163–170 (2012)

4. Zabian, A., Ibrahim, A., Al-Kalani, F.: Dynamic head cluster election algorithm for clustered Ad-Hoc Networks. J. Comput. Sci. **4**(1), 19 (2008)
5. Dhamodharavadhani, S.: A survey on clustering based routing protocols in mobile Ad Hoc Networks. In: IEEE International Conference on Soft-Computing and Networks Security (ICSNS) (2015)
6. El-Bazzal, Z., Kadoch, M., Agba, B.L., Gagnon, F., Bennani, M.: A flexible weight based clustering algorith-min mobile Ad hoc Networks. In: International Conference on Systems and Networks Communications, Tahiti (2006)
7. Adabi, S., Jabbehdari, S., Rahmani, A.M., Adabi, S.: SBCA: score based clustering algorithm for mobile Ad Hoc Networks. The 9th International Conference for Young Computer Scientists, Hunan, China (2008)
8. Yick, J., Mukherjee, B., Ghosal, D.: Wireless sensor network survey. Comput. Netw. (Elsevier) **52**(12), 2292–2330 (2008)
9. Hussein, A.H., Salem A.O.A., Yousef, S.: A flexible weighted clustering algorithm based on battery power for mobile Ad Hoc Networks. In: IEEE International Symposium on Industrial Electronics, Cambridge, UK (2008)
10. Aslam, M., Javaid, N., Rahim, A., Nazir, U., Bibi, A., Khan, Z.A.: Survey of extended LEACH-based clustering routing protocols for wireless sensor networks. IEEE 14th International Conference on High Performance Computing and Communication, Liverpool, UK (2012)
11. Li, D., Jia, X., HiaLiu: Energy efficient broadcast routing in static adhoc wireless networks. IEEE Trans. Mobile Comput. **3**(2)
12. Xing, Z., Gruenwald, L., Phang, K.K.: A robust clustering algorithm for mobile Ad Hoc Networks. Handbook of Research on Next Generation Networks and Ubiquitous Computing (2008)
13. Selvam, R.P., Palanisamy, V.: Stable and flexible weight based clustering algorithm in mobile ad hoc networks. (IJCSIT) Int. J. Comput. Sci. Info. Technol. **2**(2), 824–828 (2011)
14. Li, C., Wang, Y., Huang, F., Yang, D.: A novel enhanced weighted clustering algorithm for mobile networks. In: IEEE 5th International Conference on Wireless Communications, Networking and Mobile Computing, Beijing, China (2009)
15. Choi, W., Woo, W.: A distributed weighted clustering algorithm for mobile Ad Hoc Networks. In: Proceedings of the Advanced International Conference on Conference on Telecommunications and International Conference on Internet and Web Applications and Services (AICT/ICIW 2006), Guadelope, French Caribbean (2006)
16. Dahane, A., Berrachand N., Kechar, B.: Energy efficient and safe weighted clustering algorithm for mobile wireless sensor networks. In: The 9th International conference on future networks and communications (FNC) vol. 34, pp. 63–70 (2014)
17. Monsef, M.R., Jabbehdari, S., Safaei, F.: An efficient weight-based clustering algorithm for mobile Ad-hoc Networks. J. Comput. **3**(1), 16–20 (2011)
18. Brust, M.R., Andronache, A., Rothkugel, S.: WACA: a hierarchical weighted clustering algorithm optimized for mobile hybrid networks. In: Third International Conference on Wireless and Mobile Communications, (ICWMC), Guadeloupe, France (2007)
19. Shayesteh, M., Karimi, N.: An innovative clustering algorithm for MANETs based on cluster stability. Int. J. Model. Optim. **2**(3), 80–86 (2012)
20. Safa, H., Artail, H., Tabet, D.: A cluster-based trust-aware routing protocol for mobile ad hoc networks. Wireless Netw. **16**(4), 969–984 (Springer) (2010)
21. Heinzelman, W.R., Chandrakasan, A., Balakrishnan, H.: Energy-efficient communication protocol for wireless microsensor networks. In Proceedings of the 33rd Hawaii International Conference on System Sciences, Maui, HI, USA (2000)
22. Zhou, H., Zhang, J.: An efficient clustering algorithm for MANET based on weighted parameters. In: 8th International Symposium on Computational Intelligence and Design (ISCID), Hangzhou, China (2016)

Ensemble of Convolutional Neural Networks for Face Recognition

V. Mohanraj, S. Sibi Chakkaravarthy and V. Vaidehi

Abstract Convolutional Neural Networks (CNN) are becoming increasingly popular in large-scale image recognition, classification, localization, and detection. Existing CNN models use the single model to extract the features and the recognition accuracy of these models is not adequate for real-time applications. In order to increase the recognition accuracy, an Ensemble of Convolutional Neural Networks (ECNN) based face recognition is proposed. The proposed model addresses the challenges of facial expression, aging, low resolution, and pose variations. The proposed ECNN model outperforms the existing state of the art models such as Inception-v3, VGG16, VGG19, Xception and ResNet50 CNN models with a Rank-5 accuracy of 97.12% on Web Face dataset and 100% on YouTube face dataset.

Keywords Face recognition · CNN · Pre-trained models · Machine learning
Computer vision

1 Introduction

CNN has attracted the Computer Vision (CV) research by significantly improving the state of the art applications such as Medical Image Analysis, Face Recognition, Robotics, Self-driving, etc. [1, 3]. The key success of using CNN in CV applications is due to its scalable quantities of processing speed, power in accuracy and the massive training dataset [2]. However, deep learning techniques are far better than the

V. Mohanraj (✉) · S. Sibi Chakkaravarthy
Department of Electronics Engineering, Madras Institute of Technology, Anna University,
Chennai, India
e-mail: mohanraj@mitindia.edu

S. Sibi Chakkaravarthy
e-mail: sb.sibi@mitindia.edu

V. Vaidehi
School of Computer Science and Engineering, VIT University, Chennai, India
e-mail: vaidehi.vijayakumar@vit.ac.in

467
J. Kalita et al. (eds.), *Recent Developments in Machine Learning and Data Analytics*,
Advances in Intelligent Systems and Computing 740,
https://doi.org/10.1007/978-981-13-1280-9_43

conventional machine learning techniques. Technology giants like Google, Deep-Mind and Facebook, etc., are already making a huge stride in the CV space [4–7]. In olden days Facebook asked you to tag your friends in your photograph, but the advancement of deep learning techniques leads the ability to recognize your friends from your group photograph. Facebook algorithms are efficient enough to recognize the faces with 98% accuracy from the posted group/individual photographs [6]. The recent face recognition method by Google also uses nearly about 200 million face images which are larger than any other publicly available dataset. Hence, processing these large dataset is a complex task and requires high computational devices like Graphics Processing Units (GPU) [7, 8].

Face recognition in an uncontrolled environment is still a challenging task. The recent study on Face recognition under uncontrolled environment reveals the practical difficulties of implementing. The conventional face recognition algorithms use the low level features and shallow models to represent the faces and facial features. In this decade, many authors have proved the effectiveness of the deep learning models in effective face recognition. Recent deep learning models such as Alexnet, ZFNet, VGG, GoogleNet, Inception, etc., are effectively utilized to extract high level visual features [9].

In contrast to the above stated models this paper proposes a novel ensemble of deep learning models for an efficient face recognition [10]. An ensemble is a infinite collection of models that can be used to obtain better average predictive accuracy than using a single model in the ensemble collection [12]. This paper also addresses the existing CNN model to extract the features for face matching.

This paper is organized as follows: Chap. 2 presents the Ensemble of Convolutional Neural Networks for face recognition, Chap. 3 analyzes the performance of proposed ECNN method and Chap. 4 concludes the paper.

2 Convolution Neural Network for Feature Extraction

Convolution Neural Networks applies multiple filters on the raw input image for extracting high-level features [11]. Figure 1 shows the extraction of features, where x represents input face image and $f(x)$ shows the extracted features.

The similarity of two face images is computed with the extracted features using Euclidean distance metric algorithm [13]. Therefore, the encodings of two images of the same person are similar to each other and the encodings of two images of different persons are different. Figure 2 shows the comparison of two face images of the same person.

The VGG16 network architecture produces the probabilities for the input 1000 ImageNet class labels [14]. When treating networks as a feature extractor, the network at an arbitrary point is chopped off. Now the last layer in the network is a max pooling layer which will have the dimension of 7*7*512. Thus achieving 25,088 feature values for each face image. The above discussed process is repeated for an entire

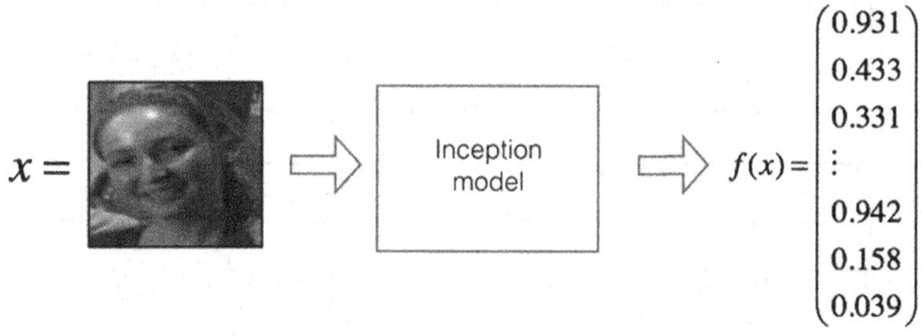

$$f(x) = \begin{pmatrix} 0.931 \\ 0.433 \\ 0.331 \\ \vdots \\ 0.942 \\ 0.158 \\ 0.039 \end{pmatrix}$$

Fig. 1 ConvNet feature extraction

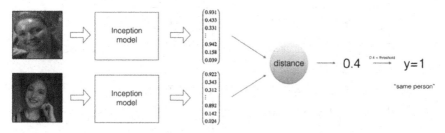

Fig. 2 Comparison of two face images

dataset resulting the total size of N images, each with 25,088 columns. The extracted feature vectors are trained using Logistic Regression classier [15]. Figure 3 shows the architecture of VGG 16 CNN model.

3 Ensemble of Convolutional Neural Networks (ECNN) for Face Recognition

The proposed ECNN method extracts the features from three different ConvNets models, namely VGG16, Incpetion-v3, and Xception [16]. The ConvNets models are used for extracting features as the weights and architectures of these models are available in public for Computer Vision research. The weights and architecture of the chosen models are loaded locally prior to the training phase [17]. The architecture of the proposed ECNN method is presented in Fig. 4 and the images in training database are resized to a fixed dimension as shown in Table 1.

After preprocessing, each of the images in the database is given to VGG16, Inception-v3, and Xception model architecture for extracting features by removing the top fully connected layers of the ConvNets model. The extracted features of the three different CNN models of a single image is concatenated and stored in a list. Further its corresponding label is stored in another list for compatibility. This process is repeated for all the images in the training database. The extracted image

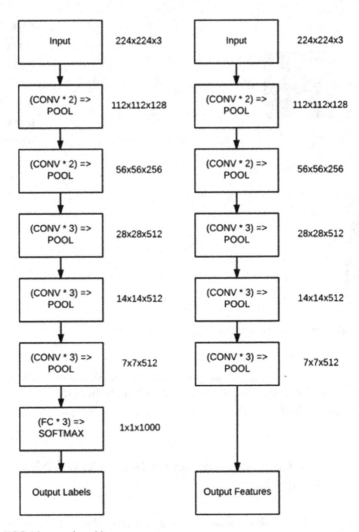

Fig. 3 VGG 16 network architecture

features and labels are stored locally in an HDF5 file format as NumPy arrays. This feature extraction process is carried out for two different benchmark Face datasets such as YouTube and WebFace.

Ensemble methods generally refer to training a "large" number of models and then combining their output predictions via voting or averaging to yield an increase in classification accuracy. Ensemble methods are specific to deep learning and Convolutional Neural Networks. Multiple networks are trained and each network return the probability for each class label. These probabilities are averaged together, and the final classification is obtained. Finally, the obtained probability values are averaged and the final classification is obtained.

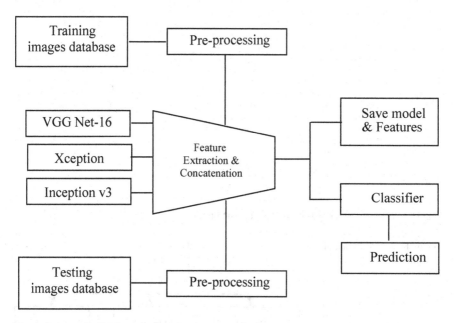

Fig. 4 Proposed ECNN architecture for face recognition

Table 1 Fixed dimension to resize the images for ECNN method

ConvNets model	Fixed dimension (for preprocessing)
VGG16	224 × 224
VGG19	224 × 224
ResNet50	224 × 224
Inception-v3	299 × 299
Xception'	299 × 299

Figure 5 shows the Random Forest with multiple decision trees. The outputs of each decision tree are averaged together to obtain the final classification.

Figure 6 shows the ensemble of neural networks consisting of multiple networks. When classifying an input image, the data point is passed to each network where it classifies the image independently of all other networks. The classification across networks are then averaged to obtain the final prediction.

The VGG16 ConvNets model accepts the input image size of 224*224 and it has 13 convolution layers with different number of filter sizes such as 64,128, 256, 512. The size of each filter is 3*3. Along with this the VGG16 network poses five max pooling and three fully connected layers. The size of the feature vector extracted for VGG16 before the last fully connected network is 4096. Inception-V3 ConvNets model accepts the input image size of 299*299, inception module extracts multi-level features by performing convolutions with different filter size such as $1 \times 1, 3 \times 3$, and 5×5 convolution. The weights for Inception V3 are smaller than VGG and ResNet,

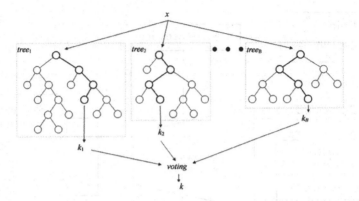

Fig. 5 Random forest consists of multiple decision trees

Fig. 6 Ensemble of neural
networks

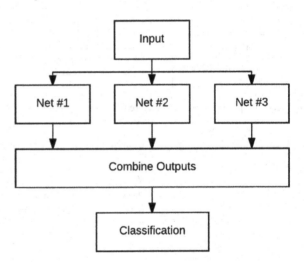

of memory size 96 MB. Xception is an extension of the Inception architecture which replaces the standard Inception modules with depth wise separable convolutions and it has a small weight memory size of 91 MB.

After extracting features from the training phase, the stored features and labels are loaded and split into training and testing data based on a parameter "train_test_split". If "train_test_split" is chosen as 0.1, then it means 90% of the overall data is used for training and 10% of the overall data is used to evaluate the trained model. Based on parameter tuning and Grid-search methodology. It is found that Logistic Regression (LR) outperformed all the other machine learning classifiers such as Random Forests (RF), Support Vector Machine (SVM) and K-Nearest Neighbors (KNN). Two performance metrics are chosen to evaluate the trained model, namely Rank-1 and Rank-5 accuracy. Rank-1 accuracy gives the accuracy of the trained model when tested with an unseen test data on the first chance. Rank-5 accuracy gives the accuracy of the trained model when tested it with an unseen data given five chances.

Algorithm

Begin
Input: ImageNet data $D_{\text{ImageNet}} = \{x_i, y_i\}_{i=1}^{m}$
Output: model M, Accuracy A, Classifier H
Input the WebFace and YouTube dataset
For feature $f = 1$ to n do

    Learn F_n based on $D_{1..m}$
    Learn H based on F_n
    Compute A based on H
    Return A & H

End for
End

4 Results and Discussion

The experimental setup for the proposed methodology is carried out using Intel Xeon processor with the NVIDIA Titanx GPU and 28 GB RAM. The proposed ensemble CNN model is implemented using Python with Keras library. The entire experiment is carried out on Windows-7 Operating System (OS) with Tensorflow as backend. Other python packages used for implementation are NumPy, SciPy, scikit-image, h5py, scikit-learn and OpenCV 2.4.10.

4.1 Dataset

Two publicly available face datasets are considered for analyzing the performance of the proposed ECNN method. The WebFace dataset contains 494,414, color images of 10,575 people with different facial expressions, illumination condition, aging, low resolution, different pose, and occlusions. Figure 7 represents the sample face images of WebFae dataset. The YouTube faces dataset contains 3425 videos of 1595 different people. Figure 8 shows the sample faces images of a YouTube face dataset.

Table 2 represents the feature vector dimension of different pre-trained CNN models. The proposed ECNN method concatenatesVGG16, Inception-V3 and Xception feature vectors which produces the final feature vector of 8192 size. Table 3 presents face recognition accuracy of WebFace dataset. From Table 2 it is observed that the proposed ECNN method performs better face recognition accuracy compared to vgg16, vgg19, inception-v3, resnet50 and xception CNN models. Table 4 shows

Fig. 7 Sample face images of WebFace dataset

Fig. 8 Sample face images of a YouTube face dataset

Table 2 Feature vector dimension of ConvNets model

ConvNets models	Feature dimension
VGG16	4096
VGG19	4096
ResNet 50	2048
Inception-v3	2048
Xception	2048

Table 3 Recognition accuracy for WebFace dataset

ConvNet Models	Accuracy	
	Rank-1	Rank-5
VGG16	69.96	88.86
VGG19	66.25	86.49
ResNet 50	29.90	55.25
Inception V3	78.81	93.91
Xception	65.82	89.02
Inception V3 + VGG16	85.44	96.25
Inception V3 + VGG19	84.01	95.58
Inception V + Xception	77.30	94.98
Inception V3 + ResNet 50	79.71	94.15
ECNN (proposed)	87.08	97.12

Table 4 Recognition accuracy of YouTube face dataset

ConvNet's Models	Accuracy	
	Rank-1	Rank-5
VGG16	99.97	99.98
VGG19	99.97	99.98
ResNet 50	99.02	99.79
Inception V3	99.93	99.97
Xception	99.92	99.98
Inception V3 + VGG16	99.97	100
Inception V3 + VGG19	99.97	100
Inception V3 + Xception	99.97	100
Inception V3 + ResNet 50	99.98	100
ECNN (proposed)	99.99	100

the recognition accuracy of a YouTube face dataset. From Table 4 it is proved that the proposed ECNN method outperforms vgg16, vgg19, inception-v3, resnet50 and xception CNN models.

4.2 Complexity Analysis

The dataset processed using any deep learning models are huge and complex. Processing such huge data will always require multiple loops to traverse the whole datasets. However, the increased number of loops will always consume a lot of time for obtaining the results. In order to reduce the number of loops and for fast processing, the present day neural model utilizes vectorization.

The proposed model can handle dataset of any size. The complexity of the proposed model is found to be in $O(m\log n)$, where m represents the total number of features and n represents the total number of stacked classifiers.

5 Conclusion

In this paper, Ensemble of Convolutional Neural Networks based feature descriptors is proposed for face recognition. Benchmarked dataset namely, YouTube and Web-Face dataset are used for the validation of the proposed model. From the experimental analysis, it is observed that the proposed ECNN model outperforms the existing ConvNets models such as VGG16, VGG19, Inception-V3, ResNet50, and Xception. In future, multiple GPUs can be used to reduce computation time of training and testing of various CNN models.

References

1. Ojala, T., Pietikäinen, M., Mäenpää, T.: Multiresolution gray scale and rotation invariant texture classification with local binary patterns. IEEE Trans. Pattern Anal. Mach. Intell. **24**(7), 971–987 (2002)
2. Lowe, D.G.: Distinctive image features from scale-invariant keypoints. Int. J. Comput. Vis. **60**(2), 91–110 (2004). https://doi.org/10.1023/b:visi.0000029664.99615.94
3. Dalal, N., Triggs, B.: Histograms of oriented gradients for human detection. IEEE Comput. Soc. Conf. Comput. Vis. Pattern Recogn. (CVPR) **1**, 886–893 (2005)
4. Yang, J., Jiang, Y.G., Hauptmann, A.G., Ngo, C.W.: Evaluating bag-of-visual words representations in scene classification. In: Proceedings of the international workshop on multimedia information retrieval (ACM), pp. 197–206 (2007)
5. Krizhevsky, I., Sutskever, I., Hinton, G.E.: Imagenet classification with deep convolutional neural networks. In: Conference on Neural Information Processing Systems (NIPS), pp. 1106–1114 (2012)
6. Zeiler, M.D., Fergus, R.: Visualizing and understanding convolutional networks. In: European Conference on Computer Vision (ECCV) (2014)
7. Simonyan, K., Zisserman, A.: Very deep convolutional networks for large-scale image recognition. In: International Conference on Learning Representations (ICLR) (2015)
8. Szegedy, C., Liu, W., Jia, Y., Sermanet, P., Reed, S., Anguelov, D., Erhan, D., Vanhoucke, V., Rabinovich, A.: Going deeper with convolutions. In: IEEE Conference on Computer Vision and Pattern Recognition (CVPR) (2015)

9. Sermanet, P., Eigen, D., Zhang, X., Mathieu, M., Fergus, R., LeCun, Y.: Overfeat: integrated recognition, localization and detection using convolutional networks. In: International Conference on Learning Representations (ICLR) (2014)

10. He, K., Zhang, X., Ren, S., Sun, J.: Deep residual learning for image recognition. In: IEEE Conference on Computer Vision and Pattern Recognition (CVPR) (2016)

11. Cao, Z., Yin, Q., Tang, X., Sun, J.: Face recognition with learning based descriptor. In: Proceedings of Computer Vision and Pattern Recognition (CVPR) (2010)

12. Li, P., Prince, S., Fu, Y., Mohammed, U., Elder, J.: Probabilistic models for inference about identity. In: IEEE Transactions on Pattern Analysis and Machine Intelligence (PAMI) (2012)

13. Berg, T., Belhumeur, P.: Tom-vs-Pete classifiers and identity preserving alignment for face verification. IN: Proceedings of British Machine Vision Conference (BMVC) (2012)

14. Schroff, F., Kalenichenko, D., Philbin, J.: Facenet: a unified embedding for face recognition and clustering. In: IEEE Conference on Computer Vision and Pattern Recognition (CVPR) (2015)

15. Taigman, Y., Yang, M., Ranzato, M., Wolf, L.: Deep-Face: closing the gap to human level performance in face verification. In: IEEE Conference on Computer Vision and Pattern Recognition (CVPR) (2014)

16. Romero, A., Ballas, N., Kahou, S.E., Chassang, A., Gatta, C., Bengio, Y.: Fitnets: hints for thin deep nets. arXiv preprint arXiv:1412.6550 (2014)

17. Ba, J., Caruana, R.: Do deep nets really need to be deep? In: Advances in Neural Information Processing Systems 27, arXiv:1312.6184 (2014)

Clustering High-Dimensional Data: A Reduction-Level Fusion of PCA and Random Projection

Raghunadh Pasunuri, Vadlamudi China Venkaiah and Amit Srivastava

Abstract Principal Component Analysis (PCA) is a very famous statistical tool for representing the data within lower dimension embedding. K-means is a prototype (centroid)-based clustering technique used in unsupervised learning tasks. Random Projection (RP) is another widely used technique for reducing the dimensionality. RP uses projection matrix to project the data into a feature space. Here, we prove the effectiveness of these methods by combining them for efficiently clustering the low as well as high-dimensional data. Our proposed algorithms works by combining Principal Component Analysis (PCA) with Random Projection (RP) to project the data into feature space, then performs K-means clustering on that reduced space (feature space). We compare the proposed algorithm's performance with simple K-means and PCA-K-means algorithms on 12 benchmark datasets. Of these, 4 are low-dimensional and 8 are high-dimensional datasets. Our proposed algorithms outperform the other methods.

Keywords Principal component analysis · Random projection · K-means
Clustering · High-dimensional data

1 Introduction

Principal Component Analysis (PCA) is a commonly used tool that reduces the data from high-dimensional to low-dimensional with maximum variance preservation [1]. The applications of PCA include Data Compression, Data Visualization, Feature

R. Pasunuri (✉) · V. C. Venkaiah
School of Computer and Information Sciences, University of Hyderabad, Hyderabad, India
e-mail: raghupasunuri@gmail.com

V. C. Venkaiah
e-mail: venkaiah@hotmail.com

A. Srivastava
ANURAG, DRDO, Hyderabad, India
e-mail: amit_srivastava@anurag.drdo.in

© Springer Nature Singapore Pte Ltd. 2019   479
J. Kalita et al. (eds.), *Recent Developments in Machine Learning and Data Analytics*,
Advances in Intelligent Systems and Computing 740,
https://doi.org/10.1007/978-981-13-1280-9_44

Extraction, and so on. The K-means [2] is a clustering method in which the data with n points given in R^d and an integer K is specified. The algorithm finds K cluster centers such that the mean squared error is minimized. It begins by initializing K random points as the cluster centers. At each iteration, the closest points are moved to the corresponding cluster, and then the new cluster centers are defined (new mean of the cluster). The total squared error is reduced in each K-means iteration. This is repeated until the algorithm converges.

Random Projection (RP) works by projecting the input space into feature space with the help of random matrices. This is simple and less error prone compared to other dimensionality reduction methods [3].

Random projections can be used for unsupervised learning as in Fern and Brodley [7]. In this work, the high-dimensional data clustering is done by using multiple random projections, and the performance is compared with single random projection and PCA for the case of EM clustering. The proposed methods outperform PCA for all the three datasets that have been considered for the study.

Deegalla and Bostrom [8] have combined RP with PCA for reducing the dimensionality of the input data, and used nearest neighbor classifier to classify the data in the low-dimensional space. The experimental results advocate that the use of PCA for dimensionality reduction is giving higher accuracy when compared to RP for almost all the datasets.

IRP K-means algorithm for high-dimensional data clustering was proposed by Cardoso and Wichert in [4]. In their study, they have taken two things into consideration for the performance analysis: one is mean squared error (MSE), and the second one is time, which can be a running time (the no. of iterations) that the clustering algorithm takes to converge. We have considered the first one for performance analysis of the proposed method.

Ding and He [11] has proved that the PCA finds clusters in the data by taking the objective function of the K-means into consideration automatically, which is also a justification for the PCA-based dimension reduction. Final Conclusion from this work is that PCA is complementary to K-means clustering.

Qi and Hughes [12] has shown that, PCA on low-dimensional random projections and on the original dataset, produces the same results as PCA with certain conditions. This is shown empirically on both synthetic as well as real-world datasets, by recovering the PCs of the original data and also center of the data.

In this paper, we have combined PCA and RP with K-means clustering to improve the clustering performance on the high-dimensional data. Here, we have done a two-step preprocessing of the input data, that is, applying PCA first to get reduced directions then apply Random Projection on the PCs. Then performing clustering on the reduced data.

The contents of this paper are organized as follows. Principal Component Analysis is described in Sect. 2. In Sect. 3, we discuss about Random Projection (RP). Section 4 describes the K-means Algorithm. Section 5 describes the proposed Algorithms. Section 6 reports and discusses the empirical results. Finally, the conclusion and the future scope are presented in Sect. 7.

2 Principal Component Analysis

Principal Component Analysis (PCA) is used to find a linear transformation of the original input data. Let the original input matrix size is d dimensions and N observations, and we want to reduce it into a D-dimensional subspace. This transformation is given by

$$Y = P^T X \tag{1}$$

where P_{dXD} is the projection matrix containing D eigenvectors corresponding to D highest eigenvalues, X_{NXd} is the mean centered matrix.

3 Random Projection

Random Projection (RP) is a simple and more often used dimensionality reduction technique in the recent times. It uses matrix multiplication to reduce the data into lower dimensional space. Generation of a Random Matrix is the key point in applying Random Projection. In the projected space, the distance between the points is preserved [3, 4].

Random Projection method works by projecting the original data X_{NXd} to a subspace of size D-dimensional ($D \ll d$) using a random orthogonal matrix P_{dXD}. The orthogonal matrix P_{dXD} is having unit length columns. Symbolically, it can be written as

$$X_{N \times D}^{RP} = X_{N \times d} P_{d \times D} \tag{2}$$

The theme of Random Projection is based on the Johnson–Lindenstrauss (JL) lemma. JL lemma [3] states that if N points in vector space of dimension d are projected onto a randomly selected subspace of dimension D, then the Euclidean distance between the points are approximately preserved. More details about JL lemma are available in [5].

4 K-Means Algorithm

K-means performs cluster analysis on low as well as high- dimensional data. It is basically an iterative algorithm which takes input as N observations and divide them across K nonoverlapping clusters. The clusters are identified by initializing K random points as centroids and iterating them over N observations. The centroids for K clusters are calculated by minimizing the error function used to discriminate a point from its cluster, in this case Euclidean distance. The lesser the error, more is the goodness of that cluster.

Let $X = \{\mathbf{x}_i, i = 1, \ldots, N\}$ be the set of N observations to be clustered into a set of K clusters, $C = \{c_k, k = 1, \ldots, K\}$, where $K \ll N$. The K-means clustering objective is to minimize the mean squared error, that is, defined as the distance (Euclidean) between the cluster mean to the points in that cluster. The mean of a cluster c_k is denoted by μ_k and is defined as

$$\mu_k = \frac{1}{N_k} \sum_{\mathbf{x}_i \in c_k} \mathbf{x}_i \tag{3}$$

where N_k is the number of observations in cluster c_k.

5 Proposed Algorithms for High-Dimensional Data Clustering

This section describes the proposed two algorithms that cluster the high- dimensional data efficiently. Most of the algorithms present in the literature are basically distance based. These algorithms which are meant for low-dimensional data clustering, cannot produce meaningful clusters because of high dimensionality and unrelated or unuseful features present [9, 10]. To make these conventional algorithms suitable for clustering high-dimensional data , first we need to project the input high-dimensional data into a low-dimensional space and then perform clustering on the projected points. PCA performs this by taking the number of principal components less than the number of input dimensions. Taking the above-mentioned issues into consideration, we have proposed two algorithms in this paper to perform clustering in high-dimensional data efficiently and effectively. The PCA is combined with K-means clustering in Algorithm 1, and we fix $D < d$ in Algorithm 1 to cluster high-dimensional data.

PCA and RP both are combinedly used as a preprocessing step before clustering is done on the reduced space in Algorithm 2, which we call it as PCA^{rp} K-means. PCA mapping is the same as the kernel-based clustering.

Given two data points p and q and the PCA defining a mapping ϕ from the input space R^d to the feature space F

$$\phi : R^d \to F. \tag{4}$$

The Euclidean distance between x and y in the input space is

$$d(p, q) = \sqrt{\|p - q\|^2}. \tag{5}$$

After the points p and q are mapped into the feature space, the Euclidean distance between $\phi(p)$ and $\phi(q)$ in the feature space becomes [13, 14]:

$$d_F(p, q) = \sqrt{\|\phi(p) - \phi(q)\|^2}.$$ (6)

Equation (5) can be substituted by Eq. (6). as we concentrate on the projected feature space not on original space. From this viewpoint, we can say that the proposed PCA-K-means is a mapping from high to low dimension, where we actually apply clustering algorithm. The PCA-K-means Algorithm consists of two steps and can be described as in Algorithm 1.

Algorithm 1 PCA-K-means Algorithm

Input: Dataset $X_{N \times d}$, number of clusters K, Reduced Dimension D
Output: cluster membership G.
begin
1: Apply PCA on the original input X
2: Perform K-means clustering in the reduced space
3: **return** G

In Algorithm 1, X represents the input data and G is the cluster membership vector. Here, we first apply PCA on the original input X, and get the reduced space. Then, we perform K-means clustering on the reduced space to get the clustering results (G). The basic idea is, instead of applying K-means clustering on the original high-dimensional data, first we reduce the dimensionality of the data into low-dimensional one, and then we perform K-means clustering in the reduced space.

Algorithm 2 PCA^{rp} K-means Algorithm

Input: Dataset $X_{N \times d}$, number of clusters K, Reduced Dimension D
Output: cluster membership G.
begin
1: Apply PCA on X to get $X^{pc}_{N \times d_1}$
2: Set a random matrix $P_{d_1 \times D}$
3: Set $X^{rp}_{N \times D} = X^{pc}_{N \times d_1} P_{d_1 \times D}$
4: Perform K-means clustering on the reduced data $X^{rp}_{N \times D}$
5: **return** G

---

where N is the number of points/patterns in the input dataset, d is the number of original features/dimensions present in the input dataset. D is the size of the reduced dimension to which the input dataset is projected and $D < d$. This algorithm starts by reducing the dimensionality of input data from d to D by applying PCA. By this, $X_{N \times d}$ becomes $X^{pc}_{N \times d_1}$. Then we generate a random projection matrix $P_{d_1 \times D}$. Then, we apply random projection on PCA- reduced dataset $X^{pc}_{N \times d_1}$ by using random projection matrix, which gives the matrix $X^{rp}_{N \times D}$. Then, we perform K-means clustering on this reduced matrix to get the clusters.

6   Experimental Study

We demonstrate the proposed algorithm's performance on 12 datasets, which contain both high dimensional and low dimensional as well. Mean squared error (MSE) is the performance measure that is considered for reporting the clustering performance. The low value of MSE means the clustering accuracy is high.

6.1   Datasets Used in Experiments

For the experimental study, we have considered eight high-dimensional and four low-dimensional datasets to evaluate the performance of the proposed algorithms. A detailed specification of the datasets is present in Table 1. The high-dimensional datasets used in the experiments are as follows: AT & T Database of Faces (formerly ORL Database), contains 400 face images of 40 persons, 10 images per each person. Global Cancer Map (GCM) dataset consists of 190 tumor samples and 90 normal tissue samples. Leukemia dataset contains 72 samples of two types: 25 acute myeloid leukemia (AML) and 47 acute lymphoblastic leukemia (ALL). Another gene expression dataset we have taken is Lung Cancer, that contains 180 samples, which are classified into malignant pleural mesothelioma (MPM) and adenocarcinoma (ADCA). Yale dataset contains 165 face images of 15 persons and 11 images per person, with a dimensionality of 1024. The number of classes or subjects is 15 in this dataset. Columbia Object Image Library (COIL20) dataset consists of 1440 images of 20 classes, 72 images per class, and the dimensionality is 1024. Colon dataset comprises of 62 samples, each sample is having 200 genes. This data can be classified into 22 normal and 40 tumor tissue samples. Prostate cancer dataset is having a total of 136 tissues: 77 tumor and 59 normal specimens.

The low-dimensional datasets used in the experiments are as follows: Iris consists of 150 iris flowers, with sepal length, sepal width, petal length and petal width as measurements, giving 150 points $x_1, x_2, \ldots, x_{150} \in R^4$. The data points are in 4 dimensions. Wine dataset contains 178 samples and 13 dimensions. ZINC is a data repository of chemical structures. We have taken 50000 samples randomly with 7 features (ZINC7) as one dataset and with 28 feature (ZINC28) as another dataset for our experimentation. See http://www.zinc.docking.org for more information.

6.2   Results and Discussion

The results we have reported here are the average of 10 independent runs. These results are shown in Table 2. The performance of the K-means is compared with the PCA-K-means and PCA-RP-K-means on low-dimensional data is shown in Table 2. From Table 2, it is evident that, PCA-K-means is better than K-means for the low-

Table 1 Specifications of datasets

S.no.	Dataset	No. of samples	No. of dimensions	No. of classes
1	AT&T Faces (ORL)	400	10304	40
2	Yale	165	1024	15
3	GCM	280	16063	2
4	Leukemia	72	7129	2
5	Lung	181	12533	2
6	COIL20	1440	1024	20
7	Colon	62	2000	2
8	Prostate	136	12600	2
9	Iris	150	4	3
10	Wine	178	13	3
11	ZINC7	50000	7	–
12	ZINC28	50000	28	–

Table 2 MSE for several datasets. Sample average over 10 runs

Dataset	Classic K-means	PCA-K-means	PCA-RP-K-means
Iris	0.69	0.077	0.441
Wine	1231	310	1185
ZINC7	309	58	136
ZINC28	99	75	99
AT and T Faces (ORL)	6.51×10^6	6.668×10^6	1.61×10^6
Yale	2.39×10^6	9.162×10^5	2.66×10^5
GCM	6.92×10^9	2.6575×10^9	0.823×10^9
Leukemia	5.02×10^9	1.4101×10^9	2.96×10^8
Lung	9.55×10^8	3.183×10^8	1.21×10^8
COIL20	16.424	17.381	4.51947
Colon	3.16×10^7	4.6×10^7	1.82×10^7
Prostate	15×10^7	14.8×10^7	7.24×10^7

dimensional datasets considered. The clustering performance of PCA-RP-K-means is much better than the clustering performance of K-means on all the low-dimensional datasets studied except ZINC28, for which both the methods are same in the performance. From Table 2, we can also say that, the clustering performance of the proposed PCA-RP-K-means is much better compared with K-means method. The PCA-K-means is showing good results for GCM, Leukemia, Yale, and Lung datasets when compared with K-means. For Prostate, COIL20 and ORL datasets, both methods performance is more or less same. For the Colon dataset, K-means (3.16×10^7)

is performing better compared to PCA-K-means (4.6×10^7). When compared with PCA-K-means, the PCA-RP-K-means is giving better performance with a 5 times improvement for Leukemia dataset, 4 times improvement for ORL and COIL20 datasets, 3 times improvement for GCM and Yale datasets, 2 times improvement for Colon, Lung and Prostate data sets. The overall performance of PCA-RP-K-means is much better than PCA-K-means method.

7 Conclusions and Future Directions

We have incorporated PCA and RP with K-means clustering for better clustering high- and low-dimensional data. By conducting an experimental study we found that K-means is giving good performance when combined with PCA than the normal K-means alone when the dimensionality is high. We have proposed two hybrid algorithms: PCA-K-means and PCA-RP-K-means. PCA when combined with Random Projection gives us good quality clusters in the reduced dimensional space. The proposed PCA-K-means and PCA-RP-K-means are outperforming the classic K-means algorithm on the given low and high-dimensional datasets. Our experimental results strongly advocate the improvement in the performance.

References

1. Jolliffe, I.: Principal component analysis. Wiley Online Library. USA
2. Lloyd, S.: Least squares quantization in PCM. IEEE Trans. Inf. Theory **28**, 129–137 (1982)
3. Johnson, W., Lindenstrauss, J.: Extensions of Lipschitz mappings into a Hilbert space. Contemp. Math. **26**, 189–206 (1984)
4. Cardoso, A., Wichert, A.: Iterative random projections for high-dimensional data clustering. Pattern Recognit. Lett. **33**, 1749–1755 (2012)
5. Dasgupta, S., Gupta, A.: An elementary proof of a theorem of Johnson and Lindenstrauss. Random Struct. Algorithms **22**, 60–65 (2003)
6. Dasgupta, S.: Experiments with random projection. In: Proceedings of the Sixteenth Conference on Uncertainity in Artificial Intelligence (UAI-2000), pp. 143–151 (2000)
7. Fern, X.Z., Brodley, C.E.: Random projection for high dimensional data clustering: a cluster ensemble approach. In: Proceedings of the Twentieth International Conference of Machine Learning (2003)
8. Deegalla, S., Bostrom, H.: Reducing high-dimensional data by principal component analysis vs. random projection for nearest neighbor classification. In: Proceedings of the 5th International Conference on Machine Learning and Applications (ICMLA), FL, pp. 245–250 (2006)
9. Bouveyron, C., Girard, S., Schmid, C.: High dimensional data clustering. Comput. Stat. Data Anal. **52**, 502–519 (2007)
10. Assent, I.: Clustering high dimensional data. Wiley Interdisc. Rev. Data Min. Knowl. Discov. **2**(4), 340–350 (2012)
11. Ding, C., He, X.: K-means clustering via principal component analysis. In: Proceedings of the 21st International Conference on Machine Learning, ACM (2004)
12. Qi, H., Hughes, S.M.: Invariance of principal components under low-dimensional random projection of the data. In: Proceedings of ICIP 2012 IEEE, pp. 937–940

13. Zhang, L., Cao, Q.: A novel ant-based clustering algorithm using the kernel method. Inf. Sci. **181**, 4672–6658 (2011)
14. Alshamiri, A.K., Singh, A., Surampudi, B.R.: Combining ELM with random projections for low and high dimensional data classification and clustering. In: Proceedings of the Fifth International Conference on Fuzzy and Neuro Computing (FANCCO), IDRBT, Hyderabad, India, pp. 89–106 (2015)

Dynamic Shifting Genetic Non-adjacent Form Elliptic Curve Diffie–Hellman Key Exchange Procedure for IoT Heterogeneous Network

M. Durairaj and K. Muthuramalingam

Abstract Maintaining an operational balance between two different types of nodes in a heterogeneous network is a vital process that affects power and security management in the network. Dynamic shifting genetic non-adjacent form elliptic curve Diffie–Hellman (DSGNECDH) key exchange procedure is introduced here for key generations and updates. As the name suggests, this procedure uses a dynamic shifting in security based on the computation power of the involving nodes and uses genetic algorithm for key updates. Non-adjacent form elliptic curve Diffie–Hellman Key exchange procedure is applied to share the keys generated by dynamic shifting genetic algorithm.

Keywords Heterogeneous network · Security and power management
Genetic security key generation · Non-adjacent form elliptic curve cryptography
Diffie–Hellman key exchange

1 Introduction

Internet of things (IoT) [1] makes it possible almost all electronic gadgets to communicate with each other. A typical IoT network can have nodes ranges from low-power simple wireless sensors to high-power supercomputers. IoT provides basic communication between the nodes but there are still a lot of improvements required here like security enhancements and power management. The nodes of a heterogeneous network are diverged in many ways from power consumption to computational speeds. Exerting a single authentication and security protocol for entire network is not recommended here. The selection and application of various protocols draw more research

---

M. Durairaj · K. Muthuramalingam (✉)
School of Computer Science and Engineering, Bharathidasan University,
Tiruchirappalli, India
e-mail: bardmuthu@gmail.com

M. Durairaj
e-mail: durairaj.bdu@gmail.com

© Springer Nature Singapore Pte Ltd. 2019
J. Kalita et al. (eds.), *Recent Developments in Machine Learning and Data Analytics*,
Advances in Intelligent Systems and Computing 740,
https://doi.org/10.1007/978-981-13-1280-9_45

deeds these days because of the tremendous growth in a number of IoT devices. Key calculation methods, selection of key sizes, key annulment and updation periods are to be maintained dynamically; thus, this process is nondeterministic in nature. Fitness function-based genetic algorithms are used to solve this kind of problems in efficient way. Hence, a new genetic algorithm-based security scheme is projected in this paper with legacy key calculation and updation procedures to fulfil the need of IoT-based heterogeneous networks. Standard network evaluation metrics like key calculation time, key updation time, throughput, IP-delay, jitter, latency, end-to-end delay, power consumption and security are measured for DSGNECDH to analyse the performance. Changes in these benchmark parameters are used to validate new security protocols.

2 Related Works

Many key generation and sharing techniques are available for heterogeneous networks. More reliable and frequently used methods are elliptic curve cryptography key generation (ECC) [2, 3], non-adjacent form elliptic curve key generation (NAF) [4] and elliptic curve-Diffie–Hellman (ECC-DH) [5, 6] key exchange procedures. Performances of these three existing procedures are measured for comparison.

2.1 Elliptic Curve Cryptography Key Generation (ECC)

The mathematical structures of elliptic curves over finite field are used to generate keys they can be used to key agreements and digital signatures. The predominant advantage of ECC key generation is the security strength of the keys. A smaller ECC key provides as much security as Galois field [7] based longer keys. Current ECC cryptographic calculations are based on an elliptic curve equation $y^2 = x^3 + ax + b$ and two main arithmetic functionalities of point addition and point doubling. In an elliptic curve set E, if two different points are taken as $P(X_p, Y_p)$ and $Q(X_q, Y_q)$, the group arithmetic addition operation can be applied to the points to calculate another point $R(X_r, Y_r)$ which should also be in the group E such that $R = P + Q$. Coordinates of point R can be calculated as $X_r = S^2 - X_p - X_q$ where $S^2 = 2X_p + X_q + X_r - X_p$. Since the point R goes to the straight line PQ, then $S = \frac{Y_r - Y_p}{X_r - X_p}$ and $Y_r = Y_p + S(X_r - X_p)$. Point doubling for a point $P(X_p, Y_p)$ in the group E, the group operator can be used to calculate the third point $R(X_r, Y_r)$ which is also in E as $P + P = 2P = R$.

$$X_r = S_2 - 2X_P \quad \text{where,} \quad S = \frac{3X_P^2 - a}{2Y_P} \text{ and } Y_r = Y_p + S(X_r - X_p)$$

Primary Curve is plotted for a cyclic group G with prime order P using generator g. Let these users are A and B. Then, A's random key: $X_A \in G \Rightarrow$ computational result $\Rightarrow y_A = g^{x_A} \bmod p$. B calculates y_B in a similar manner, and key solvation is performed as $y_B^{x_A} \bmod p$ and $y_A^{x_B} \bmod p$. Then, the shared secret key is $g^{X_A X_B} \bmod p$ with prime curve equation: $y^2 \bmod p = (x^3 + ax + b) \bmod p$ over z_P and binary curves equation $y^2 + xy = x^3 + ax^2 + b$ over $GF(2^m)$. Shared points of the elliptic curve are represented $E_P(a, b)$ with base point: $G = (x_G, y_G)$ with large prime order n. A's random Integer: $X_A < n \Rightarrow$ computed $Y_A = X_A G$ and B's random Integer: Y_B similar calculation. Then, the secret key is calculated as $K = X_A X_B G$. By this way using an elliptic curve, a set of keys are generated and shared between sender and receiver. The main advantage of using elliptic curve is keys for the security is high even though the sizes of the keys are small.

2.2 Non-adjacent Form Elliptic Curve Key Generation (NAF)

Non-adjacent form is a balanced binary representation of ECC. In NAF, integers are represented into binary with an extra value -1. NAF also ensures one or more adjacent value is 0. This is used to increase the speed of modular exponentiation and point multiplication arithmetic processes of an arbitrary group like elliptic curve.

NAF of $a \in N$ is defined as any tuple of $j \in N$ integers $a_i \in \{-1, 0, +1\}$ with $0 \le i < j$

- $a = \sum_{i=0}^{i=j-1} a_j 2^j$
- $j > 0 \rightarrow a_{j-1} \ne 0$
- $0 < i < j \rightarrow a_i a_{i-1} = 0$

For the above equations, each a has a unique non-adjacent form of the curve. That is, for $a \ne 0$, $j = \left[\log_2(3|a|)\right] - 1$. For any elliptic curve $E = \left(e_{m-1} e_{m-2} \ldots e_1 e_0\right)_2$, an NAF restriction $z = (z_{m-1} z_{M-2} \ldots z_1 z_0)_{\text{NAF}}$ can be applied. The main advantage of NAF over ECC is the speedy modular arithmetic without compromising other network characteristics. Even though NAF provides more security with less computational resources, the key calculation and updation process still perplexed for nodes with less computational powers like single channel monitoring wireless sensor nodes.

2.3 Elliptic Curve-Diffie–Hellman Key Exchange Procedure (ECC-DH)

ECC-DH is an anonymous key agreement protocol with private and public key pairs. Using the key pairs, the sender and receiver can share a secret message and they can generate new key pairs as well. The network nodes involved in the communication generate their own private and public keys. For example, if node A and node B

are involved in communication, the private key of node A is d_A and the public key $H_A = d_A G$. Similarly for node B, d_B is the private key and $H_B = d_B G$ is the public key, where G is the base point at the elliptic curve. Then, node A and B have to exchange their public keys H_A and H_B. After sharing the public keys, node A can calculate the shared secret key $S = d_A H_B$ using its own private key and public key of node B. Similarly, node B can calculate $S = d_B H_A$. Although ECC-DH uses for secured key sharing, this procedure is vulnerable to the MIM (man-in-the-middle) attack [8]. MIM attack can take place while node A and B are sharing their public keys in an insecure channel. Using these public keys H_A and H_B by solving discrete logarithm problem, an intruder can compute d_A and d_B. In heterogeneous network, ECC-DH consumes more power when the sender node and receiver node are of different types. If ECC-DH is modified to use lightweight keys to save power, then the network will be less secure because of the keys' sizes.

3 Proposed Method and Implementation

Proposed DSGNECDH is based on four major functional blocks. They are as follows:

1. Node classifications,
2. Cluster classifications,
3. Power index calculation and
4. Key generation and updation.

3.1 Node Classifications

Heterogeneous IoT network nodes are having high variance based on their operational and computational powers. In DSGNECDH, the nodes are classified into four different categories based on their type. Wireless sensor nodes (classification index γ_1), low-power computational mobile devices (classification index γ_2), smartphones and laptops (classification index γ_3) and high-power servers (classification index γ_4). Wireless sensor nodes are limited computational power nodes [9] used to gather one or more environmental sensory information. Mostly they are battery operated and designed for a limited time operation. The nature of a wireless sensor node is to collect and communicate required information to a base station with or without the support of other peer nodes wirelessly. Wireless sensor nodes are designed to work as continuous measurement and communication device which prefers insecure communications rather than secured communications. Only in some restricted area, these nodes are designed to communicate confidential data, but in general, wireless sensors are dealing with common public data. Hence, these nodes are available with entry-level security mechanisms. Environmental Earth sensing devices, seashore sensing devices, forest fire detection devices, air pollution monitors and water pollution monitors are some examples for wireless sensor network nodes [10, 11]. In DSGNECDH, wireless sensor nodes are classified as γ_1. Low-power computational mobile devices are used to communicate delicate data which requires a little more security procedures than in WSN nodes. Unlike WSN's continuity in secure streaming data, low-power computing devices are designed to work periodically with the large volume of data that should be communicated in a secured way. These devices are designed to operate for longer endurances, and their power sources are replaceable or rechargeable. A number of configurable security protocols are embedded in this type of devices. Text data, voice data and lightweight multimedia data are used in this category. Basic communication mobile phones and health monitoring devices are classified into this category, and this classification index is denoted in γ_2.

Smartphones and laptops are designed with higher computational infrastructures like multicore processors and plenty of memory. Though these devices are much powered, they are required to provide mobility. Hence, these devices are mainly designed with batteries as their power sources. Power saving mode and maximum performance mode are available in these devices to switch between performance and operational longevity. A number of highly configurable high-security communication protocols are embedded in these devices to provide customized security. These protocols are much resource consuming while comparing with γ_1 and γ_2 device category protocols. This category is represented by γ_3 in proposed method. High-power servers are used in places like data centres and they deeds a general destination hub for all the above-mentioned devices. These devices are power monsters operating with continuous multiple power sources. This category consists of multichannel communications and plays a vital role in IoT-based heterogeneous networks. γ_4 is used as the classification index for this category.

3.2 Cluster Classifications

Clusters are classified into several groups based on a number of different category nodes [12], and these classifications are represented as $\eta_1, \eta_2, \ldots, \eta_n$. A group with a large number of low computational resource nodes (γ_1 class nodes) is labelled as η_1, and a group with a large number of high computational resource nodes (class nodes) is labelled as η_n. Node group formation of the proposed method is given in Fig. 1.

Group η_1 has a collection low-power devices; hence, the security policy of this cluster is limited to use less than or equal to 64-bit numeral arithmetic which includes prime number key calculations. This lightweight security policy ensures the power saving while generating and sharing security keys. Alternatively, nodes of group η_4 are granted to use large number arithmetic (Fig. 2).

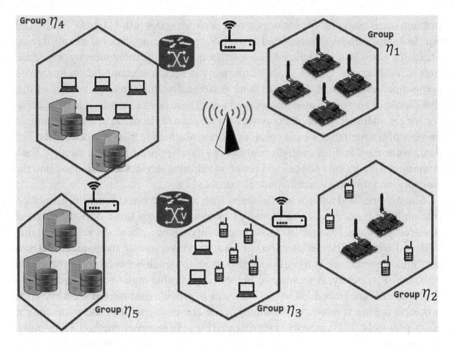

Fig. 1 Power groups in heterogeneous network

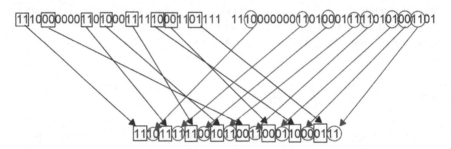

Fig. 2 Key generation using genetic process

3.3 Power Index Calculation

Power index is calculated based on the processing power and available power source of the IoT nodes involved in communication. A customized formula is introduced in DSGNECDH to calculate Power Index. While two nodes A and B need to communicate in group, the processing power of node A is represented as ϕ_A and the processing power of node as ϕ_B. Similarly, the available power source of node A and B is represented as Δ_A and Δ_B. The processing power ϕ is measured in MHz and remaining power from the source Δ in percentage (%) is based on the capacity (mAh) of node's battery power. Then, the power index φ is calculated using the following equation:

$$\varphi = \sqrt{\left(\frac{\Delta_A}{\phi_A} - \frac{\Delta B}{\phi_B}\right)^2} \tag{1}$$

3.4 Key Generation and Updation

Keys are generated using a calculated value of φ using the standard procedure NAF. Segmented sieve of Eratosthenes method is used to calculate the prime numbers which can operate at a better speed. For node categories γ_3 and γ_4, prime numbers are calculated and accumulated in the node memory. Whenever an attack or a suspicious activity is detected [13], key updation process is initialized by either node in communication. The key updation process is performed using genetic algorithm [14], and a dynamic shift process is applied based on the values of φ in different nodes and cluster groups. A fitness function $f = \frac{1}{(1+e^{-\lambda})}$ is assigned for the genetic key updation algorithm, where

$$\lambda = \sigma_1\lambda_1 + \sigma_2\lambda_2 \cdots \sigma_n\lambda_n$$

σ Crossover rate,
λ_1 Frequency test and
λ_2 Gap test.

$$\sum_{i=1}^{n}\lambda_i = 1$$

Crossovers are applied whenever a key updation is required in an active communication between two IoT nodes. Mutations are generated when a cluster head or more number of nodes generate an intruder detection signal. The number of crossovers (ν) is calculated as

$$\nu = \omega \times \mu \times \xi / n^2 \times max(\eta_x) \tag{2}$$

where

ω Crossover rate,
μ Key size,
ξ Number of keys,
n Number of nodes and
η_x Cluster classification.

Fig. 3 Intra-cluster
communication

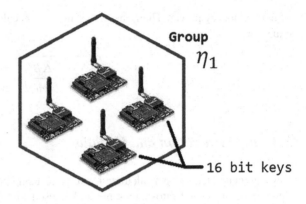

Number of mutations ρ is calculated as $\rho = \sigma \times \mu \times \xi / n^2 \times max(\eta_x)$ (3)

where

σ Mutation rate,
μ Key size,
ξ Number of keys,
n Number of nodes and
η_x Cluster classification.

Sample key updation process using genetic NAF-ECC for parent prime numbers 3761536111 and 3761536333 of node A and B is as follows:

Key 1 = 3761536111 = 11100000001101000111110001101111(Prime)
Key 2 = 3761536333 = 11100000001101000111110101001101(Prime)

Random genetic key generation
Generated key value in binary is 11101111110010110011000110000111, which is equal to 4023071111 and is also a prime number shared between the nodes. Whenever a prime number is not found as the outcome of this procedure, then the genetic procedure is repeated with another random seed. Intra-cluster communication [15] between same low-power nodes uses 16 bits keys given in Fig. 3. The key size is increased according to the cluster power index φ for increasing security.

Key sizes of inter-cluster communication are calculated based on the cluster power indexes $\varphi(\eta_x)$ and $\varphi(\eta_y)$ where x and y are different power clusters. Inter-cluster key size equation is given in Eq. 4

$$\mu = \frac{2^{4x} \cdot 2^{4y}}{2}$$ (4)

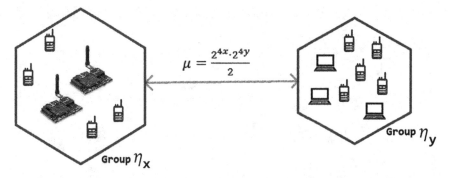

Fig. 4 Inter-cluster communication

DSGNECDH handles three network states; they are neutral, power saving and security improvement. In neutral scenario, neither power nor security is changed. By letting the communication in a default state, DSGNECDH provides uninterrupted communication between nodes. In power saving state, DSGNECDH saves power by using reduced key sizes where security is not a prime concern. In security improvement state, DSGNECDH increases the key sizes to improve security where a little more power is compromised (Fig. 4).

4 Experimental Setup

Network simulator OPNET [16] is used to simulate 1000 number of nodes randomly placed in 1000 square metre area. Network node types are configured to represent different types of nodes like servers, gateway, routers, laptops, mobile devices and IoT-based wireless sensors. All standard protocol types like IEEE 802.11 b/g/n are used in a mixed mode based on the node types. Mobility and network traffics are randomized to represent a typical real-world communication. OPNET tools sets such as node model, packet format, protocol selector, process model, network behaviour monitor and network capturer are configured and used using advanced C (AC) programming script. Visual Studio IDE [17] is used to create the user interface and to code the proposed method DSGNECDH. A computer with Intel® Core™ i5 processor, 4 GB RAM is used to run the experiment. Simulation is performed for 1 h simulation time (5 min of real time) with 12 min timestamps. OPNET simulation setup is given in Fig. 5.

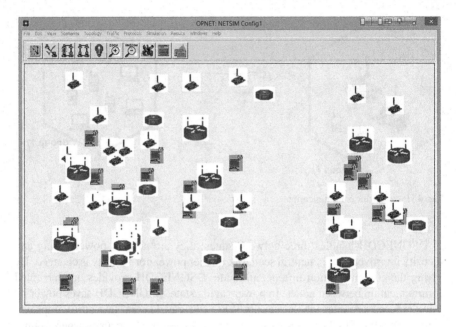

Fig. 5 OPNET simulation setup

5 Results and Analysis

Network metrics throughput, IP-delay, jitter, latency, end-to-end delay, key calculation time, key updation time, power and security levels are measured five times in a simulation hour with equal 12 min timestamps. Results of existing methods ECC, NAF-ECC, ECC-DH and proposed DSGNECDH are logged in report files, and graphs are plotted based on the results for the above network metrics.

5.1 Key Calculation Time

Key calculation time is one of the important metrics of the network that dissembles the quality of the network. Measured key calculation times for ECC, NAF, ECC-DH and DSGNECDH are given in Table 1.

Key calculation time is to be kept under manageable duration to get higher throughput. As per the observations (Table 1), DSGNECDH used lesser time to calculate keys than the other methods. DSGNECDH takes an average time of 578 mS which is lesser than other timings 934, 967 and 804 mS of ECC, NAF-ECC and ECC-DH, respectively, compared with Fig. 6. This lesser value reflects the higher efficiency of the proposed method DSGNECDH.

Table 1 Key calculation time

Key calculation time (mS)				
Time (min)	ECC	NAF-ECC	ECC-DH	DSGNECDH
12	934	961	815	578
24	942	959	801	585
36	923	955	791	572
48	942	984	815	577
60	933	977	802	579

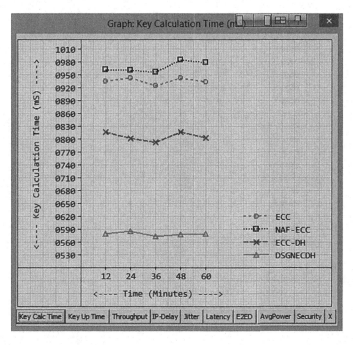

Fig. 6 Key calculation time

5.2 Key Updation Time

Keys are updated periodically to ensure the security of a communication. Time taken for updating a key also affects the performance of a network. Key updation times for methods ECC, ECC-NAF, ECC-DH and DSGNECDH are measured and tabulated as Table 2. Smaller key calculation duration represents the higher efficiency of the method. DSGNECDH consumes 243 mS on average for key updates, which is lesser than other methods in comparison. Key updation time comparison chart is given in Fig. 7.

Table 2 Key update time

Key update time (mS)				
Time (min)	ECC	NAF-ECC	ECC-DH	DSGNECDH
12	660	695	575	242
24	647	694	567	244
36	656	697	556	237
48	658	713	559	236
60	652	705	572	256

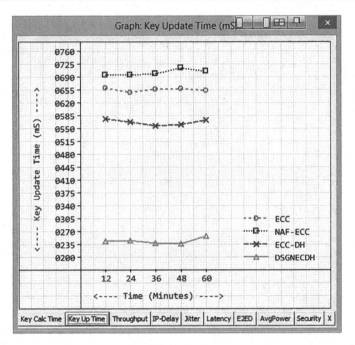

Fig. 7 Key update time

5.3 Throughput

Throughput is an essential metric to measure the quality of a network schema. The good quality network architecture provides higher throughput values. Measured throughput values are given in Table 3 and compared in Fig. 8. DSGNECDH provides a higher throughput of 1333 Mbps, which is better than the other methods. Highest throughputs of ECC, NAF-ECC and ECC-DH are 828, 816 and 1188 Mpbs in order. Observed 1333 Mbps throughput of DSGNECDH ensures higher data transmission rate and better data packet delivery ratio in a typical heterogeneous network with IoT nodes.

Table 3 Throughput

Throughput (Mbps)

Time (min)	ECC	NAF-ECC	ECC-DH	DSGNECDH
12	812	807	1185	1325
24	812	809	1186	1329
36	807	810	1178	1330
48	804	808	1188	1333
60	818	816	1174	1317

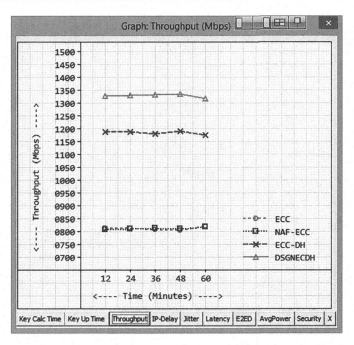

Fig. 8 Throughput

5.4 IP-Delay

IP-delay occurs while the routers are processing packet headers. Lesser IP-delay refers that the communication protocol is suitable for faster router operations. While measuring IP-delay for the methods in comparison, DSGNECDH causes the very less duration of 94.8 mS. IP-delay values are tabled in Table 4 and compared in Fig. 9.

Table 4 IP-delay

IP-delay (mS)				
Time (min)	ECC	NAF-ECC	ECC-DH	DSGNECDH
12	165	138	119	97
24	171	135	117	91
36	168	126	113	99
48	171	141	122	90
60	165	125	110	97

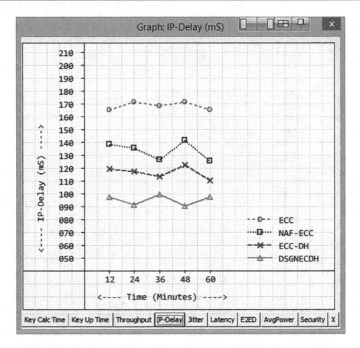

Fig. 9 IP-delay

5.5 Jitter

Jitter is a variation in delay caused by nondeterministic communication states. Complete mobility is permitted to the heterogeneous network nodes which causes nondeterministic states while some nodes travel from one cluster to another cluster. Network handover is to be performed faster to ensure an uninterrupted communication. While the jitter is maintained at a minimum level, then the quality of the network is higher. While measuring the jitter values of various methods, DSGNECDH has a very low jitter duration of 41.6 mS on average. Jitter values are given in Table 5. Jitter for the methods ECC, NAF-ECC, ECC-DH and DSGNECDH is compared in Fig. 10.

Table 5 Jitter

Jitter (mS)				
Time (min)	ECC	NAF-ECC	ECC-DH	DSGNECDH
12	64	61	50	44
24	63	64	53	42
36	70	56	59	41
48	61	55	49	41
60	67	56	52	40

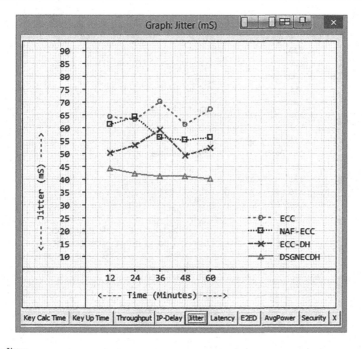

Fig. 10 Jitter

5.6 *Latency*

Latency is a time delay between data input timing and node's response timing. A quality network protocol makes a node to serve an input data communication request quickly. Reduced latency improves the network quality. Latency for ECC, NAF-ECC, ECC-DH and DSGNECDH is given in Table 6. It is observed that DSGNECDH has lesser latency values than the other methods in the comparison. Average latency for ECC, NAF-ECC, ECC-DH and DSGNECDH is 258.2, 231.2, 201 and 154.8 mS, respectively. Latency values are compared in Fig. 11.

Table 6 Latency

Latency (mS)				
Time (min)	ECC	NAF-ECC	ECC-DH	DSGNECDH
12	256	228	207	159
24	263	225	202	150
36	251	233	200	146
48	264	236	198	155
60	257	234	198	164

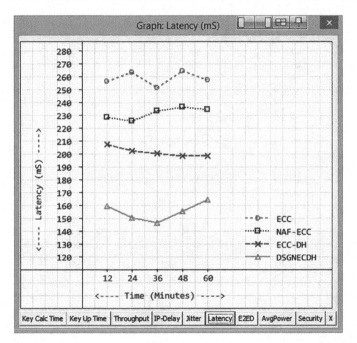

Fig. 11 Latency

5.7 End-to-End Delay

The transmission time taken for a data packet to pass through source to destination is called end-to-end delay. This delay includes IP-delay, jitter, latency and system delay. For a good quality network infrastructure, end-to-end delay must be maintained with lower level.

A communication protocol's performance can be evaluated by measuring end-to-end delay. The lesser value of end-to-end delay assigns the higher quality of the protocol. As per the observations, ECC, NAF-ECC, ECC-DH and DSGNECDH get an average end-to-end delay of 506.4, 439.2, 383.8 and 308 mS in order. End-to-end

Table 7 End-to-end delay

End-to-end delay (mS)				
Time (min)	ECC	NAF-ECC	ECC-DH	DSGNECDH
12	503	450	393	326
24	521	431	393	289
36	499	424	379	306
48	511	458	388	309
60	498	433	366	310

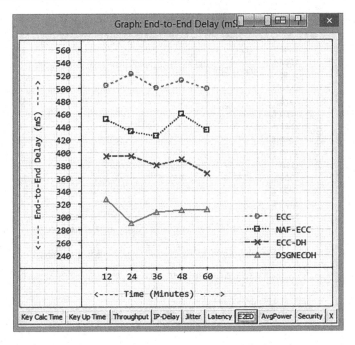

Fig. 12 End-to-end delay

delay values for ECC, NAF-ECC, ECC-DH and DSGNECDH are given in Table 7. End-to-end delay comparison is given in Fig. 12.

5.8 Power Consumption

A large number of IoT nodes in a heterogeneous network are battery powered and thus have a limited power source. To extend the mobility and lifetime, they should consume less power without degrading their performances. Optimal usage of power by communication protocol in a network helps to improve node's performance. Power

Table 8 Average power consumption

Average power consumption (mW)				
Time (min)	ECC	NAF-ECC	ECC-DH	DSGNECDH
12	1387	1550	1078	780
24	1388	1556	1058	764
36	1379	1583	1072	756
48	1386	1587	1065	788
60	1369	1567	1071	754

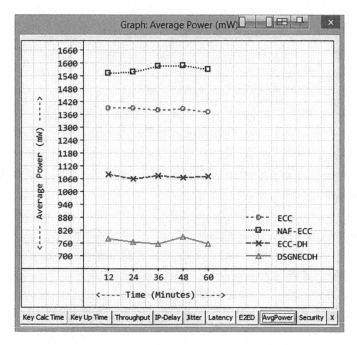

Fig. 13 Power consumption

consumption methods ECC, NAF-ECC, ECC-DH and DSGNECDH are measured and tabulated in Table 8, which is to compare and identify good quality protocol. DSGNECDH consumes lesser power during all the five checkpoints. Even its highest power consumption of 788 mW is less than the lowest power consumption of nearest performance protocol ECC-DH which consumes 1058 mW. Power measurements are given in Table 8 (Fig. 13).

5.9 Security

Security is a vital metric of modern networks. There are more possibilities for intrusion in a heterogeneous network. Providing security without affecting the other parameters is one of the important characteristics of a communication procedure. While measuring security levels of ECC, NAF-ECC, ECC-DH and DSGNECDH, they scored security levels of 87, 90, 94 and 96% on average. Measured security levels are given in Table 9, and a comparison chart is provided as Fig. 14.

Table 9 Security level

Security (%)

Time (min)	ECC	NAF-ECC	ECC-DH	DSGNECDH
12	86	89	92	95
24	85	90	94	96
36	85	89	94	95
48	85	90	93	95
60	87	89	94	95

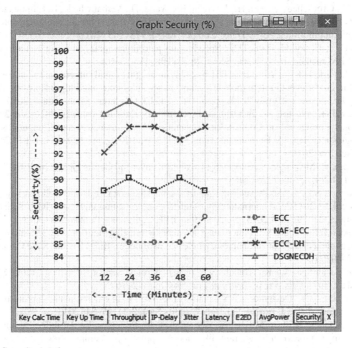

Fig. 14 Security level

6 Conclusion

Providing security and power optimization without botching other parameters of a heterogeneous network is a challenging job and not handled well by existing methods in many environments. To provide a better security and optimized power utilization, a new method DSGNECDH is proposed, and its performance betterment is testified based on the simulation results. By providing a lucid bridge between two different node types for security key management, DSGNECDH is recommended as a more legible procedure to use in heterogeneous IoT networks.

References

1. Lee, G.M., Crespi, N., Choi, J.K., Boussard, M.: Internet of things. In: Emmanuel, B., Crespi, N., Magedanz, T. (eds.) Evolution of Telecommunication Services, vol. 7768, pp. 257–282. Springer, LNCS (2013)
2. Azarderakhsh, R., Järvinen, K.U., Mozaffari-Kermani, M.: Efficient algorithm and architecture for elliptic curve cryptography for extremely constrained secure applications. IEEE Trans. Circuits Syst. I Regul. Pap. **61**(4), 1144–1155 (2014)
3. Liu, Z., Großschädl, J., Zhi, H., Järvinen, K., Wang, H., Verbauwhede, I.: Elliptic curve cryptography with efficiently computable endomorphisms and its hardware implementations for the internet of things. IEEE Trans. Comput. **66**(5), 773–785 (2017)
4. Liu, C.L., Horng, G., Tsai, D.S.: Speeding up pairing computation using non-adjacent form and ELM method. Int. J. Netw. Secur. **18**(1), 108–115 (2016)
5. Seo, J., Park, J., Kim, Y.J.: An ECDH-based light-weight mutual authentication scheme on local SIP. In: 2015 Seventh International Conference on Ubiquitous and Future Networks (ICUFN), Sapporo, Japan, 7–10 July 2015, pp. 871–873. IEEE Conference Publication (2015)
6. Kodali, R.K., Naikoti, A.: ECDH based security model for IoT using ESP8266. In: 2016 International Conference on Control, Instrumentation, Communication and Computational Technologies (ICCICCT), Kumaracoil, India, 16–17 Dec 2016, pp. 629–633. IEEE Conference Publication (2016)
7. Charpin, P., Mesnager, S., Sarkar, S.: Involutions over the Galois Field F2n. IEEE Trans. Inf. Theory **62**(4), 2266–2276 (2016)
8. Khader, A.S., Lai, D.: Preventing man-in-the-middle attack in Diffie-Hellman key exchange protocol. In: 2015 22nd International Conference on Telecommunications (ICT), Sydney, NSW, Australia, 27–29 April 2015, pp. 204–208. IEEE Conference Publication (2015)
9. Aktakka, E.E., Najafi, K.: A micro inertial energy harvesting platform with self-supplied power management circuit for autonomous wireless sensor nodes. IEEE J. Solid-State Circuits **49**(9), 2017–2029 (2014)
10. Khan, S., Pathan, A.S.K., Alrajeh, N.A. (eds.): Wireless Sensor Networks: Current Status and Future Trends. CRC Press, Taylor & Francis Group (2016)
11. Lazarescu, M.T.: Design of a WSN platform for long-term environmental monitoring for IoT applications. IEEE J. Emerg. Sel. Top. Circuits Syst. **3**(1), 45–54 (2013)
12. Liu, L., Chen, X., Liu, M., Jia, Y., Zhong, J., Gao, R., Zhao, Y.: An influence power-based clustering approach with page rank-like model. Appl. Softw. Comput. **40**(C), pp. 17–32 (2016) (Elsevier Science Publisher)
13. Pawar, S.N., Bichkar, R.S.: Genetic algorithm with variable length chromosomes for network intrusion detection. Int. J. Autom. Comput **12**(3), 337–342 (2015). Springer
14. Sastry, K., Goldberg, D.E., Kendall, G.: Genetic algorithms. In: Burke, E.K., Kendall, G. (eds.) Search Methodologies, pp. 93–117. Springer, Heidelberg (2013)

15. Xu, L., O'Grady, M.J.G., O'Hare, M.P.: Reliable multihop intra-cluster communication for wireless sensor networks. In: 2014 International Conference on Computing, Networking and Communications (ICNC), Honolulu, HI, USA, 3–6 Feb 2014, pp. 858–863. IEEE Conference Publication (2014)

16. Yang, S., He, R., Wang, Y., Li, S., Lin, B.: OPNET-based modeling and simulations on routing protocols in VANETs with IEEE 802.11p. In: 2014 2nd International Conference on Systems and Informatics (ICSAI), Shangai, China, 15–17 Nov 2014, pp. 536–541. IEEE Conference Publication (2014)

17. Amann, S., Proksch, S., Nadi, S., Mezini, M.: A study of visual studio usage in practice. In: 2016 23rd International Conference on Software Analysis, Evolution, and Reengineering (SANER), Suita, Japan, pp. 124–134 (2016)

Geo-Statistical Modelling of Remote Sensing Data for Forest Carbon Estimation—A Case Study on Sikkim, Himalayas

Pradeep Kumar, Ratika Pradhan and Mrinal Kanti Ghose

Abstract The mountainous topography coupled with the high biodiversity throws up the range of challenges in remote sensing of forests for forest carbon assessment. The correlations between vegetation indices or grey-level co-occurrence matrix (GLCM) metrics and forest carbon cannot be reliably used in the estimation of forest carbon in biodiversity-rich mountainous topography of Sikkim Himalayas due to geometric, spectral and radiometric distortions. In this paper, an attempt has been made to apply the geo-statistical modelling of aboveground terrestrial vegetation carbon in Sikkim using products derived from remotely sensed satellite data, GIS and ground sampling data. The paper makes use of universal kriging as an interpolation method that makes use of semivariogram models for spatial autocorrelation to make a prediction of forest carbon at unsampled locations. The errors associated with prediction have been estimated. Remote sensing derived forest type and forest density maps have been used as cokriging parameters. Cross-validation technique has been used to evaluate the interpolation results. The analysis reveals that the total aboveground vegetation carbon stored in the forests of Sikkim is about 29.46 million tonnes. Importantly, the montane wet temperate forests contain the maximum forest carbon of 8.46 million tonnes, while the least carbon (1.03 million tonnes) is stored in the moist alpine scrub.

Keywords Sikkim · Forest carbon · Geo-statistical modelling · Semivariogram Kriging · Remote sensing · GIS

P. Kumar (✉)
Forests, Environment and Wildlife Management Department,
Government of Sikkim, Gangtok 737102, India
e-mail: pradeepifs@gmail.com

R. Pradhan
Department of Computer Applications, Sikkim Manipal Institute of Technology,
Majitar 737136, Sikkim, India
e-mail: ratika.p@smit.smu.edu.in

M. K. Ghose
Department of Computer Applications, Sikkim University, Tadong,
Gangtok 737102, Sikkim, India
e-mail: mkghose@cus.ac.in

© Springer Nature Singapore Pte Ltd. 2019
J. Kalita et al. (eds.), *Recent Developments in Machine Learning and Data Analytics*,
Advances in Intelligent Systems and Computing 740,
https://doi.org/10.1007/978-981-13-1280-9_46

1 Introduction

With the climate change posing menacingly serious challenges to humanity, the forest carbon as a bulwark against climate change is gaining increasing importance in international efforts to contain climate change. As a major proportion of about 23% of earth's forests, lies in mountains [2]), it poses a big challenge in terms of inaccessibility for ground inventory-based assessment of forest carbon. Remote sensing can play a vital role to a great extent in surmounting the issues of inaccessibility and assist in scaling up spatial distribution. However, the mountainous topography coupled with the high biodiversity constraints the application of remote sensing of forests for forest carbon assessment [5, 7]. The studies reveal that the correlations between vegetation indices or grey-level co-occurrence matrix (GLCM) and forest carbon are poor in such areas, and thus cannot be reliably used in the estimation of forest carbon in Sikkim Himalayas [6]. In such situations, remote sensing can be used for deriving other factors, on which the forest carbon density depends and then based on field samples, geo-statistical methods can be used for analysing and predicting the forest carbon distribution.

The spatial variability in the forest carbon consists of three components, viz., deterministic and stochastic processes and measurement errors. The geo-statistical models attempt to explain the natural spatial pattern and its variability in the biomass due to physical processes up to a certain level by interpolating values where sample plots have not been laid. The associated error measures of uncertainty can also be modelled. The more the model is realistic, the better are the interpolated values and the associated uncertainties are accurate and represent the real distribution of forest carbon.

2 Objectives

The vast expanses of forest areas make it absolutely impractical to measure each and every tree for carbon. It is wiser from the economical and practical point of view to measure only representative trees on certain sample plots and to be used to model the distribution of the rest of the unsampled forest for estimating the forest carbon. The parameters for the models for whole area can be captured through remote sensing and GIS. With this background, the objectives of the present study are as follows:

- Geo-statistical modelling of aboveground terrestrial vegetation carbon in Sikkim using products derived from satellite remote sensing, GIS and ground sampling.
- Generation of geospatial data of the aboveground terrestrial vegetation carbon in Sikkim.

3　Materials and Methods

3.1　Study Area

Sikkim, a mountainous state of India in the Eastern Himalayas, extending approximately 114 km from North to South and 64 km from East to West, between 27° 00′ 46″ and 28° 07′ 48″ N latitude and 88° 00′ 58″ and 88° 55′ 25″ E longitude, as shown in Fig. 1, is taken as the study area for the present study. It encompasses a great altitudinal compression ranging from 300 to 8585 m above mean sea level. Forestry is the major land use in the state covering around 47.80% of the total geographical area [8]. Because of the great altitudinal compression, the forests of Sikkim exhibit tremendous biological diversity.

3.2　Methodology

3.2.1　Study Design and Data Collection

For establishing the correlation between the actual field based carbon and the image parameters, actual ground sample plots of 0.1 ha (31.61 m × 31.61 m) during 2009–2010 have been considered and the girth at breast height (1.37 m) for tree species falling within the plot is measured. Based on the extent of area covered by different NDVI ranges, the proportionate numbers of plots are randomly selected in the areas with corresponding NDVI ranges. Based on the physical and financial constraints, a total number of 55 sites have been selected. The locations of the sites are determined using handheld global positioning system (GPS). The sites are divided into four quadrates (NE, SE, SW and NW), and four sample plots from each site at about 75–90 m away from centre have been laid as shown in Fig. 2.

Each site has a cluster of four plots. The distribution of sites for plot clusters is shown in Fig. 3. A total of 207 plots are laid. The total trees measured are 1088 in East Sikkim, 1526 in South Sikkim, 9321 in West Sikkim and 6063 in North Sikkim.

3.2.2　Estimation of Plot Biomass and Carbon

The collected data as stated above are processed to estimate the biomass and carbon. As the present study is based on optical remote sensing data, the aboveground tree biomasses including woody biomass and the leaf biomass of all trees (i.e. all dia classes) have been considered for carbon estimation of the trees. The site-specific volume equations developed by Forest Survey of India (FSI) have been used for calculating the woody volume of trees for each sample plot. The carbon contents of the tree have been obtained by multiplying the tree biomass with percentage carbon content of the species.

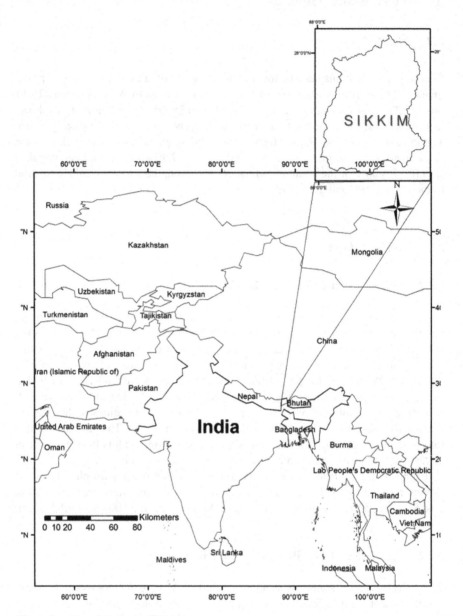

Fig. 1 Location of study site (Sikkim)

Fig. 2 Layout of sample plots

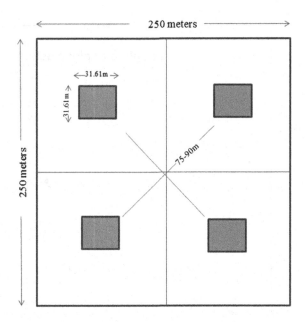

3.2.3 Forest-Type Mapping

The Champion and Seth's [1] classification method is adapted for the classification of forests, and the forest-type mapping by FSI on the basis of GIS framework using remote sensing data has been adopted. For the present study, the forest types are regrouped as below and the forest-type distribution is shown in Fig. 4:

- Tropical moist deciduous forests,
- Subtropical broadleaved hill forests,
- Montane moist wet temperate forests,
- Himalayan moist temperate forests,
- Subalpine forests and
- Moist alpine forests.

3.2.4 Forest Density Mapping

Forest density mapping is aimed at finding the canopy density which has the bearing on carbon density of forests. The forest density mapping has been carried out by digital interpretation followed by ground truthing of remotely sensed IRS P6 LISS III satellite images. For this, the Forest Report 2009, wherein forests have been classified into four categories namely dense forest (canopy density > 70%), moderately dense forest (canopy density 40–70%), open forest (canopy density 10–40%) and scrub (canopy density < 10%), has been considered as the datasets. Forest density distribution in Sikkim is shown in Fig. 5.

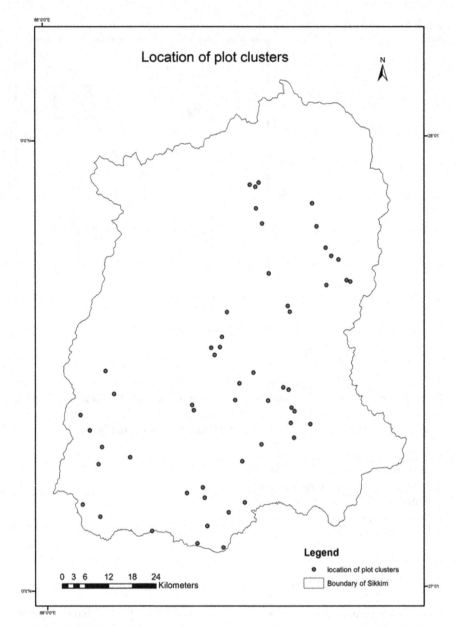

Fig. 3 Location of sample plot clusters

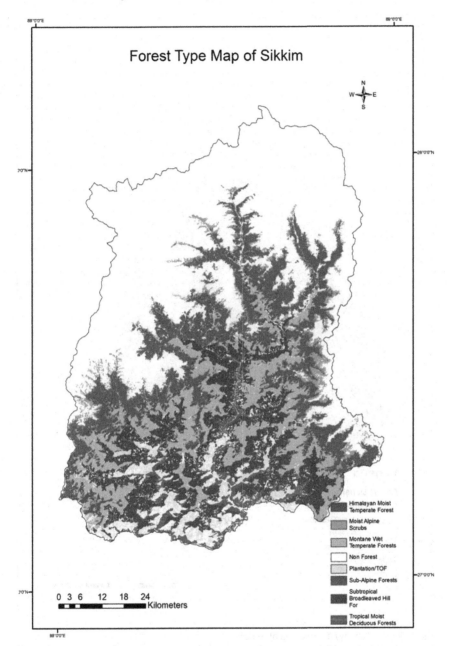

Fig. 4 Forest-type map of Sikkim

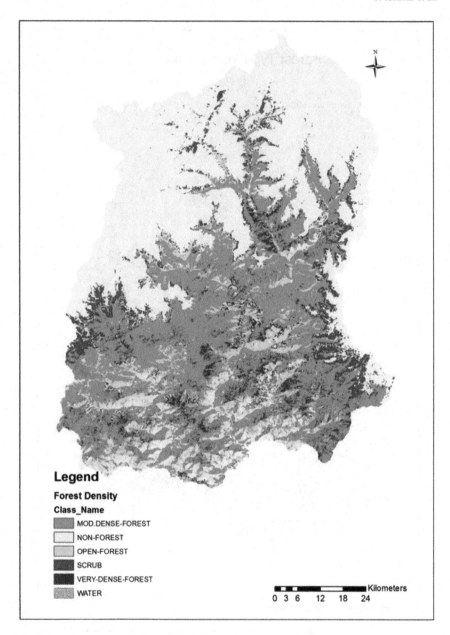

Fig. 5 Forest (canopy) density map of Sikkim

3.3 Exploratory Spatial Data Analysis

It is important to conduct the exploratory spatial data analysis to find out the suitable geo-statistical approach for forest carbon density prediction at unsampled locations. To gain insights into the data and to find out the most appropriate methods and suitable parameters, the following spatial data analysis tools have been used.

3.3.1 Histogram

Histogram examines the summary statistics and distribution of forest carbon across the plots. The histogram for the current dataset of forest carbon in sample plots is shown in Fig. 6.

3.3.2 Normal QQ Plots

The normal QQ (Quantile–Quantile) plots give an idea about the univariate normality of the distribution in dataset. For the current dataset of forest carbon in sample plots, the normal QQ plot is shown in Fig. 7 where the points are deviating from the 45° reference line.

From Figs. 6 and 7, it is evident that any geo-statistical approach which presupposes the normality of the data distribution cannot be applied in the present case.

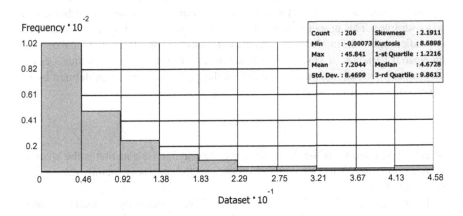

Fig. 6 Histogram of plot carbon

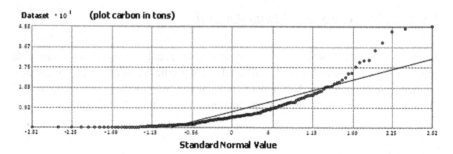

Fig. 7 Normal QQ plot

3.4 Geo-Statistical Modelling

Geo-statistical modelling assumes that some of the spatial variation in the forest carbon distribution can be attributed to and modelled by spatial autocorrelation and it is possible to explicitly model the spatial autocorrelation. Spatial prediction or spatial interpolation techniques can be used to predict the value of forest carbon at unsampled locations. For making a prediction at a point, all the samples are not considered, rather a subset of samples known as window is considered so that those samples which are quite far away and have no influence are not unnecessarily considered. The size of window, the number of samples and their configuration affect the prediction surface.

Semivariogram is a function that represents the variation in the sample values against the distance between samples. When the data is autocorrelated, the semivariogram will show low variance for nearby samples and high variance for relatively far away samples. This property makes the semivariogram fit an important component of prediction for unsampled locations in interpolation methods. In statistical domain, the spatial variation of environmental variable of forest carbon or biomass can be viewed as information signal consisting of the three components as defined by the universal model of variation [3]

$$Z(s) = Z * (s) + \varepsilon'(s) + \varepsilon'' \tag{1}$$

where $Z^*(s)$ constitutes the deterministic component, $\varepsilon'(s)$ constitutes the spatially correlated random component and ε'' constitutes the pure noise mainly due to measurement error.

Using Eq. (1), on the field measurements at different sampling intensities, it can be estimated as to how much of variation is due to deterministic component attributable to environmental factors, how much is due to spatial autocorrelation and how much is due to uncorrelated noise or measurement errors. Processing the data may require removal of trends and declustering to account for preferential sampling. In methods like kriging, spatial structure in dataset is modelled using semivariogram or covariance function. Search strategy defines the number of data points to be used for predicting the forest carbon value at unsampled location and the spatial configuration of data points. The predicted value and the associated uncertainty are dependent on these two factors.

3.5　Spatial Interpolation of Forest Carbon Through Kriging

Kriging is a collection of interpolation method that makes use of semivariogram models for spatial autocorrelation to make a prediction and the associated errors. Kriging is based on two aspects, viz., spatial autocorrelation estimated through semivariogram or covariance function estimation and the prediction by making use of generalised linear regression technique. There are a number of kriging models like simple, ordinary or universal.

Kis [4] has proposed that when the data does not meet the requirement of second-order stationarity, i.e. the correlation between any two values of plot forest carbon is not dependent solely on their relative position in forest and the mean and variance are not the same in the entire forest area, the universal kriging is to be preferred [4]. Accordingly, the universal kriging for spatial interpolation has been used for the present study. Forest type and forest density have been used as cokriging datasets. The model has been optimised and four sectors with 45° offset have been taken. For search neighbourhood, five neighbours are included.

Kriging like other interpolation models also considers that the forest carbon in sample plots which are closer to each other will be more alike than those situated far away. This relationship has been explored through empirical semivariogram. The semivariogram modelling has been carried out to determine the best fit for the model that will pass through the points in the semivariogram. This line is fitted in such a way that the weighted squared difference between each point is least, so that the autocorrelation in the data can be quantified. Lag size is the distance which is used for grouping a large number of combinations of the location of sample pairs. The lag size in the semivariogram in the current study has been found to the tune of 4448 m, the range is 31,091 m and the number of bins taken is 12. The output kriging raster cell size is kept at 358.44 m. The semivariogram fitting is shown in Fig. 8. The red points show every pair of location of sample plots.

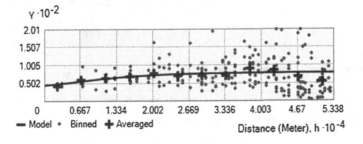

Fig. 8 Semivariogram fitting of plot carbon data

Fig. 9 Forest carbon density distribution map of Sikkim

Table 1 Forest carbon in different forest types of Sikkim

Sl no.	Forest type	Area (in ha)	Total carbon (in tonnes)
1	Tropical Moist Deciduous Forests	16,345	1,293,846
2	Subtropical Broadleaved Hill Forest	84,552	8,073,852
3	Montane Wet Temperate Forests	86,332	8,457,205
4	Himalayan Moist Temperate Forest	21,101	2,066,943
5	Subalpine Forests	90,570	7,636,050
6	Moist Alpine Scrubs	15,602	1,029,736
7	Plantation/TOF	9117	899,066
	Total	323,619	29,456,698

The semivariogram fitting generated the equations for kriging. The matrices and vectors for kriging equation depend upon the measured location and predicted location. The kriging weights of each measured value in the searching neighbourhood are dependent on these matrices and vectors. Based on above semivariogram fitting during cokriging, the geospatial layers of forest carbon density prediction have been converted to raster images. The prediction map for non-forest areas is corrected by masking any predictions in non-forest areas. The forest carbon distribution obtained by using the geo-statistical modelling is shown in Fig. 9.

It is observed from Fig. 9 that the continuum of distribution of forest carbon density is from high to low, where each cell has a predicted value of carbon. For finding the distribution of forest carbon across forest types, the values of forest carbon for all the pixels falling in a particular forest type are summed. The results of forest carbon distribution across the forest types are shown in Table 1.

Thus, the total aboveground vegetation (trees only) carbon stored in the forests of Sikkim is estimated to be 29,456,698 tonnes or 29.46 million tonnes. It is further

Fig. 10 Distribution of forest carbon among different forest types

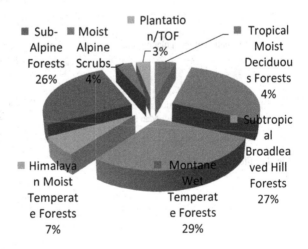

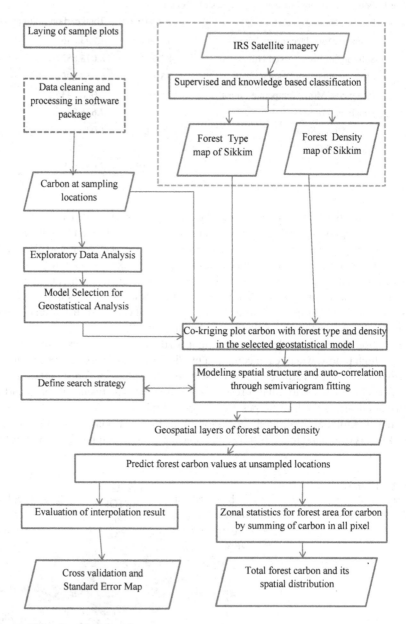

Fig. 11 Flow chart of methodology

Fig. 12 Scatterplot of predicted values versus measured values

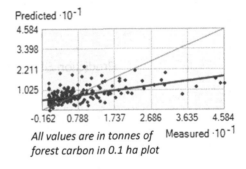

All values are in tonnes of forest carbon in 0.1 ha plot

observed that the montane wet temperate forests contain the maximum forest carbon of 8.46 million tonnes, while the least carbon (1.03 million tonnes) is stored in the moist alpine scrub. The pie chart of distribution of forest carbon among different forest types is shown in Fig. 10.

The summary of methodology followed is given in the form of flowchart in Fig. 11.

3.5.1   Evaluation of Interpolation Results

The cross-validation method of fitting the prediction of forest carbon against the actual distribution as measured through the sample plots has been used to evaluate the interpolation results. Cross-validation sequentially omits one data and then calculates the value of that data based on the remaining data. Finally, a comparison of the measured and predicted values is made to calculate the errors in terms of root mean square error.

In the present case, the mean prediction error is 0.0643 tonnes in 0.1 ha plot area, which is nearer to 0 implying that the predictions are unbiased. For the plot sizes of 0.1, the root mean square error is found to be 7.09 tonnes and the average standard error is found to be 7.21 tonnes, which infers that the model is appropriate. Standard error here refers to the standard deviation of the prediction for any individual point.

A scatter plot of predicted values versus measured values is given in Fig. 12. These are expected to fall around 1:1 line. But the slope is less than 1. This essentially is due to the property of interpolation method based on kriging which under predicts large values and overpredicts small values. The line which fits the scatter is shown in blue.

The prediction uncertainty at each location in forest carbon density distribution is measured by the prediction standard error. It can be said that the true value of forest carbon at any point will lie inside the interval of predicted value ±2 times the standard error at least for 95% of the times. The prediction standard error map is shown in Fig. 13.

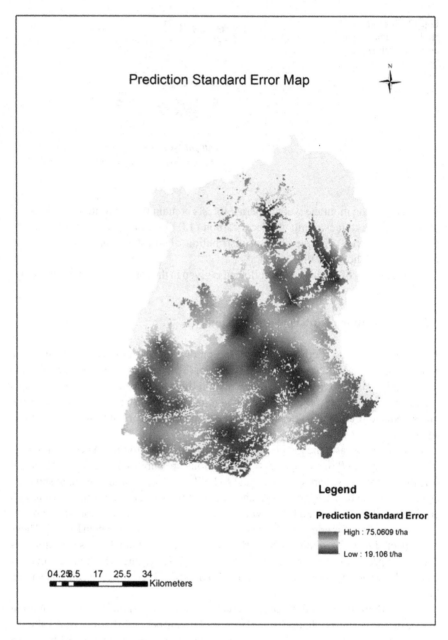

Fig. 13 Prediction standard error map

3.6 Conclusions

The proposed geo-statistical model reveals that the total aboveground vegetation carbon stored in the forests of Sikkim is 29.46 million tonnes. Montane wet temperate forests contain the maximum forest carbon of 8.46 million tonnes, while the least carbon (1.03 million tonnes) is stored in moist alpine scrub. Accuracy can be further improved by dividing forest types into more classes and also having more canopy density classes subject to the capabilities of the remote sensing data to capture these differences.

From Figs. 6 and 7, it can be seen that the data distribution is skewed to the left, i.e. there are more number of plots in low forest carbon density areas. The bounding box, where samples have been laid, covers only some part of Sikkim and not the whole of Sikkim. As the interpolation cannot be done for points outside the bounding box, the predictions outside the bounding box have been extrapolated that may lead to a higher degree of uncertainty.

References

1. Champion, H.G., Seth, S.: A Revised Survey of the Forest Types of India. Manager of Publications, Delhi, India (1968)
2. FAO, Price, M.F., Schweiz (eds.): Mountain Forests in a Changing World: Realizing Values, Addressing Challenges [International year of forests 2011]. FAO, Rome (2011)
3. Hengl, T.: A Practical Guide to Geostatistical Mapping, vol. 52. Hengl (2009)
4. Kiš, I.M.: Comparison of Ordinary and Universal Kriging interpolation techniques on a depth variable (a case of linear spatial trend), case study of the Šandrovac Field. Rudarsko-Geološko-Naftni Zbornik **31**(2):41–58. (2016). https://doi.org/10.17794/rgnzbornik.v31i2.3862
5. Kumar, P., Ghose, M.K.: Remote sensing based forest carbon assessment in mountains: the challenges of Terrain. Asian J. Geoinform. **16**(4) (2017a)
6. Kumar, P., Ghose, M.K.: Remote sensing-derived spectral vegetation indices and forest carbon: testing the validity of models in mountainous terrain covered with high biodiversity. Curr. Sci. **112**(10):2043. (2017b). https://doi.org/10.18520/cs/v112/i10/2043-2050
7. Lu, D.: The potential and challenge of remote sensing-based biomass estimation. Int. J. Remote Sens. **27**(7), 1297–1328 (2006). https://doi.org/10.1080/01431160500486732
8. SFR: India State of Forest Report 2015. Forest Survey of India Dehradun, India (2015)

Author Index

Printed in the United States
by Bookmasters

Printed in the United States
By Bookmasters